SCIENCE IN THE BEDROOM

SCIENCE
IN THE
BEDROOM

*A History of
Sex Research*

VERN L. BULLOUGH

BasicBooks
A Division of HarperCollinsPublishers

Designed by Ellen Levine

Library of Congress Cataloging-in-Publication Data
Bullough, Vern L.
Science in the bedroom : a history of sex research / Vern Bullough.
p.   cm.
Includes bibliographical references and index.
ISBN 0–465–03020–3 (cloth)
ISBN 0–465–07259–3 (paper)
1. Sexology—Research—History   I. Title.
HQ60.B85   1994   93–46396
306.7'072—dc20   CIP

95 96 97 98 ❖/RRD 9 8 7 6 5 4 3 2 1

For BONNIE BULLOUGH,
my wife, my frequent collaborator,
and always my helpful colleague

Contents

ACKNOWLEDGMENTS   viii

Introduction   1

1   Sex Research and Assumptions: From the Greeks to the
    Nineteenth Century   9

2   Homosexuality and Other Factors in Bringing about
    Sex Research   34

3   Hirschfeld, Ellis, and Freud   61

4   The American Experience   92

5   Endocrinology Research and Changing Attitudes   120

6   From Freud to Biology to Kinsey   148

7   From Statistics to Sexology   172

8   The Matter of Gender: Masculinity, Femininity, and
    Cross-Gender Behavior   210

9   Other Voices, Other Views   241

10 Problems of an Emerging Science   273

NOTES   301
INDEX   361

ACKNOWLEDGMENTS

Though this work has been under way for many years, Susan Rabiner of Basic Books was the crystallizing force in finally getting it done. The whole staff at Basic Books deserves praise, but I would also like to single out Jane Judge, who was project editor, and Candace Levy, copyeditor. The index was prepared by Liz Cunningham.

INTRODUCTION

In terms of observable human anatomy it would seem that sex is a simple thing. A male has a penis and testes and produces sperm, while a female has a vagina, uterus, and breasts and menstruates. These simple distinctions have probably always been evident to humans. Note that the information about the female reproductive organs was much less complete than that about the male, because the female organs are less visible. Also much of the extant data have been derived from male sources. Still, it was assumed that when the male put his semen (seed) into the female's vagina there was a possibility that the woman would then proceed to nurture the seed until it grew into a baby.

Although there probably has never been a time when males did not know they had something to do with procreation, if only because the animal examples around them were so obvious, they also knew that not every act of sexual intercourse led to a pregnancy. That is, intercourse might be a necessary cause of pregnancy, but it was not always believed to be sufficient in itself. The reasons why pregnancy did not always occur were fully understood only in the twentieth century. The issue was further complicated by the fact that some males were impotent or sterile and some females did not get pregnant no matter how many times they had intercourse. What was known, moreover, raised further questions and even problems. Sexual intercourse, for example, was a source of pleasure whether or not a pregnancy resulted. Even touching the penis or clitoris could give a pleasurable sensation as could mutual masturbation between two individuals regardless of sex. This knowledge, early in human history, led to discussions over whether it was natural for an individual to touch himself or herself or for two persons of the same sex to lie together. Different cultures had different ideas about these phenomena. Various cultures also had their own answers to the question of why some acts of intercourse resulted in male babies and others in female babies. This remains a question for which we have no simple explanation. There was also a great deal of concern about the occasional appearance of babies with ambiguous genitalia, hermaphrodites if you will. What was the cause of this? Again different cultures gave different answers, some going so far as to claim that such offspring were monsters. Humans also wondered what the menses were and what their purpose was.

We have answers to these and many other questions from the ancient Egyptians as well as from early civilizations that appeared in the Tigris-Euphrates Valley, the Indus Valley, and China. More than two thousand years later, information was collected by the classic Greeks, and this became the basis of much of Western knowledge about human sexuality. These Greek ideas remained more or less dominant until the nineteenth century, when advances in science allowed us to ask new kinds of questions about sex and sexual behavior. The lack of what might be called scientific answers, however, did not prevent people in the past from attempting to frame answers to questions about human sexuality. These answers varied from culture to culture and changed through time, but generally they were a combination of observations, mythology, morals, and magic. Even when investigators came up with more rational explanations for some aspects of human sexual activity, they continued to rely on tradition for most answers, simply because the science of the time was not sophisticated enough to give more complete explanations.

Further complicating any attempt to come up with scientific explanations about sexuality is the fact that sexual behavior, unlike many other aspects of human activity, traditionally has been regarded more as a matter of morals than physiology or psychology. In the Western world, this has meant that attitudes toward sex have been dominated by the assumptions of the Judeo-Christian tradition, particularly as this was interpreted by Christian writers from the third to the sixth centuries. These early Christians set the patterns of Christian dogma and are collectively known as the "Fathers of the Church." Sexual matters were points of major concern to these men and continued to be important for later generations.

For example, sexually related issues made up more than a quarter of medieval Roman Catholic canon law, and continue to do so. Particularly influential in setting Western Christian ideas was St. Augustine (354–430), who held that the ideal Christian life was one of celibacy. Fortunately, he recognized that not all people could live a celibate life, and he reluctantly allowed that God had countenanced marriage as well. According to Augustine, however, the only purpose of sex within marriage was for procreation, and sexual intercourse should only be undertaken with that purpose in mind. All other sexual activities were sinful. Augustine went further and taught that the only proper position in sexual intercourse was with the woman on the bottom and that the only proper act was one in which the penis penetrated the vagina. He absolutely condemned the use of any other orifice (such as oral-genital sex).[1] He also condemned masturbation either done alone or with others.

In fact, it was the knowledge of sex that constituted the original sin that occurred in the Garden of Eden. Augustine, who also set Christian doctrine on this, held that the sin of Adam and Eve is transmitted from parents to children through the sexual act, which, by virtue of the lust that accompanies it, is inherently sinful. It was only through baptism that the taint of this original sin was washed away. The result of the writings of St. Augustine and other Church Fathers was to make Western Christianity what I have called a sex-negative culture. Even though the Western tradition has always included significant minorities such as Jews as well as Christian groups that do not subscribe to the Augustinian theology, it was the Augustinian interpretation of sex that tended to dominate public discourse until the twentieth century.

As a result, research into sex was a somewhat doubtful enterprise, because those who engaged in it were regarded as not quite respectable. This attitude continued to exist even during my own lifetime. A good illustration of this mind-set is the case of William Masters. As he was finishing medical school at the University of Rochester, he began seriously to think of devoting his life to research in sex. He turned for advice to Professor George W. Corner, who, as will be explored later in this book, was a significant figure in American sex research. Corner advised him not to start out as a sex researcher but to establish his reputation as a scientist in another field. After Masters made his reputation, was in his late thirties or early forties, was happily married, and had a family, then and only then could he begin to investigate aspects of human sexuality without damaging his scientific reputation. I received similar advice, and I gradually and gingerly entered into serious sex research only after gaining tenure at a university. Whether such hesitation was necessary in my case is unclear, but the perception that it was tended to affect the nature of sex research during much of the twentieth century.

Initially, I was interested in studying the kinds of individuals who could be regarded as societal barrier breakers because of their willingness to investigate sexual topics. Thus this book seems occasionally to digress from the theories and findings to give brief biographies of the different personalities who have engaged in sex research. I was particularly interested in their coping mechanisms, and though these varied, some general patterns emerge. One way that many of these individuals coped was simply by denying that they were sex researchers. Instead, they called themselves endocrinologists, urologists, gynecologists, or some other less stigmatizing term, even though their primary interest was in sex. Another way many seemed to cope was to distance themselves from their subjects by labeling them with words such as *deviant, pathological,* and *perverse,* emphasizing in the process that although it was necessary for society to know about the existence of such individuals, most people, including the researcher, were not like them.

Still others, and many of the early American researchers are in this category, were concerned by the evils wrought on innocent wives and children from sexually transmitted diseases, which the investigators believed resulted from contact with prostitutes. They came to believe that the answer lay in sex education and thus emphasized the dangers of premarital and extramarital sexuality and the need to practice abstinence except within marriage. It was only with the failure of these early efforts that some researchers turned to science to gain a better understanding of human sexuality, although for many of them science was seen as a partner in their moral crusade against sexually transmitted diseases.

Some of these individuals, however, entered into the studies of sexuality to understand themselves and to clear up their own as well as the public's misconceptions. Among the early sex researchers were homosexuals, who were interested in demonstrating that homosexuality was not a perversion. Feminists were determined to eliminate much of the erroneous information about women put forth by the male establishment, which tended to hold that women were inferior to males. Some of these individuals remain propagandists for a particular viewpoint, but many moved beyond their initial preoccupation to become major scientific investigators.

The personality of the researchers is only part of the question, however. Another aspect of the development of sex research is why the last part of the nineteenth century began to see the development of what Paul Robinson called the modernization of sex.[2] This is a subject that has been little explored. The pioneer in the field was Edward M. Brecher, who was far more interested in the events surrounding Alfred Kinsey and his successors than in an attempt to explain how modern research developed.[3] This is a question that I have spent many years trying to answer, and this book is the result of my inquiry.

Part of the answer is simply the growth of specialization in science, which has been a fact of life for more than one hundred years. If opthalmologists could specialize in the study of the eye, then others could begin to specialize in urology and gynecology. These medical specialists, however, dealt with sex problems only indirectly. There were a growing number of scientists who felt that the sex organs needed to be studied but that sexual relationships and sexual behaviors also needed to be analyzed. They believed that investigation should not be limited to the medical arena but should draw from a number of the emerging disciplines in the biological and social sciences.

Within the scientific community itself, interest in sexual matters had become widespread, in part because of the theories of Charles Darwin (1809–82). Darwin believed that sexual selection was a key to evolution. Sex coupling, the bringing together the sexual elements of two organisms who had been exposed to different environments, led to more vigorous offspring than did self-fertilization. Building on this insight, some of Darwin's followers went so far as to hold that sex existed for the good of the species. This misinterpretation of Darwin became the dominant view, albeit an erroneous one, until the middle of the twentieth century, and proved a strong impetus for studies in sexuality.[4] It was also a major factor in removing some of the stigma from the study of sex.

Other researchers seized on the concept of sexual selection to explain any number of factors. Sir Francis Galton (1822–1911), a cousin of Darwin's, believed that sexual selection could prove to be a means for improving the human race. He coined the term *eugenics* in 1885, and though he himself concentrated on urging judicious marriages and large families for the wealthy and gifted (terms he used more or less synonymously), others went much further and argued that the high birth rate of the poor and less intellectually endowed was a threat to civilization. One result was a growing concern with attempting to better plan family size through various birth control methods. Though other factors also affected the birth control movement, as will be described later in the book, the attempt to limit the growth of "inferiors" and extend the numbers of "superiors" cannot be overlooked.

Some scholars who agreed with Darwin about the importance of sexual selection in evolution came to believe in the existence of degeneracy caused by a defect in an individual's heredity. The resulting degenerate strains—for the defect was believed to be both progressive and inherent—involved nervous illness, physical weakness, and deviant behavior. For some writers, almost any departure from proper, conventional behavior was regarded as a sign of degeneracy. Included in this was any variation of the norms of sexual behavior, and the result was the growth of the concept of the sexual degenerate.

Cesare Lombroso (1836–1909) used Darwinian assumptions to bolster his argument that sexual deviates were on a lower stage of the evolutionary ladder than normal, heterosexual individuals. Lombroso and his allies accepted the belief that animal life had evolved from lower forms, but they went further, arguing that life had progressed sexually from a hermaphroditic or self-fertilizing stage to a higher monosexual stage. Just as life itself had evolved, so had human species, and as humans had progressed from primitive society to higher levels of civilization, they had outgrown robbery, murder, promiscuity, and perversion, or at least the most civilized among humans had done so. Because, however, a child had to repeat the progression of the species to become civilized, it was understandable that those with a defective heredity would become criminals, deviants, or mental defectives. It was also understandable why sexual behavior common among primitive groups or observed among animals would be unacceptable in higher, civilized societies.[5]

Inevitably, another factor in the development of sex research, as hinted at above, was the efforts of many of those labeled as deviate, defective, or criminal to find alternative answers. As is discussed later, one of the major elements in the emergence of the German sexological movement was the effort of homosexuals to escape the stigmatizing labels applied to them by such people as Lombroso.

A principal reason that the theories of Lombroso and others had become so well known was because of public concern about the perceived problems of the growth of the modern city. The nineteenth century saw the enormous growth of cities as individuals moved from an agricultural way of life to an industrialized and commercialized one. Vast numbers of rural inhabitants made their way to the expanding metropolitan areas of the nineteenth century, which led not only to rapid population growth but to wider disparities among the various segments of the population as well as to the demand for greater law enforcement.

Problems that would have been overlooked or gone unreported in a rural community were accentuated by the concentration of population. For example, hundreds, perhaps thousands of books were written on prostitution in the nineteenth century as authorities struggled with the problems of how to deal with it. Similarly, previously isolated homosexuals, who might have felt themselves to be unique, found others like themselves in the crowded cities. Inevitably, as they formed groups, they came to the attention of the police and the courts, who lacked understanding and precedent about how to deal with all sorts of sexual behavior. Much of this behavior had previously been ignored or had been regarded as something that only an isolated individual had done. One of the founders of modern sex research, Richard von Krafft-Ebing, took as his task the exploration of a variety of sexual behaviors to assist the courts, and he did so as one convinced that sexual pathology was a real threat to society. Some of his contemporaries even felt that the sex drive itself held potential danger, and Sigmund Freud, while not quite going this far, emphasized the importance of the ego and superego in controlling such a primeval force.

The beginning of what is called the first wave of feminism also affected the development of sex research. Much of the interpretation of sexual matters had been from the male point of view, and in general, such a view held that women were regarded as inferior males turned inside out. Some of the early sex researchers were women, and as indicated in the following chapters, they sought successfully to challenge many of the male stereotypes and assumptions. The feminists were also active in campaigns for the abolition of prostitution, for access to birth control information, and for other issues, which tended to encourage new kinds of sex research. Many of the early sex researchers were also interested in marriage reform, since they felt women were often prisoners in loveless marriages. The result was an outpouring of marriage manuals and of new findings about human sexuality as well as the organization of groups to bring about reform. Such activity also led to the splintering of sex researchers into the so-called pure scientists whose goal was to publish their findings about human sexuality and the so-called reformers who wanted to lobby to implement some of their findings.

All of these factors tended to come together in the last few decades of the nineteenth century and were carried over with greater impetus into the twentieth. Ideas about sexuality changed, and the traditional Christian norms accepted since St. Augustine came under attack and in many cases were discarded. Sex research itself broadened into an interdisciplinary field that depended on expertise in biology, psychology, and sociology; historical and cross-cultural perspectives; and information provided by professionals in medicine, law, nursing, religion, and others.

I wrote this book to explain the development of sex research, to help a wider audience understand what sexology is all about, and to describe how changes in pubic attitude occurred. It is a subject to which I have devoted much of the past decade, and it is my fervent hope that my analysis will prove helpful. Though I concentrate on the last 150 years, research in the nineteenth century did not begin in a vacuum. It too had a history, which at least must be summarized, and a mind-set, which must be examined. This is where the first chapter begins, after which I turn to the more recent developments and controversies.

1

SEX RESEARCH AND ASSUMPTIONS

FROM THE GREEKS TO THE NINETEENTH CENTURY

*G*enerally, as indicated in the introduction, much of the early information about sex was gathered from animal observations. Interestingly, however, observations about sexual activity among plants was ignored, perhaps because with few exceptions, such as the date palm, it was not very obvious. One result of this early lack of attention to plants has been that in Western culture flowering plants such as the rose have been regarded as symbols of chastity and coupling animals as symbols of concupiscence. There have even been religious groups, such as the Albigensians, who prohibited their adherents from eating any product of sexual union but allowed them to eat vegetables, fruits, and fish, because they believed these forms of life did not result from sexual union.

All of this is a way of emphasizing that Western concepts of sexuality up to the nineteenth century represented a hodgepodge of ideas and contributors. Historically, the most influential premodern author on sexual activity among animals and humans was the fourth-century B.C. Greek philosopher Aristotle, whose *History of Animals*, *Parts of Animals*, and *Generation of Animals* can be regarded as the foundation not only of Western zoology but also of Western sexology. So great was his influence that almost anything attributed to him was believed, with the result that his name was attached to books that had little to do with what he said. For example, the most widely used source of information about sex in the English-speaking world from the seventeenth through the nineteenth centuries was known as *Aristotle's*

Masterpiece, although only fragments of the information and misinformation it served up can be traced to Aristotle.[1]

Sex was a key to Aristotle's classification scheme, because on this basis he assigned animals to one of three groups: those that reproduced (1) by sexual means, (2) by asexual means, and (3) by spontaneous generation. This last category included a number of lower animals such as fleas, mosquitoes, and flies, which he believed were produced out of putrefying substances. Among the shellfish, he tried to differentiate those that reproduced through bud formation from those that came from self-generation. His data were a combination of acute observation and folklore. He held that the hermit crab grew spontaneously out of the soil and slime and found its way into unoccupied shells, shifting to ever larger shells as it grew. Usually, however, he came down on the side of spontaneous generation only when he could find no other explanation, as in the case of the eel. He reported that eels lacked milt or spawn and when dissected no passages for such secretions could be found. Adding to the puzzle was the fact that eels seemed to appear spontaneously after a rainfall, while they never appeared in stagnant pools, even in times of drought. This led him to conclude that eels were derived from the "earth's guts" and grew spontaneously in mud on sustenance from rainwater.

Some of Aristotle's descriptions of reproduction showed remarkable skills of observation. For example, he wrote that in mating of the octopus, two octopuses swam about, intertwining mouths and tentacles until they fit closely together. Then one octopus rested its so-called head against the ground and spread its tentacles; the two sexes then brought their suckers into mutual connection. He added that some asserted that "the male has a kind of penis in one of his tentacles, the one in which are the largest suckers, and they further assert that the organ is tendinous in character, growing attached right up to the middle of the tentacle, and that the latter enables it to enter the nostril or funnel of the female."[2]

This is a more or less accurate description of mating in which the male uses his hectostylus (a tentacle serving as the arm of procreation) to remove a semen cartridge from his own mantle and then places it in the female's. Because Aristotle's classification scheme seemed so all-comprehensive, even though it was a mixture of masterful insights and popular superstition, there was a reluctance to challenge his conclusions. Most damaging was his belief in spontaneous generation, and it was not until this concept was lain to rest that it was possible to understand fully the importance of sexuality. In the seventeenth century, the Italian Francesco Redi demonstrated that maggots were not born from putrescent material but came from fly eggs. Still, though Redi demonstrated the source of maggots, he was unwilling to mount a frontal assault on the theory of spontaneous generation, perhaps because the Christian theology, based on Aristotelian assumptions, supported such a belief to justify the concept of original sin. Put simply, the people of the time could not visualize how Adam and Eve could have lived happily in Paradise before their expulsion if they were plagued by the earthly ills brought on by tapeworms, roundworms, and other parasites. Yet, according to the Bible, God had created all plant and animal forms before he created humans. To answer this apparent contradiction, the theology of the Church held that parasites came about not with Creation but as a consequence of original sin through spontaneous generation. Redi was not about to confront the Church about its belief in spontaneous generation but was content to modify Christian doctrine in some particulars. Others continued such modification until Louis Pasteur's discovery in the nineteenth century that fermentation was caused by the existence of minute organisms led most of the scientific community not only to abandon the Aristotelian notions but in fact for the most part to ignore questions about the origin of life and simply describe the way various forms of life reproduced. In this way, a conflict with religion was avoided, although gradually religious ideas changed.

Long before Pasteur's discovery, some investigators had even begun to speculate that plants might also have a form of sexual reproduction. One of the early explorations of this possibility appears in the writings of the seventeenth-century Englishman Nehemiah Grew. In his *Anatomy of Plants*, published in 1685, he postulated that the flowers on plants were sexual organs, the pistils were the female organs, the stamens were the male ones, and pollen was the male seed. Since the flowers he observed encompassed elements of both sexes, Grew also held that plants were hermaphrodites. This belief, at least for some plants, was challenged by his German contemporary Rudolph Jacob Camerarius in his 1694 essay *Letter on the Sex Life of Plants*. Camerarius had observed that an isolated fruit-bearing mulberry tree produced only empty, sterile seed vessels. He wondered if this might be because it had not been fertilized, and this led him to hypothesize that there might well be male and female plants. He began experimenting with a plant popularly known as dog's mercury (*Mercurialis perennis*), which he had observed in two different variations. Some of the plants had only stamens in their flowers but bore no seeds or fruits, while others, lacking stamens in the flowers, bore fruit. Isolating the fruit-bearing from the pollen-producing plants, he demonstrated that although seed vessels still appeared on the former, they were sterile.

Though this further strengthened the idea of male and female elements in plants, the basic question remained of how they made contact with each other. In the eighteenth century, Josef Gottlieb Kölreuter through his study of the stamens in hibiscus plants found that both wind and insects were

important in the fertilization process and that fertilization was only successful when it took place between plants belonging to the same species.

Continuing and expanding on this research was another German, Christian Konrad Sprengel, who published *The Newly Revealed Mystery of Nature
in the Structure and Fertilization of Flowers* in 1795. Convinced that God had
put everything in nature for a purpose, Sprengel set out to determine what
useful purposes the different parts and properties of flowers had been created
to serve. He concluded that color in flowers was to attract insects and that
certain colors attracted certain insect forms. He then observed that the
flower was adapted not only to the general conditions of its own life but also
to those of the insects it wanted to attract. He argued that even though some
flowers appeared to be hermaphrodites and had both stamen and pistil, a differential development cycle made it impossible for the flower to be fertilized
by its own pollen. Instead, fertilization was by pollen conveyed by insects
from other flowers. This led him to philosophize that nature did not desire a
flower to be fertilized by its own pollen.[3] His whole approach made his findings appear as nature's demonstration of the wonders that God had worked.

If investigators had difficulty in fully understanding the importance of sexual reproduction in lesser forms of life, they had similar problems in trying to
explain how conception took place not only in humans but in most animal
species. Again, the basis for many of the Western cultural notions about conception and sexual reproduction came from the Greeks, and Aristotle was a
major figure. But Aristotle did more than try to explain human reproduction;
his philosophical assumptions about reproduction led him to posit different
roles for men and women. This gave a political connotation to ideas about
sex and reproduction from which we have not yet fully escaped.[4]

Aristotle believed that the male was the major factor in reproduction,
although he did grant that the female supplied the matter for shaping. "If,
then, the male, stands for the effective and active, and the female considered as female, for the passive, it follows that what the female would contribute to the semen of the male would not be semen but material for
semen to work upon."[5] In his defense, Aristotle presented the role of the
female somewhat more favorably than did some of his contemporaries, who
simply regarded the woman as a vessel designed to carry the male seed to
fruition.[6] There was, moreover, a strong minority of Greek writers who went
so far as to give equal credit to the female as the male, although their views
never became as influential as those of Aristotle. A good example is the
writer of the Hippocratic work *On Generation*, who stated that two seeds
were involved in conception, the male contributing semen and the female,
vaginal secretions.[7] The second-century medical writer Galen also held the
two-seed doctrine, claiming both the male and female seeds had coagulative

power and receptive capacity for coagulation but that one was stronger in the male and the other in the female.

Although some later writers also adopted the two-seed doctrine, Aristotle's views triumphed in the West, because they had been adopted and advocated by the eleventh-century Arabic writer Avicenna, who was a key transmitter of Aristotelian ideas to medieval Europe. Avicenna, who also incorporated some of concepts of Galen, held that the male agent was equivalent to the clotting agent of milk and the female to the coagulum.[8] This was the explanation adopted by St. Albertus Magnus in the thirteenth century. Though Albertus used the term *female semen*, he made it clear that it could only be called semen in an equivocal sense and that the male contributed the essential material for generation.[9] In this belief, he was supported by his pupil St. Thomas Aquinas, who held that the female generative power was imperfect compared with the male.[10]

Better understanding of the process depended on a more effective knowledge of the human body. In this, the anatomical work of Andreas Vesalius in the sixteenth century proved important, as did the more specialized studies on female anatomy by Gabriele Falloppio, his younger contemporary. Falloppio described the clitoris as well as the tubes that bear his name. Falloppio was also important in the study of male anatomy; he described the arteria profunda of the penis, which led to a better understanding of how the organ became erect.

Even as knowledge about the anatomy of the reproductive organs of males increased, those of the female, particularly the uterus, remained linked with mystery and superstition. Plato, the teacher of Aristotle, for example, had popularized the belief that an inactive uterus caused female hysteria. He wrote that when the uterus, an indwelling creature desirous of childbearing, "remains barren too long after puberty, it is distressed and sorely disturbed, and straying about in the body and cutting off the passages of the breath, it impedes respiration and brings the sufferers into extreme anguish and provokes all manners of disease besides."[11] Fortunately, most medical writers rejected such notions. Soranus of Ephesus, the second-century writer on gynecology, held that although the uterus was conceived by some as having an independent existence, similar to that of an animal, it did not. Such misunderstandings arose, he explained, because the uterus actually did have some characteristics similar to those of an animal, as it would contract when cooling agents were applied and relax when warming ones were.[12] Galen also opposed the theory of the wandering womb, emphasizing that it was impossible for it to jump over the stomach to touch the diaphragm, but nonetheless, he believed that the womb desired to be pregnant and that the only solution for many female complaints was intercourse and pregnancy. This was because the uterus produced a secretion similar to the male semen and that the retention of this substance in the uterus led to the spoiling and corruption of the blood. This in turn led to a cooling of the body and an irritation of the nerves, and eventually resulted in hysteria.[13]

Although Vesalius and other anatomists of the sixteenth century finally put to rest the concept of the wandering uterus, there was still considerable misunderstanding of female physiology and exactly how reproduction came about. The first effective challenge to the overwhelming importance of the male came from William Harvey, whose *Anatomical Exercitations Concerning the Generation of Living Creatures* was published in Latin in 1651 and translated into English in 1653. Although best known for his demonstration of the circulation of the blood, Harvey spent much of his life studying generation. He observed reproduction and gestation in all types of animals, with particular emphasis on the day-to-day development of the chick embryo. He also dissected the uteri of deer at various stages during mating and pregnancy. This research led him to highlight the significance of the egg in generation. Some hint of the role of the egg was earlier advanced by Hieronymus Fabricus of Aquapendente, a student of Fallopius, but it was Harvey who pushed the oviparous theory to its logical conclusion.

Harvey could find no evidence that the seminal mass of the cock entered into or even touched the hen's eggs during their formation. He did find that for a time a hen could continue producing fertile eggs after all detectable traces of semen had vanished from her body. To Harvey, this seemed to offer solid evidence that the contribution of the cock's semen to generation was indirect and incorporeal; it simply conferred a certain fecundity on the hen and then played no further role in the actual generation of the egg or chick. Once endowed with this fecundity, the hen could, entirely on her own, produce fertile eggs.

To explain how this happened, Harvey compared the process to the spread of disease by contagion as it was then understood. Exposure to a sick individual, he held, could engender within a second individual an internal principle that subsequently reproduced in him or her the same specific disease. For Harvey, fecundity was transmitted to the female in the same way by the male semen. The role of the hen and cock is not to produce a chick but to create a fertile egg that subsequently gives rise to a chick through its own innate powers. He believed he had demonstrated this by his dissections of deer uteri, in which he observed with his own eyes (this was before the development of the microscope) that it was long after the disappearance of semen in the uterus that the first evidence of conception appeared. This led him to formulate his famous dictum that "an egg is the common primoridum of all animals," implying in effect that the role of parents in generation was indirect. They produced a fertile egg or conceptus or seed, and this subsequently produced a new animal or plant through innate vegetative powers.[14]

Other researchers came forth with supporting conclusions, particularly Marcello Malpighi, whose study of chick embryos was published in 1672. At about the same time Regnier de Graaf observed changes taking place in rabbits' ovaries in the first days after fertilization and concluded that similar changes probably took place not only in the rabbit doe but in the human female.

Ovists, however, ran into a temporary roadblock created by the studies of Anton van Leeuwenhoek (1632–1723), who made observations with a simple, early microscope that were not equaled until the more powerful instruments of the nineteenth century allowed others to confirm in detail what he had said. Leeuwenhoek became interested in male semen after Johan Ham, a medical student, consulted him about microscopic creatures that he had observed in the semen of a patient suffering from nocturnal emissions. Ham brought a glass bottle containing the semen to Leeuwenhoek for confirmation. Leeuwenhoek observed the microscopic creatures, noting that they were different from other *animalcula* (microscopic animals) he had observed; all such creatures in the semen had round bodies with tails five or six times as long as their bodies. As he studied them further, he reported they made swimming movements similar to those of an eel and that such movements gradually slowed and ultimately ceased, although the *animalcula* remained clearly recognizable.

To make certain these little animals were not the result of any sickness, Leeuwenhoek and Ham examined the semen of "healthy" males and observed the same kind of creatures. They estimated there must be a thousand or more in the space that could be occupied by a grain of sand. As additional data were gathered under a variety of conditions, it was found that the *animalcula* died with 24 hours if kept in cold temperatures, but they survived several days if kept in warm conditions. Leeuwenhoek called these creatures *spermatozoa*.

Almost immediately after the results of his discoveries were published in the proceedings of the British Royal Society in 1678, all kinds of findings began to be reported. One observer noted seeing a miniature horse in the semen of a horse, and another, a miniature donkey in the semen of a donkey. Still others believed they could distinguish male and female sperm, and at least one researcher reported that he saw male and female sperm copulating and then giving birth to little sperm. The effect of the discovery, in spite of the ludicrous tales, was to reassert—at least in the minds of many—the supremacy of the male in reproduction.[15]

One of the results of the observations of eggs and spermatozoa was to emphasize that sex was universal. This belief was strongly supported in the eighteenth century by the great Swedish biologist Carolus Linnaeus (1707–78), who based his classification scheme of all living things on their means of reproduction. For him, *vivum omne ex ovo*, or everything living comes from the egg—although in the case of plants, the egg was in the form of a seed. So strong were his beliefs that he wrote that there could be no deviation from this general law of nature. As Colin Milnes of the Linnaean Society in England said in 1771: "No being owes its formation to chance, and all are probably produced by a similar mode of generation, depending on the concourse of the two sexes, the minutest insect as well as the elephant, the smallest moss, as well as the most stately and elevated oak."[16]

Linnaeus, however, spoke too soon, and in the early years of the nineteenth century, his classification system came under challenge as it became apparent that there were asexual spore-bearing forms of life. But a more serious problem was the difficulty in identifying what was involved in fertilization. A minority of investigators even went so far as to argue that the so-called spermatozoa seen under the microscope were simply parasites, while another group held they were separate organisms. Moreover, even the advocates of sperm as the key were as yet unable to explain how sperm entered the male body or how they were manufactured. Answering these questions was impossible until the development of the cell theory by Matthias Jakob Schleiden and Theodor Schwann. In 1838, Schleiden concluded from his intense microscopic observations that plants were not a single living thing but an aggregate of individual, self-contained organic molecules. Uncertain of what he had discovered, he discussed it with his friend Schwann, who replied that he had seen similar structures in animal membranes. Their discovery resulted in the foundation of cytology (the study of the structure and function of cells) and the realization of the complexity of life.

One of their early discoveries was that although unicellular organisms normally reproduced by budding and division they also produced sexually by a process described as autogamy, conjugation, and copulation. Schleiden and Schwann speculated that although the budding process resulted in tremendous population growth it also resulted in extreme inbreeding, a difficulty that could be overcome by occasional autogamy. Some forms of life, they found, alternated between sexual and asexual propagation, as did *Plasmodium*, the agent that causes malaria. In humans, *Plasmodia* reproduce by dividing, but in the intestine of the mosquito, they reproduce sexually. Schleiden also emphasized the importance of pollen in the fertilization process of plants. The combination of the observations of Schleiden and Schwann gave new importance to sex in reproduction, a concept that Darwin, as mentioned in the introduction, made all-important.

But what kind of process led to fertilization? The existence of a mammalian egg had been discovered in 1827, but it was not until 1875 that Oscar Hertwig observed the moment of fertilization in the sea urchin, one of the so-called higher animals. Hertwig chose the sea urchin because its eggs are transparent, occur in large number, and develop rapidly. This made it possible for him to observe the coming together of two nuclei in the egg, an event he interpreted as fertilization. Not until four years later did Herman Fol actually observe the spermatozoon penetrate the egg and contribute the second nuclei.

Further light was thrown on the process with discovery of chromosomes in 1873, a finding that Eduard van Beneden used to demonstrate that fertilization resulted in joining together two half sets of chromosomes to form one full set. This phenomenon, it was postulated, also took place in human beings, although the ova of the human female were not observed until the twentieth century.[17]

This rather simplified summary of the developing biological understanding of the nature of sex and reproduction seems to imply an ever-steady trend toward greater and more accurate knowledge. Missing from this account are the numerous dead ends and erroneous leads that were followed and that for a time challenged what is now the perceived wisdom. Moreover, the account itself cheats a little by carrying ideas and findings into the early twentieth century, somewhat beyond the starting place of this book. Issues of debate were still not settled in the nineteenth century, and throughout much of the century, there was considerable debate over the reasons why sexual reproduction occurred in the first place. Interestingly, much of the debate centered on the role of the female, generally to the disadvantage of women. There was a tendency, for example, to associate the ova with asexual cells that began the reproductive process only after being triggered by the sperm, though it was not yet clear how or why. This explanation, minus the asexual aspect, is not far from what we accept today. It resulted, however, in many "authorities" emphasizing that sex was simply the means of procreation used by higher organisms that had, by the division labor, placed the task of procreation into the hands (or, more apposite, ovaries) of a special individual, the female. Some saw this as both sociological and biological proof that the female existed solely to bear and raise offspring. In a sense, this is as accurate as arguing that the only purpose of the male is to fertilize the egg. This latter perspective, however, was not what many of the nineteenth-century interpreters argued. Rather, they held—following traditional societal attitudes toward women—that the subordinate status of women was a law of nature; to argue otherwise was to threaten the social and biological fabric of the species.[18]

This example of arguing for a biological justification for the subordination of women emphasizes just how much attitudes and beliefs about human sex and gender were based on much more than a physical explanation. Rather, such attitudes depended heavily on psychological, sociological, cultural, and historical factors. Inevitably, each new finding about sexuality in nature took on a social and political context when it was applied to humans. Generally, data were interpreted to strengthen traditional Christian ideas about sexuality, which as indicated in the introduction, was deeply influenced by the sex-negative ideas of St. Augustine. As the assumptions about sex and reproduction on which religious authority had based its moral teachings were undermined, commentators on human sexuality seized on science to give a new foundation for traditional morality. If anything, this tended to strengthen negative attitudes toward sex, because those serving up advice could argue that they gave more accurate guidance than their predecessors, since they included the new findings of science. Moreover, as secular knowledge grew in the eighteenth and nineteenth centuries, there was a corresponding growth of faith in science itself.

Although there had been some challenges within the Christian community to Augustinian ideas about sex and although Protestants, in general, and Puritans, in particular, looked on sex within marriage as a source of pleasure, all segments of the Christian community continued to believe that marriage was a prerequisite for sexual intercourse.[19] It was in this mind-set that discussion about human sexuality took place. Increasingly, however, sexual discussion was dominated by the medical community, in part because most of the discoveries recounted so far in this chapter were made by physicians. Engaging in research, however, was limited to a few individuals, whereas giving advice to patients about sexual matters was something that every physician was expected to do. The practicing physician, moreover, was in a difficult situation when patients came to him (and all physicians up to the nineteenth century were males) for treatment of sexually related problems. Even with the observations and findings of a Harvey or a Leeuwenhoek, it was difficult to explain to a patient complaining about an inability to conceive why this might be the case. The state of knowledge was simply not up to this problem. Instead, the perceived necessity to offer a diagnosis and make a prognosis encouraged the growth of what might be called *medical systems*. In simple terms, these systems were organized medical theories that were all-inclusive explanations of why illness occurred and what a physician could do to treat it. System makers had existed since medicine emerged as a profession, but the systems of the seventeenth and eighteenth centuries became ever more elaborate and continued to have adherents well into the nineteenth century. Moreover, many of the system makers, influenced by their own philosophical and religious beliefs, held that sexual activity in itself was potentially dangerous. As a result, illness and pathology replaced sin in the discussion of sexual activities.

One of the most influential system makers was the great eighteenth-century Dutch clinician Hermann Boerhaave (1668–1738), who in his *Institutiones Medicae* wrote that the rash expenditure of semen brought "on a lassitude, a feebleness, a weakening of motion, fits, wasting, dryness, fevers, aching of the cerebral membranes, obscuring of the senses, and above all the eyes, a decay of the spinal chord, a fatuity, and other like evils."[20]

Boerhaave's observations on the dangers of sexual activity fit well with some of the new medical theories such as vitalism, a theory based on the works of Georg Ernst Stahl (1660–1734) and others. Vitalism emphasized a unity of body and soul, symbolized by the *anima*, which protected the body from the deterioration to which it tended. When the movements representing normal life were altered by the body or its organs, disease supervened. Disease was thus little more than the tendency of the *anima* (or of nature) to reestablish the normal order of tonic movements as quickly and efficiently as possible. A contemporary of Stahl, Friedrich Hoffman (1660–1742) held that the living organism was composed of fibers having a special characteristic *tonus*, namely the ability to contract and dilate. This process was regulated by the nervous system, headquartered in the brain. When the *tonus* was normal, the body was healthy, but every modification of the *tonus* brought with it a disturbance of health. For him, individuals who practiced masturbation gradually damaged their memory because of the strain on the nervous system.[21]

Several other theorists, including John Brown (1735–88) and Théophile de Bordeu (1772–76), built on this foundation. Brown's medical philosophy was based on his own experience with gout, a disease from which he had suffered most of his life. In his preface to his *Elements of Medicine*, he wrote that after failing to cure himself with traditional treatment, he sought other remedies, and eventually arrived at the conclusion that "debility was the cause of his disorders and that the remedy was to be sought in strengthening measures." To banish his gout, he had to strengthen himself, avoid debilitating foods, and treat himself with wine and opium. Whether his gout was cured or not remains doubtful, but from his experience, he built a medical philosophy known as Brunonianism. Basic to his teachings was the notion of excitability, defined as the essential distinction between the living and the dead. The seat of excitability was in the nervous system, and all bodily states were explained by the relationship between debility and excitement. Too little stimulation was bad, but excessive stimulation could be worse because it could lead to debility by exhausting the excitability. Excitability was compared to fire: If there was not enough air (insufficient excitement), the fire would smolder and go out; but under a forced draft (too much excitement), the fire would burn excessively, become exhausted, and also go out. From these assumptions, he concluded there were two kinds of diseases: those arising from excessive excitement (sthenic) and those from deficient excitement (asthenic). Too much stimulation carried an asthenic ailment into a sthenic one. For example, mutual contact of the sexes as in kissing or even being in each other's presence, gave an impetuosity to the nerves. This nervous condition could be relieved by sexual intercourse, but in giving temporary relief, it could also result in the release of far too much turbulent energy; and if carried to excess, this too caused difficulty. Bordeu ended up with the same conclusions as Brown. Bordeu, however, maintained that the lymphatic glands as well as the nervous system had vital activity, and secretions—including semen—drained the vital essences that resided in every part of the body.[22]

Particularly influential in emphasizing the dangers of sexual activities were the writings of S. A. D. Tissot (1728–87). Tissot believed that physical bodies suffered a continual waste, and unless this loss was periodically restored, death would result. Though much of the waste lost through natural processes such as urination and bowel movements could be restored by ingesting food, even with an adequate diet, the body could not fully restore the waste that resulted from blood loss, diarrhea, and sexual activity. Particularly dangerous was the loss of seminal emission or vaginal discharges in the female.

Semen, he explained, was extremely important to the well-being of the male. As evidence of this, he pointed out that its appearance coincided with the growth of the beard and the thickening of the muscles, something that did not happen if a male was castrated. Because semen was so important to the development of the male, it seemed obvious to him that any excessive loss of it would weaken the male. Although he agreed that some loss was necessary for reproduction to replenish the human race, the real danger was excessive intercourse. Even more dangerous was the loss of semen brought about by nonprocreative sex or through unnatural means—practices that he called *onanism*. Though onanism is sometimes simply equated with masturbation, Tissot used the term in a broader sense to include all nonprocreative sex from homosexuality to masturbation to the use of the wrong orifice to the use of a contraceptive. He listed a number of dangers that could result from the excessive waste of semen, including (1) cloudiness of ideas and sometimes even madness; (2) a decay of bodily powers, resulting in coughs, fevers, and consumption (tuberculosis); (3) acute pains in the head, rheumatic pains, and an aching numbness; (4) pimples of the face, suppurating blisters on the nose, breast, and thighs, and painful itching; (5) eventual weakness of the power of generation as indicated by impotence, premature ejaculation, gonorrhea, priapism, and tumors in the bladders; and (6) disordering of the intestines, resulting in constipation, hemorrhoids, and so forth.

Onanism affected females in the same ways as males, but not surprisingly, the effects were ever so much more severe. In addition to suffering from the same problems as did men, women also were likely to have severe cramps, ulceration of the cervix, uterine tremors, incurable jaundice, and hysterical fits. Moreover, masturbation in the female often resulted in mutual clitoral manipulation, and this resulted in women loving other women with as much fondness and jealousy as they did men, a practice that lowered them from the high status of womanhood to the level of the most lascivious, vicious brutes.

Even worse than masturbation in adults was the existence of the practice in youth who had not yet attained puberty. Here, it tended to destroy the mental faculties by putting too great a strain on the nervous system.[23]

It is easy for late-twentieth-century commentators to make fun of Tissot and others who wrote in a similar vein. One of the reasons earlier writers attributed so many dangers to sexual activity is that the sequelae of sexually transmitted diseases, particularly the third stage of syphilis, were accredited to sexual acts in general. It was not until the nineteenth century that all the stages of syphilis were discovered and explained, and there could be a more rational explanation of what Tissot attributed to excessive sexual activity. Actually, there was probably even some kind of rough correlation between what Tissot attributed to sexual activity and the sequelae of syphilis. Moreover, those most likely to get syphilis were those who had a variety of sexual partners and so, in Tissot's terms, were visualized as losing abnormal amounts of their body energy through waste. Even after the course of syphilis was theoretically worked out by Philip Ricord (1800–1889), there was still confusion, because it was not until the isolation of the spirochete in 1905 that Ricord's hypotheses could be proven.[24]

Tissot's ideas, in one form or another, proved extremely influential through much of the nineteenth and even into the twentieth century. For a time it seemed that each writer tried to outdo the other in portraying the dangers of sex. In the United States, Tissot's ideas were popularized by Benjamin Rush (1745–1813), probably the most significant American physician at the end of the eighteenth century; he is remembered today as a signer of the Declaration of Independence. Rush, who had studied in Edinburgh, returned to America to introduce a variation of John Brown's medical ideas. All disease could be reduced to one basic causal model: either the diminution or increase of nervous energy. Because sexual factors were a major cause of excitement, Rush taught that careless indulgence in sex would lead to everything from impotence to epilepsy to death.

Rush, however, did not entirely discount the positive aspects of sex, stressing that abnormal restraint in sexual matters was also dangerous, because it might produce "tremors, a flushing of the face, sighing, nocturnal pollution, hysteria, hypochondriasis, and in women, the furor uterinus."[25]

One did not have to be a physician to trumpet the dangers of sex. Sylvester Graham (1794–1851), a health reformer perhaps best remembered for his advocacy of unbolted wheat, or graham flour, and commemorated today in the graham cracker, taught that his contemporaries suffered from an increasing incidence of debility, skin and lung diseases, headaches, nervousness, and weakness of the brain, much of which resulted from sexual excess. This was because the human body was a composite of animal and organic life, both of which were controlled by a network of nerves.[26] Because reproduction, almost alone of the body's functions, involved both an animal and organic component, sexual activity was seen as putting a unique strain on the body. Moreover, it was not simply the sex act itself that posed dangers and eventually resulted in insanity but lascivious thoughts and the contemplation of sexual activities also could be harmful. Individuals, therefore, had to be ever alert to the dangers of sexual desire. Sexual feelings could even be aroused by eating dishes that were highly seasoned and "overrich" as well as by eating too much meat.[27] One result was the growth of fears about the dangers of untrammeled sexuality.[28]

In fact, the dangers of sex became part of the arsenal of all the faddist reformers of the last part of the nineteenth century, including John Harvey Kellogg, whose Battle Creek Sanitarium introduced new breakfast foods to the world. He gave a long list of signs that distinguished a person who was masturbating. Listed among the signs was every form of conduct found in teenagers: fickleness, bashfulness, unnatural boldness, lassitude, and capricious appetite. Masturbation also, according to Kellogg, resulted in acne, paleness, shifty eyes, the use of tobacco and profanity, bed wetting, and fingernail biting. As masturbation continued, the ultimate dangers were terrible to behold, because the nervous shock resulting from the exercise of the sexual organs was the most profound to which the nervous system was subject. Even those who engaged in sex for procreative purposes had to limit their activities or else insanity would result.[29]

But why, if sexual activity was so harmful, had not generations of individuals become insane in the past? Those concerned with chronicling the great dangers of sex had an answer. It was the growing complexities of modern civilization and the higher evolutionary development of humanity that had made sexual activity so much more dangerous now than it had ever been in the past. In fact, many nineteenth-century physicians taught that the whole of modern society suffered from neurasthenia, a deficiency of nerves. This was a new disease first discovered by the American George M. Beard (1838–83), and it was symptomatic of the growing complexities of modern civilization as well as the higher evolutionary development of humanity. It was particularly widespread and dangerous to the educated, brainy workers in society who represented a higher stage on the evolutionary scale than the less-advanced social classes. The theory had wide appeal, because it seized on the new scientific concepts of evolution to justify class and race consciousness. According to Beard, the chief cause of nervous exhaustion was sexual intercourse, and it was important for individuals to regulate and guard against any unnecessary (nonprocreative) sexual activity.[30] For a time, Beard's ideas were extremely influential in both America and Europe.

Predictably, the result of these theories was the reinforcement by science of the traditional Western hostility to sex, which had been such an influential factor in Christianity. Writers of most popular sex manuals, whether physicians, clergy, teachers, or reformers, seized on new scientific evidence to mount a crusade against sex, hitting with sledgehammer force the horrible dangers of "unnatural sex" (in other words, all nonprocreative sex, including the use of any form of contraception). The word *masturbation*, not *onanism*, often became a codeword for all sorts of unnatural sexual activity, from the use of contraceptives to homosexuality.[31] Unnatural sex was regarded as ten times worse than simple illicit intercourse between an unmarried man and woman, because at least children might result from heterosexual fornication.[32] It not only was described as criminal and pernicious[33] but also eventually came to be described consistently as a disease, and sometimes even a contagious one.[34]

Unnatural sex was worse than almost any other disease for it constantly drained off the vital body fluids and gradually took away life itself.[35] Every loss of semen was regarded as equivalent to the loss of 4 ounces of blood,[36] and although the body could eventually replace this loss, it took time to recuperate. The only thing worse than seminal loss in the male was masturbation and unnatural sex in the female.[37] Though not all health professionals subscribed to such exaggerated ideas, many felt it was much less dangerous to exaggerate than to leave the patient and the public in ignorance.[38] Others were fearful of the ostracism that might result if they said anything contrary to the dominant view about sex and so simply remained silent. Silence, however, did not always save them, because they were sometimes attacked by other physicians, who alleged that their silence led to wasted lives.[39] Often the attack was led by respectable and honored physicians. Abraham Jacobi (1830–1919), for example, the founder of pediatrics in the United States,

was simply reflecting some of the best medical opinion when he blamed
infantile paralysis and infantile rheumatism on masturbatory practices,
which had weakened the body's resistance.[40] Masturbation also encouraged
the formation of "morbid attachments" for persons of the same sex[41] and
led to pederasty and child abuse.[42]

Some commentators attempted to define masturbation as including
coitus interruptus, oral-genital contact, pederasty, bestiality, mutual masturbation, coitus interfemora, and self-pollution.[43] Others simply used the
term without defining it. One of the dangers posed by the masturbator,
according to G. Stanley Hall (1846–1924), a founder of American psychology and an early writer on adolescence, was that such a person seduced others into experimenting with masturbation, and once established, masturbation was the major cause of "one or more of the morbid forms of sex
perversion."[44]

In the new conservatism on sexual issues, any unusual position in intercourse or any action besides that designed to impregnate the female could
be labeled as masturbation and righteously condemned. One physician
reported the case history of a young woman whose husband "had the fatal
habit of applying the tongue and lips to his wife's genitals to provoke in her
a venereal orgasm." This had resulted in "gastralgia" and "constant exhaustion" in the woman. So concerned was the physician treating her that he felt
obliged to warn her that her very life would be in danger unless her husband
ceased this foul practice. Moreover, if her husband continued such practices,
he would probably get cancer of the tongue. It was not just the dangers of
cunnilingus with which he was concerned but the fact that such practices
resulted in orgasm and this could only lead to even greater afflictions.[45]
In case his warnings were not enough to stop them from engaging in such
practice, he added that any child born to them would have "perverted
instincts."[46]

Some writers were bothered about the possibility not only that women
could have orgasms but also that women might find sex pleasurable. In their
minds, sex, though clearly essential to procreation, should only be engaged
in for this purpose. Those able to otherwise preserve marital continence
would, it was claimed, be rewarded "by a sound constitution, an approving
mind, and the applause of the deserving and considerate."[47] Even those
unable to be totally continent but who still managed to engage only moder-
ately in sexual activity would elevate the noblest faculties of their minds.[48]
In no case was it ever possible for chastity to be excessive.[49] Married couples
interested in sex only for procreation were advised to limit their attempts at
coition to "one indulgence to each lunar month." This was all the "best
health of the parties can require," and it was sufficient to ensure pregnancy

if that was God's wish.[50] It was held by many to be a law of God that no animal, let alone a human, should "use the reproductive powers and organs for any other purpose than simple procreation."[51] The more individuals were able to confine their sexual indulgence to the generation and development of offspring, the nearer they came to fulfilling the "supreme law" of being and the nearer they came to being true Christians.[52]

In the United States, medical practitioners, most of whom thought of themselves as Christians, saw as part of their duty the education of the public to realize that God had designed "intercourse of the sexes" for the production of offspring and no other reason.[53] Though it could not be denied that the male received pleasure in doing his duty to beget children, couples were warned about seeking or prolonging pleasures. This was because, among other things, there were "undeniable instances where children begotten in the moment of intoxication remained stupid and idiots during their whole life."[54] Logic was also marshaled to demonstrate that conception during time of ill-humor, bodily indisposition, or too much nervous strain would also affect the resulting offspring.[55]

Women especially had to be careful not to enjoy sex, because they were maternal, rather than sexual, creatures. Only the diseased female had an "excessive animal passion."[56] Advice, however, was of necessity somewhat contradictory, because some writers, following older medical texts, believed that a woman would not get pregnant unless she had an orgasm during intercourse. Generally, however, this idea was ignored or downplayed in most nineteenth-century medical texts, although fortunately, it remained a popular belief. Certainly any woman who participated in the enjoyment of sex while she was pregnant endangered her role as a mother and could bring about a miscarriage.[57] In fact, any couple engaging in intercourse during pregnancy reduced the constitutional vigor of the fetus and predisposed it to debilitating diseases.[58] Moreover, women were particularly susceptible to such debilitating diseases. A standard textbook on gynecology published in 1888 had a concluding chapter titled "Gynecology as Related to Insanity in Women." The author wrote:

I take it for granted that all will agree that insanity is often caused by diseases of the procreative organs, and on the other hand, that mental derangement frequently disturbs the functions of other organs of the body, and modifies diseased action in them. Either may be primary and causative, or secondary and resultant. In the literature of the past, we find the gynecologist pushing his claims so far as to lead a junior in medicine to believe that if the sexual organs of the women were preserved in health, insanity would seldom occur among them.[59]

Women, in fact, were believed by some to dislike sexual intercourse. The well-known English physician William Acton taught that women had no desire for sex at all and only engaged in sex with their husbands to have children or to keep their husbands from straying.[60] He admitted that some low and vulgar women might not only enjoy sex but seek out sexual partners, but this was something no proper wife and mother would ever do. For many of the writers on sex, a female who enjoyed sex was either a sick creature or somehow an imperfect woman. Interestingly, many women writers seemed to agree, perhaps because they saw that one of the few ways they could liberate their sex from the continual burdens of pregnancy and lactation was to deemphasize female sexuality. E. B. Duffy, a married woman, wrote that real women should regard all adult men not as lovers but as stepsons for whom they had a mother's "tenderness."[61]

Almost all advice writers recommended that intercourse between a married couple cease when the women entered menopause, because this was nature's way of putting a "cessation" to the female's sexual functions.[62] Those men and women who persisted in engaging in sexual intercourse after this time would find themselves exhausted and dangerously prone to illness.[63] Thus in "well regulated lives," the sexual passions become less and less imperious, diminishing gradually until—at an average age of forty-five for woman and fifty-five for men—they were but rarely awakened and seldom satisfied.[64]

Perhaps the best summation of all these negative ideas about sex was by Elizabeth Osgood Goodrich Willard, who coined the term *sexology*. She held that the sexual orgasm was more debilitating to the system than a hard day's work. She regarded sex as more or less a loathsome thing and was unhappy that people were generated under a system that was so easily abused. She argued that humankind must stop the waste of energy

> through the sexual organs, if we would have health and strength of body. Just as sure as that the excessive abuse of the sexual organs destroy their power and use, producing inflammation, disease, and corruption, just so sure is it that a less amount of abuse in the same relative proportion, injures the parental function of the organs, and impairs the health and strength of the whole system. Abnormal action is abuse.[65]

With such attitudes toward the dangers of sexual abuse, there was a determined effort to find preventatives for those unable to control their own sexuality. Some physicians perforated the foreskin of the penis and inserted a ring or cut the foreskin with jagged scissors to lessen possibilities of male pleasure. Others applied ointments that would make the genitals of both sexes tender to the touch, and still others applied hot irons to the inner thighs. In some cases, clitoridectomies were performed on females, and in a few cases, amputation of the penis was attempted to prevent masturbation in males. Castration was occasionally recommended. Most popular, however, were mechanical devices that the interested could purchase; large numbers of these are listed under the category of medical appliances in the U.S. Patent Office records, including various kinds of devices with metal teeth designed to prevent erection in the male and various kinds of guards to be worn around the genitalia of the female. There were special devices for patients in mental institutions, including a unique pair of gloves that prevented the patient from touching his or her genitals. There was even a device to prevent bedcovers from coming into contact with sensitive areas.[66]

The system makers were attempting to use scientific knowledge to preserve the status quo of traditional attitudes toward not only sexual issues but sex or gender roles as well. Women were the weaker sex, set apart by biology to bear children as evidenced by their menses. American physicians, in particular, emphasized that menstruation by itself made women special creatures, something the medical men could state with impunity because so little was know about it. Though there was a growing belief that ovulation and menstruation were connected,[67] there was uncertainty as to whether the onset of the menses marked ovulation or whether it came afterward. Even as late as the 1890s, when the first experimental work leading to the understanding of human hormones was taking place, American physicians were still discussing the question of whether the ovaries triggered menstruation, whether the uterus was an independent organ and performed the menstrual function without external aid, or whether the fallopian tubes were responsible for the monthly flow.[68]

In 1861, E. F. W. Pflüger (1829–1910) demonstrated that menstruation did not take place in women whose ovaries had been removed, a finding that reinforced the ovarian theory but did not end the debate over the physiology of menstruation, as Pflüger hypothesized that there was a mechanical stimulus of the nerves by the growing follicle that was responsible for congestion and menstrual bleeding. This led him to argue that menstruation and ovulation occurred simultaneously.[69] It was not until the twentieth century, when the physiology of hormones was better known, that the timing of ovulation and the stimuli involved were fully understood (see chapter 5). In the meantime, Pflüger's theory that nervous stimulation triggered menstruation was widely accepted.

The theory itself was not unreasonable, based on observations of the influence of stress and tension on menstrual irregularity. But many were not content simply to see this as a possible explanation of the menses. Instead, they used it to erect new theories about the nature and purpose of the female. Leaders in the theorizing were the men who seemed to be most threatened by the changing relationships between the sexes that were taking place in the nineteenth century. One indicator of this change was the successful demand by women to enter into college and universities, something that had been forbidden to them before. Similar demands and changes were occurring in other parts of the Western world, and this new generation of educated women sought to enter the all-male professions. Medical schools found themselves under attack for failure to admit women, and a few women such as Elizabeth Blackwell managed to receive medical training, something that the rank and file of the medical profession opposed, some of them with great hostility. It was in this setting that Edward H. Clarke (1820–77), a physician at Harvard Medical College opposed to the intrusion of women into his profession, set out to demonstrate why menstrual disabilities should make women ineligible for higher education.

In 1873, Clarke wrote that although women undoubtedly had the right to do anything of which they were physically capable, the physiology of being female put natural limits on their opportunities, including going to college. He explained this by stating that while the male developed steadily and gradually from birth to manhood, the female, at puberty, had a sudden and unique period of growth when the development of the reproductive system took place.[70] This, he said (following Pflüger), involved special demands on the female nervous system, because it had to work not only on developing the brain, as it did in males, but also on developing the reproductive organs. This made the female different from the male, whose nervous system could concentrate solely on intellectual development. He then argued that because the nervous system could not do "two things well at the same time" it was important for the female between the ages of twelve and twenty to concentrate most of her energy on developing her reproductive system. This implied that females should not devote much time to higher education, because if they did so, the signals from the developing organs of reproduction would be ignored in favor of those coming from the overactive brain.

There was more to his argument than this brief summary, but Clarke eventually concluded that women who concentrated on education rather than the development of their reproductive system underwent mental changes. Not possessing the physical attributes of men, they tended to lose their maternal instincts and become coarse and forceful, with the result that a new class of sexless humans analogous to eunuchs was appearing among women. To solve this alarming problem, he recommended strict separation of the sexes during education, particularly after elementary school. He urged that female schools provide periodic rest times for students during their menstrual periods. The young women would also have shorter study hours, because they were by nature weak and less able to cope.

> A girl cannot spend more than four, or, in occasional instances, five hours of force daily upon her studies, and leave sufficient margin for the general physical growth that she must make.... If she puts as much force into her brain education as a boy, the brain or the special apparatus [i.e., the reproductive system] will suffer.[71]

Inevitably, he "reluctantly" concluded that women could not be admitted to Harvard Medical School or any intellectual competition with males not only for their own protection but for the preservation of the human race.

Although it seems obvious to today's reader that Clarke was using his own biases and prejudices to construct a theory of female inferiority, it was not so evident to his contemporaries who subscribed to the dispassionate truth of science.[72] Even those physicians who challenged the association of menstruation with development of the nervous system, simply replaced it with other theories that emphasized female inferiority and instability. This was the case of John Goodman's theory of the "menstrual wave,"[73] which was used by George J. Englemann (1847–1903) in his presidential address before the American Gynecological Society in 1900 to urge that schools for girls should heed the "instability and susceptibility of the girl during the functional waves which permeate her entire being" by providing rest during the menstrual periods.[74]

Fearful that Clarke and others might be correct, educational institutions that had already admitted women, such as the University of Wisconsin, tried to protect the femininity of its women students in 1877 by officially noting as Wisconsin did that "every physiologist is well aware that at stated times, nature makes a great demand upon the energies of early womanhood and that at these times great caution must be exercised lest injury be done." Although education for women was to be desired, "it is better that the future matrons of the state should be without a University training than that it should be produced at the fearful expense of ruined health; better that the future mothers of the state should be robust, hearty, healthy women, than that, by over study, they entail upon their descendants the germs of disease."[75] Perhaps it was because of the fears aroused in young women as a result of such ideas that a number of women were pioneers in American sex research. It was only by undermining the standard assumptions about sex that women could argue that they were not victims of a natural disability.

Because modern sex research began in this setting, it is thus not surprising that one of the problems it had to overcome was this mind-set. Somehow, the would-be sex researcher had to find some justification for challenging the prevailing belief pattern without threatening his or her professional image. In a sense, much of the sexual advice literature of the nineteenth century might be regarded as a reaction to the challenges of industrialization and urbanization. It was a kind of a mythologizing of a past that had never existed combined with the hope and expectation that if the tendency to sexual immorality so ever present was kept in check, the future would be better. Though it claimed science as the source of its conclusions, it was essentially a new kind of moralizing in which sex had replaced the devil and become the symbol of evil. As indicated in the introduction, the growing urbanization posed new problems and made the observers more conscious of the sins of the flesh than was evidenced in rural villages and towns of earlier years. The growing concentration of population permitted what might have been an isolated case in the village to become the conduct of numerous individuals in the city. Certainly, it increased the numbers of prostitutes, but it also made it easier for homosexual individuals to find each other and to realize that they were not alone in the world.

Industrialization also led to vast displacement of people who lacked the family support and group censorship that had existed in a more encapsulated universe. The growing cities were not at first centers for family life; instead they attracted new immigrants, primarily male, who were at the height of their sexual drives and who were much less observant of the intellectualized prohibitions against sex than females. Moreover, it was a time of a double standard, and even St. Augustine, that most antisexual person of the early Church fathers, had argued that prostitution was a necessary evil; just as there had to be sewers to carry of the filth of the city so there had to be prostitutes to carry away the evils of sex, thereby protecting the good women and preventing those who turned to the prostitutes from turning to each other. These problems were not new; they had existed in the Rome of St. Augustine's time, had appeared in fifteenth-century Florence and Venice, and had existed in eighteenth-century London and Paris, but they became much more widespread and universal in the nineteenth century. Courts of law were called on to deal with various kinds of sex behaviors that they previously had ignored and about which they knew little. Judges and political authorities wanted and needed information, and they turned to the medical profession for assistance. Modern sex research then began with what the people of the time called "perversions" and "degeneracy," which, as emphasized, could be interpreted as almost all sexual behavior. It was believed that if science could throw light on these actions, they could be eliminated. There was a consciousness that modern society was somehow different from the past and that new solutions might be needed.

One of the earliest examples in the sex field of the attempt to use scientific data was in the area of prostitution. At least in some countries, government administrators and legislative bodies, motivated by a new sense of responsibility for the well-being of the citizenry, not only in protecting the women in the community but in attempting to limit the spread of sexually transmitted diseases, had rationalized the control of prostitution to a remarkable degree. In Paris, they established municipal guidelines, registration systems, and venereal disease examinations, all under the direction of the *police des moeurs.* Physicians became involved in the program rather unobtrusively at first, from an administrative decision in 1802 to provide facilities to examine public prostitutes for venereal disease; shortly after this, such examinations were made mandatory. What this Paris legislation did was link the old system of toleration of the necessary evil with an organized method of disease control, one widely copied by other European governmental jurisdictions.[76]

It was not enough, however, simply to establish new regulations or institutions; there was also concern with how they were working. The physician was the natural person to give answers, since both the growing concern over public health issues and the long-established tradition of medical writing on sexuality gave the physician the status of expert in a society that was often ambivalent about open discussions of sexual matters. Such an ambivalence, however, did not mean that the public was uninterested in gaining information about sex, which seemed to undermine public morality.[77]

Prostitution especially needed objective study, and it was the investigations of the physician Jean Baptiste Parent-Duchâtelet (1790–1836) into the life of prostitutes in Paris that might be regarded as the pioneer endeavor of modern social science research into sexual topics. Parent-Duchâtelet was the leading public hygienist of the first part of the nineteenth century, a member of the Paris Conseil Général de Salubrité and a founding editor and contributor to the *Annales d'Hygiène Publique et de Médecine Légale.* Though primarily an expert on sewage and waste disposal, Parent-Duchâtelet conducted a study of prostitution that, when posthumously published, made his reputation. He initiated his study in an effort to go beyond traditional medicine and, in his words, bring the methods of science to the study of people.[78]

To this end, he gathered information about the 3,558 registered prostitutes of Paris (he excluded the clandestine, part-time, and amateur prostitutes). He found that the inscribed prostitute was in her late teens or early twenties, illiterate, poor, probably illegitimate or from a broken family, and likely to have regarded herself as a prostitute for a relatively brief period. She was also willing to leave prostitution if something better turned up, a finding that has more or less consistently appeared in research into prostitution in Western culture for more than 150 years. Today, this type of research would probably be undertaken by a sociologist, not a physician, but sociology and the social sciences had not yet emerged as independent or specialized fields of study, and the physician was in a position to explore such topics, because he or she was the expert on sexual matters.[79] Sex, after all, was a part of anatomy and physiology, and though physicians had always mentioned the subject in their writings, they had shared expertise in the past with priests and canon lawyers. In the new age of science, however, it was the physician as scientist that the public turned to for data.

Many physicians responded, such as the American William Sanger later did, because they were concerned with public health issues, particularly the problem of sexually transmitted diseases. Although it was perhaps somewhat stigmatizing for the male physician who studied prostitutes, it was not a subject that would cause anyone to doubt his masculinity. This was not the case with homosexuality, yet it was homosexuality more than prostitution that consolidated the position of the medical expert. The major development in this field, however, took place in Germany, and it was as a result of the aftermath of these studies that modern sexology can be said to have begun.

Members of the German-speaking medical community had as early as the eighteenth century demonstrated interest in the link between physical disorders and sexual problems. This interest was on a somewhat different basis from Tissot's, because the German physicians were more interested in classifying than Tissot was.[80] This meant that a major problem these early investigators faced was defining the kind of sexual behavior with which they were concerned. This was particularly important for those engaged in forensic medicine, because they were called on to advise the courts. Johann Ludwig Casper (1796–1864), whose writings on medical forensics were widely read and translated, described the difficulties he encountered with the term *paederastia*, often used in medical works and in the courts to describe same-sex activities. Casper held that the such usage was erroneous, because same-sex desires and practices existed between adult individuals and not just between adults and children. Similarly, he felt the word *sodomy*, often used as a catchall term in the courts and in medical writings, should be limited to bestiality.[81] Although he continued to hold that more precision was needed for physicians to make accurate diagnoses, he failed to include any precise diagnoses in the various editions of his handbook on forensic medicine, probably because he himself felt inadequate to do so.[82]

Physician writers themselves were handicapped by their lack of knowledge about the variations in sexual behavior, and most of them who wrote about it were what might be called antiquarians of sexuality. They mainly assembled a number of different cases about various sexual phenomena, primarily from history and some from their own practice. To explain such behavior, many in the middle of the nineteenth century turned to the philosophical views of Arthur Schopenhauer (1788–1860), who held a deterministic view of human action and motivation. Whatever a person did or does was necessarily an expression of his or her inner will and thus was fixed and unalterable. For Schopenhauer, a wrongful sexual act was an act in which a person by expressing his or her own will denied or inhibited the will of another. Justice consisted of refraining from such injurious acts. What these early German sexologists did not do was any of the kind of analysis associated with the best medical practice of the time. Still, even the Schopenhauer approach established a receptivity to research in the area of sexuality.

The final impetus for more intense medical investigations came from individuals in the German homosexual community. Though homosexual activity and desire have, I believe, existed in every society and time period, the concept of homosexuality is a nineteenth-century one. The acceptance of the notion implies that homosexuals are different from others, that homosexuality is "possessed" by some people and not others,[83] a notion that fit in with the ideas of Schopenhauer.

Many of these men and women who were attracted to members of the same sex ended up in law courts for one reason or another, while others made their appearance in the physician's office.

Both the state and physicians wanted answers on how to deal with such individuals. In this respect, the German-speaking areas were probably no different from other areas of Europe or America. However, Germany was in the process of unification under Prussian hegemony, and Prussian laws on same-sex activity were different from those in other areas of Germany. Much of western Germany had been influenced by the changes in the civil law wrought by the Napoleonic code, which said nothing about same-sex relationships per se but concentrated on such things as age of consent and use of force to define illegal conduct.[84] The Prussian code, on the other hand, made sodomy a crime, and this was interpreted to include same-sex activities. This fear of a possible change in legal status through the incorporation of their home areas into Prussia encouraged some homosexuals to be very vocal. The major figure in focusing this concern was Karl Heinrich Ulrichs (1825–95), perhaps the modern world's first "self proclaimed homosexual."[85] The result of his efforts and its implications for sex research are the subjects of the next chapter.

2

HOMOSEXUALITY AND OTHER FACTORS IN BRINGING ABOUT SEX RESEARCH

KARL HEINRICH ULRICHS

Karl Heinrich Ulrichs, the seminal figure in bringing about new thinking and research on same-sex love, was born August 28, 1825, in Westerfeld, East Friesland, in Hannover. He came from a long line of Lutheran pastors, and both his mother's brother and father served as such, and later one of his sisters married a pastor. His father was an architect in the service of the royal Hannoverian government and died as a result of an accident on a construction site when Ulrichs was ten. Ulrichs attended the University of Göttingen for two years, after which he transferred to the University of Berlin. In 1847, he took his examination to become a civil servant in Hannover and began to advance up the ranks until he rather abruptly resigned in 1854, probably over a homosexual incident.

As he became more conscious of his homosexuality, he, like many homosexuals both before and after him, began an investigation of same-sex attraction in an attempt to find answers. He first tried to explain it in terms of animal magnetism, basing his ideas on the earlier, and generally discarded, theories of Friedrich Anton Mesmer (1733–1815). The more he studied, the more important research into the topic became for him. In February 1862, he wrote an autobiographical statement that he deposited under seal in the Freies Deutsches Hochstift für Wissenschaften, Künste, und Allgemeine Bildung (Free German Foundation for Science, Art, and General Culture) in Frankfurt, to which he belonged. In the statement, which has survived, he emphasized the necessity of researching and propagandizing on same-sex love. The necessary first step for him was to make his homosexuality known, and in June of that same year, he told his sister Ulrike that he was attracted to other males, a statement to which she initially reacted with considerable hostility. Undeterred by her reaction, he determined he had to campaign for homosexual rights.

As one of his first efforts, he decided he had to come to the defense of an acquaintance, Johann Baptist von Schweitzer, who had been arrested on a morals charge. Two women had reported to the authorities that they had overheard Schweitzer propositioning an unidentified fourteen-year-old boy in a castle garden. Although the boy never came forward, Schweitzer was convicted on the basis of the testimony of the two women and sentenced to two weeks in jail.* Ulrichs, immediately following the arrest and without asking Schweitzer, began planning a defense, but it was not used. This failure only emphasized to Ulrichs that homosexuals had to identify themselves publicly. Though his sister was still urging him to change his ways by seeking the help of God, he felt it essential to announce his homosexuality to his family. In September 1862, he sent another letter to his sister, asking her to circulate it to other family members. In it he defended his homosexuality as natural and said that because God had given him his same-sex drive, he had the "right to satisfy it."[1] He also began work on the first of his monographs on same-sex love, which was published in 1864. He eventually wrote a total of twelve booklets. The first five were written under his pseudonym, "Numa Numantius," but the later ones, beginning in 1868, appeared under his own name as he publicly emerged from the closet in which homosexuals had hidden. The twelfth booklet appeared in 1879.[2] Ulrichs's timing was fortuitous because many in the medical-scientific community were looking for explanations about sexual variation that went beyond the biblical sin model or that could better explain differences than the Tissot masturbation thesis.

ULRICHS'S THEORY AND ACTIVITIES

Ulrichs's writings, which he distributed to various professionals, served as the foundation for researchers who were striving to understand homosexuality. Ulrichs put forth a mishmash of mythological, literary, historical, physiological, and other data as well as his own personal beliefs and experience to explain same-sex love and attraction. In all his writings, he argued that what eventually came to be called homosexuality in men was due to a strong fem-

* In spite of his conviction, which might well have been politically motivated, because he was associated with left-wing politics, Schweitzer continued to be active in politics. He later became a leading Social Democrat as well as a prominent writer of popular comedies for the stage.

inine element in such men, and this element had been present in them since their birth. Such a person he eventually came to call an *urning*, a term he derived from the speech of Pausanias in Plato's *Symposium:*

> For we all know that love is inseparable from Aphrodite, and if there were only one Aphrodite there would be only one Love; but as there are two goddesses there must be two Loves. And am I not right in asserting that there are two goddesses? The elder one, having no mother, who is called the heavenly Aphrodite—she is the daughter of Uranus; the younger, who is the daughter of Zeus and Dione—her we call common.... The Love which is the offspring of the common Aphrodite ... is apt to be of women.... But the offspring of the heavenly Aphrodite is derived from a mother in whose birth the female has no part.... Those who are inspired by this love turn to the male.[3]

According to Greek mythology, life had begun when Gaea (Earth) and Uranus (Heaven) burst out of the silver egg formed in the divine ether. The two elements copulated, giving birth to Cronos and other Titans. Uranus, however, hated children, and as they were born, he confined them in Tartarus, a place beneath the earth, as far below Hades as heaven was above. Eventually, they rebelled, and Uranus was castrated and dethroned by Cronos, the father of Zeus, Poseidon, and Hades, among others. One of Uranus's children was Aphrodite, who sprang to life from the foam gathering around his limbs and thus was not of woman born.[4]

Ulrichs assumed that all urnings were like himself and that any love that is directed toward a man, even though it is by another man, is necessarily a woman's love. This suggested to him that the source of such a feeling in a male must be traced to a strong feminine component in that man. He argued that just as the hermaphrodite was a creature of God and part of nature, so were urnings. He held that the sexes in utero were the same up to a certain stage of development, after which a threefold division took place into male, female, and urning (or urningin, the female counterpart), this last group being made up of individuals who had the physical features of one sex but whose sexual instinct failed to correspond to their sexual organs. The result was an inversion of sexual desires. Ulrichs also believed that the line of differentiation between males and females had been overemphasized; as proof, he pointed out that normal males had rudimentary breasts and normal females a rudimentary penis. Many people, not only hermaphrodites, failed to develop along expected lines, and so it seems easily understandable to him, and he thought it should be to others, why a person might have the body of one sex and the soul of another.

Though his relatives had tried to discourage him from disseminating his ideas publicly, he explained,

> I believe that I owe it to my poor and, from my standpoint, innocently persecuted comrades-in-destiny. I shared my idea with several of them and they think the publication an urgent necessity. For my part, too, I feel the need finally to present openly a justification of myself against all the humiliations that have been laid onto me up to now, against which I do not know what else to set.[5]

Ulrichs had ambitious plans to follow up the publication of his first five booklets with an organization for urnings and even a periodical devoted to urnings. He petitioned the congress on German law held in Graz in 1865, urging the abrogation of the articles in the Prussian code dealing with homosexuality. This petition had been rejected by the organizers as not suitable for consideration by the congress.

His activities were interrupted by the Prussian invasion and annexation of Hannover in 1866. Ulrichs spoke out publicly against this action and was twice imprisoned. Exiled from Hannover on his release from prison in 1867, Ulrichs went to Munich to resume his earlier fight. Undeterred by his 1865 rejection, he again petitioned to be heard in the 1867 Munich congress on German law. Though his request was denied, he was allowed to protest formally the exclusion of his proposal at the closing meeting of the General Assembly on Thursday, August 29. He began by emphasizing the need for legal uniformity, and to this end, he urged the German jurists to follow the Napoleonic code in dealing with a specific class of persons "to which many of the greatest and noblest intellects of our and other nations have belonged." At this point, his speech was interrupted by catcalls, and he left the platform without formally presenting his plan. In the aftermath of the incident, however, Ulrichs dropped the pseudonym of Numa Numantius and turned to using his own name, an action that makes him the first out-of-the-closet homosexual man in modern history.[6] His attempts at the German legal congresses to gain legal recognition for homosexuals are today also regarded by homosexuals as marking the beginning of the public movement for homosexual emancipation.[7]

INFLUENCE ON THE MEDICAL COMMUNITY

Ulrichs sought allies everywhere, particularly in the medical community. In 1867, he believed he had found just such an ally in Richard von Krafft-Ebing, who had already written on the need for scientific investigation of those sexual behaviors currently being dealt with in the German courts.[8] Encouraged by the tone of Krafft-Ebing's writing, Ulrichs sent him his own publications on homosexuality, and these helped focus Krafft-Ebing's developing concepts of homosexuality and other stigmatized behaviors. Some indications of the influence of Ulrichs on Krafft-Ebing is indicated in a letter Krafft-Ebing wrote to him on January 29, 1879:

> The study of your writing on love between men interested me in the highest degree . . . since you . . . for the first time spoke openly about these matters. From that day on, when—I believe it was in 1866—you sent me your writing, I have devoted my full attention to this phenomenon which at the time was as puzzling to me as it was interesting; it was the knowledge of your writings alone which led to my studies in this highly important field.[9]

Also influenced by Ulrichs's writing was Carl Westphal (1833–90), the physician usually given credit for putting the study of stigmatized sexual expression on a "scientific" basis with his 1869 article in the *Archiv für Psychiatrie und Nervenkrankheiten*.[10] Westphal described at length two cases: The first was of a young woman who from her earliest years liked to dress as a boy and engage in boys' games and who found herself attracted only to women; the second was of a man who wanted to wear women's clothes and act the part of a woman. In attempting to give a diagnostic category to these cases, Westphal coined the phrase *Konträre Sexualempfindung*, usually translated as "contrary sexual feeling." In spite of the new term, Westphal based his assumptions on Ulrichs's theories. He cited the early work of Ulrichs and knew the term *urning*. Westphal also agreed with Ulrichs that such an "abnormality" could be congenital, not acquired, and that in such cases, it should not be termed a *vice*. Still, Westphal emphasized that it was possible only in a few isolated cases to be a homosexual without pathological symptoms, although his illustration of these cases allowed more exceptions than might appear on the surface. He stated, for example, that though there are pathological stealing, murder, and sexual aberrations, there are also nonpathological cases of theft, murder, and variant sexual behavior. In fact, he predicted that if section 143 of the Prussian legal code condemning homosexuality was ever repealed and the "ghost" of imprisonment removed as a threat, many more homosexuals would come to the physician's office for treatment. Here, in his mind, was where they should go instead of to prison. Westphal's work marks the beginning of the medicalization of homosexuality, and though the writings of Ulrichs had served as a basis for information and conceptualization, scientists such as Westphal fit those ideas into their own illness-oriented approach.

Ulrichs eventually came to regard Krafft-Ebing and Westphal as opponents, holding that their observations had come primarily from work with individuals in lunatic asylums or jails and that they had never seen healthy urnings or urningins. Still, both essentially adopted the core of Ulrichs's theory, namely, the congenital nature of at least some forms of homosexuality.

It was, however, not Ulrichs's term, or variants of it, or even Westphal's concept of contrary sexual instinct that carried the day but, rather, one developed by another self-appointed defender of homosexuality: the German-Hungarian writer Karl Maria Benkert (1824–82), who in 1847 was formally authorized to use the noble name of his family—Károly Mária Kertbeny. It is as Kertbeny that he is most generally known. The first appearance of the words *homosexuality* and *heterosexuality* appear in the draft of a private letter Kertbeny wrote to Ulrichs on May 6, 1868.[11] From 1869 to 1875, Kertbeny lived in Berlin, and while there he also published two anonymous pamphlets encouraging the repeal of section 143 of the Prussian legal code and, more important, opposing the adoption of it as section 152 by the North German Confederation. In these pamphlets, he used the terms *Die Gleichgeschlechtlichen* ("those of the same sex") and *Der Gleichgeschlechtlicher Akt* ("the same-sex act") as well as the word *homosexual*, which he coined to distinguish such individuals from those who were heterosexual.[12] Kertbeny was not particularly interested in what caused homosexuality but in getting legal barriers against homosexuality removed. He wrote, "As much as we pride ourselves that ours is a time when science rules, when no riddle of nature goes unsolved, we must in shame admit precisely in regard to the apparent riddle of nature that scientific research, with a prudery held to only here, has up to now not once come near the subject."[13]

Kertbeny proved no more successful than Ulrichs in influencing German lawmakers, and after the unification of Germany in 1871, the offending paragraph became section 175 of the German imperial legal code. But he was successful in another regard. The term *homosexuality*, with the Greek prefix grafted onto the Latin root word, was adopted and popularized by Krafft-Ebing, and as a consequence, the term was distinguished from other activities and behaviors with which it had previously been lumped, including bestiality, sodomy, and pederasty. For a time, Westphal's term *contrary sexual feeling* was used by some medical and scientific writers, including Albert Moll, but that term was eventually dropped because of its imprecision. Though Ulrichs's terms were more precise, his nomenclature had become increasingly complicated as he coined new words to describe a number of forms of behavior. His terms were also associated with certain assumptions about same-sex love with which a researcher might feel uncom-

fortable. Other terms were also used, such as Havelock Ellis's *sexual inversion*, and for a time, the term *third sex* was popular; but it was *homosexuality*
that was adopted as the medical term, primarily because of the influence of
Krafft-Ebing.

KRAFFT-EBING

The most significant medical writer on sex of the last part of the nineteenth
century was Richard von Krafft-Ebing (1840–1902). He was the oldest of
four children born into the aristocratic Krafft-Ebing family of Mannheim,
and he had the hereditary title of Freiherr ("baron"). His mother was the
daughter of a renowned Heidelberg lawyer, and when Krafft-Ebing attended
the University of Heidelberg, he lived with his maternal grandparents. It was
while living there that he developed an intense interest in criminal cases
involving "deviant" sexual behavior, a subject of great interest to his grand-
father as well. To pursue this interest, Krafft-Ebing turned to medicine and
became a psychiatrist, or *alienist*—a nineteenth-century term used to describe
those treating mental patients. At the age of thirty-two, he was appointed
professor of psychiatry at Strasbourg and subsequently occupied similar
positions at Graz and, in 1889, at Vienna. He died on December 22, 1902,
near Graz.[14]

Edward M. Brecher, who wrote a popular study of modern sex research,
viewed Krafft-Ebing as an unmitigated disaster, a psychiatrist who, "without
a shred of evidence," compared a lust murderer with a fetishist who wore
white kid gloves or high-heeled shoes. This is a harsh and unjustified condemnation of Krafft-Ebing, who tried his best to keep up with the latest
research in sex.[15]

Krafft-Ebing, however, was very much a man of his own time. He accepted
at face value the widely acknowledged belief of the dangers of masturbation
and thought it to be a source of mental illness and sexual pathology. Still, as
a scientist, he also tried to incorporate into his system the latest findings. To
do so, he combined several prevailing nineteenth-century theories: the idea
that disease was caused by the physical nervous system, the belief that there
were often hereditary defects in this system (hence some forms of homosexuality), and the concept that degeneracy can result from overstressing the
system through such activities as masturbation. He distinguished between
innate and acquired perversions, but eventually held that even acquired perversions could exist only when there were hereditary weaknesses in the nervous system, such as epilepsy. He argued that there was a strong association
of an intense sexual instinct with epilepsy, which often led to sexual perver-

sity during or following an attack. Though he accepted this association, he also emphasized that those who believed the epileptic element was present in all cases of "peculiarity" of sexual life were wrong.[16] In fact, shortly before his death, he wrote an article in the *Jahrbuch für Sexuelle Zwischenstufen* in which he said that homosexuality was not a manifestation of degeneracy or pathology but could occur in otherwise normal subjects.[17]

Krafft-Ebing, in short, was struggling to bring greater understanding to the field of human sexuality but was very much a prisoner of his own cultural assumptions. So, it might be added, were most of the other scientists of the time. Charles Darwin, for example, also subscribed to the idea that sex was part of the nervous system.[18]

To his credit, Krafft-Ebing early recognized the importance of the sex drive. Sexuality for him was the "most important factor in social existence, the strongest incentive to the exertion of strength and acquisition of property, to the foundation of a home, and the awakening of altruistic feeling, first for a person, then for the offspring, and in a wider sense for all humanity."[19]

Still, he saw the essential purpose of sex as reproduction and believed all sexual activities lacking this ultimate purpose were "unnatural practices" and a perversion of the sexual instinct, even though such perversions often resulted from inborn characteristics. Moreover, even though sex was important, civilization had been made possible only by the tempering of lust through altruism and restraint. Religion, law, education, and morality had given civilized people the means to restrain their passions, yet both men and women were always in danger of sinking from the clear height of pure and chaste love into the mire of a common sensuality. To maintain morality, men and women had to fight a constant struggle with natural impulses: "Only characters endowed with strong wills are able to completely emancipate themselves from sensuality and share in the pure love from which springs the noblest joys of human life."[20] It was on just such assumptions that Freud laid the foundation of psychoanalysis, arguing that in the struggle to emancipate themselves from sensuality, men and women repressed their sexual drives rather than coming to terms with them.

To illustrate his theories, Krafft-Ebing presented a number of clinical case studies, 238 of them by the twelfth edition of his *Psychopathia Sexualis*.[21] Homosexuality was just one of four broad categories of sexual variation that Krafft-Ebing discussed at length and to which he gave names that are still used. The others were fetishism, sadism, and masochism, and his attempts to analyze them were similar. If someone had given a name to a phenomenon, as Kertbeny had to homosexuality, and it seemed a good term to him, he kept it. He read extensively on what others had to say, collected his case studies, and arrived at his conclusions.

Fetishism, for example, was a term that had initially been coined by the French psychologist Alfred Binet (1857–1911).[22] Krafft-Ebing adopted the term but not all of Binet's assumptions. He went on to describe the behavior as the peculiar or unreasonable fascination and sexual meaning that objects, or their parts or simply their oddities, had to some individuals because of their association to something else. Much of such association came through masturbatory fantasies, a causal factor in which he originally strongly believed, but he later proved willing to modify this belief somewhat.

When he could find no other suitable term, Krafft-Ebing coined his own. The term *sadism* he borrowed from the attitudes expressed in the novels of the Marquis de Sade (Donatien Alphonse François de Sade, 1740–1814).[23] Krafft-Ebing defined it as an act of sexual arousal (including orgasm) produced by inflicting pain. *Masochism*, he defined as opposite of sadism, namely the desire to suffer pain and be subjected to force. He again turned to a literary figure for his term, this time the writings of the Austrian Leopold von Sacher-Masoch (1836–95), who was a historian, dramatist, and novelist. Sacher-Masoch's fictional writings became stereotypes, almost always featuring a woman in furs (he had a fetish for fur) who, with a whip, symbolic of lust, scourged her male lover for his animal lusts. In his classic *Venus in Pelz*, Wanda and Gregor are, respectively, the active and passive participants in flagellation,[24] the reversal of traditional character traits, because Krafft-Ebing held that sadism was a pathological intensification of the masculine character and masochism a pathological degeneration of the female character.

Krafft-Ebing also gave brief expositions on satyriasis, nymphomania, necrophilia, incest, and pedophilia, but these topics were primarily discussed in terms of the legal implications. The significance of Krafft-Ebing to the study of sexual behavior was the large number of variant acts that he brought out into the open and made the subject of public discussion. This was contrary to his expressed intentions, namely helping his fellow physicians to cope better with the sexual problems of their patients and to aid the courts in dealing with various forms of sexual behavior. In fact, when he found that not only professionals were reading his book but the general public as well, he tried to make the book more "scientific" (and obscure?) by using more technical language and putting the specific descriptions of sexual acts into Latin. Even with these changes, he was often condemned for delving into human sexuality as a scientist and physician. The typical attitude, even among his colleagues, was expressed in an 1893 editorial in the *British Medical Journal* about an English translation of *Psychopathia Sexualis*:

We have taken some time to consider whether we should notice this book or not, and have, in the end, decided that the importance of the subject and the position of the author render it necessary to refer to it. It is, we believe, unique in the fullness with which the subject has been treated, but we question whether it need to have been translated. Anyone wishing to study the subject might just as well have gone to the original, and some may be disposed to go even further and regret that the whole had not been written in Latin, and thus veiled in the decent obscurity of a dead language. There are many morally disgusting subjects which have to be studied by the doctor and by the jurist, but the less such subjects are brought before the public the better.[25]

OTHER MEDICAL RESEARCHERS

Once Westphal and particularly Krafft-Ebing established the importance of the medical study of sexuality, other physicians, encouraged by their example, turned to the study of sex in spite of the prudishness of many of their professional colleagues. They usually argued it was essential to investigate sexual behavior to deal with patient problems. Although most of them continued to look on much of sexual behavior as pathological, not all of them agreed with Krafft-Ebing's explanation. A major disagreement arose between Krafft-Ebing and Binet, who, using fetishism as an example, argued that the medical appeal to the hereditary nature of perversions ignored the question of how such behaviors had been acquired by a given patient's ancestors. Drawing on the case histories of Westphal and Krafft-Ebing as well as those of some of the French alienists, such as Jean-Martin Charcot (1825–93), Binet emphasized early childhood experiences. Though he recognized that each child might be affected differently by the same experience, he held that there was a congenital "morbid state" in those adults adversely affected by a childhood experience.[26] For him the major forms of sexual pathology— from homosexuality to fetishism—were specifically determined by chance events, and a fetishist could have become a homosexual given exposure to a different determining event.

This theory, known as *associationism*, was later abandoned by Binet, who went on to develop the intelligence scale and become a key figure in functional psychology, a movement that attempted to interpret phenomena with reference to the part they played in the life of the organism rather than describe or analyze the facts of experience or behavior. Nonetheless, associationism had considerable influence on some of the teachers of Freud as well as others.[27] For example, Albert von Schrenck-Notzing (1872–1919), who adopted some of the theory, claimed that he could cure some homosexuality by hypnotism and suggestion therapy. At the First International Congress on Hypnotism held in Paris in 1889, he reported on a homosexual from a tainted family whom he had successfully treated in forty-five hypnotic sessions over a four-month period.[28] Von Schrenck-Notzing continued his experiments, and three years later he reported on some seventy other cases of homosexuals and patients with other stigmatized sexual behaviors; he claimed he had been able either to cure them or to reduce their urges. He then held that if homosexuality could be cured by such external influences, then it as well as other "pathological" sexual behaviors might be acquired through influences such as those posited by Binet.[29] Others built on this concept, emphasizing that precocious childhood sexual experiences, even though forgotten, may persist in the unconscious and ultimately form the psychological foundation of adult sexual experience.[30]

Krafft-Ebing came to believe that the associationist theory had some plausibility, particularly for fetishism. He thought, however, that it was unable to explain how an accidental childhood sexual association, even in the most precocious individual, could by itself lead to sadism or masochism and stated that in most cases acquired sexual anomalies were rare. This nature-versus-nurture argument was also tied into Darwinian evolutionary views, and in this sense, Krafft-Ebing was seen as a subscriber to Darwin's evolutionary theories, and he held to the belief in the transmission of marked moral qualities.[31]

Ellis summarized the developments in sexology up to Krafft-Ebing's time:

> It was during the second half of the nineteenth century, when a new biological conception, under the inspiration of Darwin, was slowly permeating medicine, that the idea of infantile and youthful "perversion" began to be undermined; on the one hand the new scientific study of sex, started by the pioneering work of Krafft-Ebing at the end of the third quarter of the century, showed how common are such so called "perversions" in early life while, on the other hand, the conception of evolution began to make it clear that we must not apply developed adult standards to undeveloped creatures, what is natural at one stage not necessarily being natural at the previous stage.[32]

The gradual acceptance of evolution, however, led to a growing acceptance of what Frank Sulloway, a biographer of Freud, called the biogenetic views of homosexuality. Many of the theorists in this area were Americans who emphasized the original bisexuality of the ancestors of humans.[33] One of those who put it most clearly was James G. Kiernan (1852–1923), the medical superintendent of the Chicago County Asylum for the Insane. He wrote:

> The original bisexuality of the ancestors of the race, shown in rudimentary female organs of the male, could not fail to occasion functional, if not organic, reversions when mental or physical manifestations were interfered with by the disease or congenital defect. . . . It seems certain that a femininely functionating brain can occupy a male body, and vice versa. Males may be born with female external genitals and vice versa. The lowest animals are bisexual and the various types of hermaphrodism are more or less complete reversions to the ancestral type.[34]

Going further was another American physician, G. Frank Lydston (1857–1923), who held that the worst types of perversions could be blamed on either maldevelopment or arrested development, with the most severe sexual aberrations making an appearance before the commencement of sexual differentiation.[35] In a sense, these Americans were reflecting what Cesare Lombroso had earlier proclaimed, namely a biogenetic evolutionary view of criminality and deviance. Lombroso held that the "criminal type" had been born, not made, and that the criminal was an atavistic being who reproduced in his or her person some of the instinct of primitive humanity and inferior animals.[36] These American physicians, however, were not adopting the atavistic view of Lombroso but instead a developmental view that could be influenced by in utero developments. In this, they were more likely to follow Ulrichs than Lombroso. This view was summed up by Krafft-Ebing shortly before his death:

> In view of the realization that contrary sexuality is a congenital anomaly, that it represents a disturbance in the evolution of the sexual life towards monosexuality and of normal psychical and somatic development in relationship to the kind of reproductive glands [possessed by the individual], it is no longer possible to maintain the idea of [degenerate] "disease" in this connection. . . .
>
>   Not infrequently one runs across neuropathic and psychopathic predispositions among homosexuals, for example, constitutional neurasthenia and hysteria . . . which may lead to the most severe aberrations of the sexual impulse. And yet one can prove that, relatively speaking, heterosexuals are apt to be much more depraved than homosexuals.[37]

THE THREE ALBERTS: MOLL, EULENBURG, AND VON SCHRENCK-NOTZING

ALBERT MOLL

Supporting Krafft-Ebing in his changing view was Albert Moll (1862–1939), a Berlin physician who paved the way for a number of Berlin investigators into the field of sex. In 1891, Moll published *Die Konträre Sexualempfindung*, in which, following Krafft-Ebing, he differentiated between innate and acquired homosexuality. Much earlier than Krafft-Ebing, he put an emphasis on the innate, arguing that because all biological organs and functions are susceptible to variations and anomalies, there is no reason why the sexual instinct should be any different. He described the homosexual as "a stepchild of nature."[38] Moll was highly critical of the belief that early sexual activity was an important correlate of later perversion. He questioned the dangers of masturbation, emphasizing that mutual masturbation was often practiced in childhood by individuals who showed no signs of inversion.[39]

Moll broke new paths by comparing normal and abnormal sexual developments side by side because he argued that it was the failure of sexologists to study normal sexuality that led to their disagreements about abnormal forms. Moll posited that two major instincts were involved in the sexual impulse: *Detumescenztrieb* ("detumescence drive") and *Contrectationtrieb* ("drive to touch, fondle, or kiss the sexual object").[40] The former was fundamentally individualistic, while the latter was social. Moll theorized that primeval organisms, which reproduced asexually by fission or budding, possessed only the instinct of detumescence but that higher organisms, which reproduced by conjugation or sex, have acquired the instinct of contrectation as well. In sexually reproducing animals, the instinct of detumescence was no longer sufficient to guarantee reproduction and had to be supplemented by the instinct of contrectation, closely resembling what we call love. Love itself was thus a byproduct of the evolution of sex. Either impulse could emerge first, but both impulses emerged well before puberty in sexually normal individuals. He wanted to "destroy" the belief "that physical puberty is a necessary preliminary condition for the sexual inclination of male and female. On the contrary, as has already been mentioned several times, the psychic element, in a number of cases, may develop much earlier than physical puberty."[41] He admitted, however, that it was often difficult to draw the line between sexual and social feelings before puberty, although even in extreme cases, the love of the child for its mother may always be distinguished from the sexual love of the child for another.

While shame, jealousy, and preferential expressions of love are all affected by the child's sexuality,[42] children may also have erections and begin masturbating as early as the first or second year of life, because both detumescence and contrectation impulses may emerge precociously in one and the same child. He adopted, with modifications, the two-stage theory of sexual development put forth by Max Dessoir (1867–1947). Dessoir had called the first stage the undifferentiated stage; this appeared in girls between the ages of twelve and fourteen and in boys between thirteen and fifteen. During the first stage, the sexual instinct could express itself in either a heterosexual or a homosexual manner or even turn toward animals, but this was followed by the second stage—differentiation—in which heterosexual relationships become the exclusive goal of the libido. Some individuals, however, remain in an embryonic state and continue to express homosexual, bisexual, or other inclinations as adults.[43] Moll insisted that the undifferentiated stage went back into early childhood, and this might well include same-sex love, but at puberty, when differentiation took place, most selected their love object from members of the opposite sex.[44] The failure to develop normally he felt was due to congenital weakness or susceptibility to various forms of perversions, although even the details of these perversions might be influenced by habits formed early in life so that each might respond to his or her congenital weakness in a slightly different way.[45] These concepts eventually led him to place more stress on homosexuality as an illness. By 1911, he was offering a cure for acquired homosexuality through association therapy, in which he replaced same-sex associations with those of the opposite sex.[46] Part of his changing opinion might have been influenced by his opposition to Hirschfeld; he eventually became the major opponent of Hirschfeld, as recounted later in this book. In fact, in spite of his contributions to sexual theory and the high regard in which he was held by Ellis and others, particularly for his early work, Moll did not get along well with the later generation of sex researchers. He was critical not only of Hirschfeld but of Freud. It might well be that his disagreements with Freud later caused him to be ignored by most English-speaking sex researchers, because Freudian ideas became so dominant in the English-speaking world in the first half of the twentieth century.

ALBERT EULENBURG

Albert Eulenburg, another prominent physician, first wrote about sex in 1895 but became more deeply involved as a significant figure in sex research only because of criticism of his twenty-six-volume *Real Encyclopädie für Medizin*.[47] This was because his multivolume comprehensive overview of

medicine was singled out by Hermann Rohleder in 1901 as an example of
the way in which the medical profession still managed to ignore sexual
issues.[48] Eulenburg responded to these criticisms by demonstrating his
expertise on such matters through publishing a series of studies on sadism
and masochism, although he used the term *algolagnia*,[49] a term put forth by
another German sex researcher, von Schrenck-Notzing.[50]

ALBERT VON SCHRENCK-NOTZING

Von Schrenck-Notzing is perhaps typical of this growing group of physician researchers in emphasizing that without careful study of the circumstances attending the development of sexual anomalies, the physician would
never be in a position to give treatment. He held that behaviors such as
homosexuality were not inborn but resulted from suggestion and that the
individual with such behavior would be open to a similar cure by hypnosis.
He claimed to have been able to cure a number of homosexuals through this
method, but he also said that he occasionally had to turn to more direct
methods to force a new direction in the behavior of the "invert." One of his
cures was to take a patient off to a brothel, where he persuaded the prostitutes to bring all their erotic arts in an effort to achieve heterosexual desire.

AMBIVALENCE OF RESEARCHERS

Generally, the physician researcher approached sexual topics gingerly, conscious of societal ambivalence about sexuality, although not always of their
own prejudices. Sex was something to be controlled and not liberated, and
the physician had to be aware of this; in fact it might well have been the fear
of unbridled sex that led some of them to class so many sexual activities as
pathological. The use of such stigmatizing terms emphasized that although
they investigated sexual problems, they themselves were very conscious of
the evils of some of the conduct they wrote about. Krafft-Ebing, for example, reflects the ambivalence of his own southern German Catholic upbring-
ing in his discussions of sex. For him, unbridled passion resembled a "volcano which scorches and eats up everything, or an abyss wherein everything
is walled up—honor, property, health. [By establishing monogamous marriages and reinforcing religious bonds] the Christian peoples obtained a spiritual and material pre-eminence over other peoples, particularly those of
Islam."[51] Krafft-Ebing was simply expressing the classical Christian dualism
between the spirit and the flesh. The flesh, of which sex was a part, was
weak, always threatening to pull men and women from contemplation of

higher things of the spirit into the gutter and sewers of fleshly existence.[52] The mind or spirit, however, is capable of controlling and should control the desires of the physical body, and the community is better off if the volcano of unbridled passion is dormant. Krafft-Ebing stressed the need for scientific description of various sexual pathologies over the need for new theory, because to connect other impulses and voices of the soul with sexual drives was to enlarge the sphere of the latter and detract from the magnificence of the former. Indeed, when greeted with the news of Freud's 1896 attempt to track the origins of hysteria to childhood sexual development, Krafft-Ebing responded that Freud's work sounded "like a scientific fairy tale."[53]

In fact, the growing possibilities of a confrontation between traditional religious ideology and the new findings about sex emphasize another important factor in the growth of sex research: the need for an attitude about sex that could go beyond the traditional Christian dualistic ones that had influenced Tissot and his followers and that continued to be dominant in most nineteenth-century studies. The key missing ingredient was a willingness to accept sexuality, not just procreation, as a fact of life; a willingness to look on sex as a vital physical force that was capable of doing more good than harm; and a willingness to see it as one of life's pleasures. Krafft-Ebing had struggled to come to terms with a need for change but had not quite succeeded. Though there was a growing middle class willing to accept pleasure as an important element in their lives, the medical community as a whole either saw no need to challenge or were unwilling to challenge traditional ideology.[54]

EUGENICS

There was a growing demand for better information about sexuality that was not so much concerned about changes in laws about homosexuality as in a better understanding of the relationships between the sexes. The British eugenicist Karl Pearson (1857–1936), for example, called for a new "science of sexualogy" to emerge to help society decide what the status of women should be:

> Not until the historical researches . . . [and] anthropological studies
> . . . have been supplemented by careful investigation of the sanitary
> and social effects of past stages of sex-development, not until we have
> ample statistics of the medico-social results of the various regular and
> morbid forms of sex-relationship, will it be possible to lay the founda-
> tions of a science of sexualogy. Without such a science we cannot

safely determine whither the emancipation of women is leading us, or what is the true answer which must be given to the woman's question.[55]

Eugenicists in fact served as a political pressure group, and many of the British and the Americans who wrote about sex at the turn of the century had some connection with the eugenics movement. Unfortunately, the eugenicists were often zealots for a certain elitist viewpoint about the status and conditions of the poor, disabled, and minorities of the world, and though this background is sometimes difficult for sexologists to acknowledge, it is nonetheless there. The eugenicist represented the attitudes of a large segment of middle-class professionals, and their concerns helped bring discussion of sexuality out into the open.

The movement had been founded by Sir Francis Galton, a major figure in nineteenth-century science. Galton, through his studies of gifted individuals, came to believe that heredity played an essential part in the development of individuals of unusual competence (that is, geniuses). To encourage both study and policy development, he founded what he called the science of eugenics—the study of forces under social control that enhance or impair the inborn qualities of future generations. The purpose of eugenics was a deliberate attempt to increase from one generation to another the proportion of individuals with better-than-average intellectual endowment.[56]

Pearson not only was the founder of the twentieth-century science of statistics but was Galton's disciple and advocate. Galton wanted to concentrate on the elite, but Pearson decried the high birth rate of the poor, which he felt was a threat to civilization. He believed that it was essential that the "higher races" supplant the "lower."[57] Although the English Eugenic Society, founded by Galton, eventually opposed Pearson's racist views, large sections of the eugenics movement continued to hold racist and antipoverty views, and the American eugenics movement, founded in 1905, initially adopted Pearson's view wholeheartedly.[58]

As a group, the American eugenicists believed that the "white race" was superior to other races and that within the white race the Nordic white was superior to other whites. It was also assumed that upper-class people had superior hereditary qualities that justified their being the ruling class. To document this assumption, eugenicists gathered all possible evidence supporting their interpretation, including the results of intelligence tests, which had been introduced in the early 1900s by Binet. In spite of the opposition of Binet himself to what he regarded as a misuse of his tests, the eugenicists held that such tests measured the innate, genetic intelligence of individuals. On the basis of such tests, eugenicists classified all people whose IQs gave them a mental age of twelve as feebleminded or morons, without regard to the educational backgrounds or deprived environments that might have led to such test results. Criminality was considered a concomitant of feeblemindedness. Insane, idiotic, imbecilic, feebleminded, and epileptic persons at the urging of the eugenicists were often sterilized—either voluntarily or involuntarily—and so in some cases were habitual criminals, "moral perverts," and others deemed socially undesirable.[59]

Although the development of genetics undermined many of the simplistic assumptions of the eugenicists, such views were not vanquished. They were particularly important in sex education in the United States, and many of the books dealing with sex in the United States as late as the 1940s were published by eugenic groups or the Eugenics Press. The eugenicists were particularly concerned with what they regarded as sexual perversion, which they equated with certain races and peoples, and so they seized on Ulrichs's ideas of genetic causes to include "sexual perverts" in their category of inferiors.

Though Galton's purpose had been to encourage the "better people" to reproduce more, the eugenicists also mounted campaigns to prevent the "lower elements"—that is the poor, blacks, immigrants, and so forth—from producing so many children. Much of the early contraceptive movement inevitably became involved with the eugenicist movement, even though individuals such as Margaret Sanger did not agree with them. Although most of the racial and class overtones have long since been removed from the current generation of Planned Parenthood programs, lingering suspicions about family planning still exist among some of the more militant representatives of minorities. The eugenicists generally were interested in contraceptives not so much for the upper classes but for the poor. It should be emphasized, however, that there were many other individuals and groups working for better contraceptives and more effective sex education who had entirely different viewpoints. Many of the early feminists, for example, were concerned with helping the overburdened mother who did not want to have additional children, and a number of people were interested in helping women in general overcome what was called their biological destiny.

Still, the activities of the eugenics movements in promoting certain sexual ideas as well as the efforts of Ulrichs to legitimatize homosexuality emphasize that much of the research about sex was not conducted by disinterested, dispassionate scientists. Sexual activity, as was stated in the introduction, had all kinds of political overtones, and in looking at developments in sex research, it is important to look at the agendas the individual researchers had and just how much such agendas influenced their findings. Fortunately, there were competing agendas; the eugenicists, for example, were countered somewhat by both feminists and radicals, and those trying to advance homosexual rights were opposed by the heterosexual moralists. Occasionally, the competing groups came together, but it is important and essential in any discussion of human sexuality to try to separate the rhetoric from the data.

PORNOGRAPHY

This separation was more difficult in the last part of the nineteenth century than it had been earlier or than it became later, because much writing about sex had come to be classified as pornography by the state. Though unofficial censorship had existed since books began to be widely circulated and official Church censorship had existed since the Catholic Index had been established at the end of the sixteenth century, the state itself generally did not get involved until the nineteenth century. The first laws against pornography in England were passed in 1853 and then supplemented in 1857 by Lord Campbell's Act. The laws gave magistrates the power to order the destruction of books and prints if, in their opinion, publication would amount to a "misdemeanor proper to be prosecuted as such."[60] The meaning of what constituted pornography was further extended by the so-called *Hicklin* decision in 1868, in which Sir Alexander Cockburn wrote that the test of obscenity was whether the "tendency of the matter as obscenity is to deprive and corrupt those whose minds are open to such immoral influences and into whose hands a publication of this sort may fall."[61] Unfortunately, such a decision potentially meant that whether or not something was pornographic would depend on whether a child could read it or, in Victorian times, whether a woman perchance might see it.

Almost immediately, pamphlets giving birth control information came under attack as pornographic, as did many treatises on sex, including for a time those written by Ellis. In United States, the person identified most with the new repression of information about sex was Anthony Comstock (1844–1915), who, with his supporters, not only managed to lobby through Congress a law governing the depositing of "obscene materials" in the U.S. mail but also got himself appointed a special agent of the post office. Not every publication on sex was obscene to Comstock, because those "properly" condemning sexual activity were not censored; however, serious discussions of contraception, prostitution, and other facets of sex that were aimed at the general public were regarded by him as obscene. Often quite detailed descriptions of such things as masturbation were approved by him if they were negative enough.

Although *Comstockery*, as the movement he was identified with came to be called in the United States, put serious difficulties in the path of any open discussion of sex, the United States in many ways remained more liberal in its publication policies than England, providing such literature did not enter the mails or through certain ports of entry, in particular New York City, where Comstock was headquartered. Comstockery was more than Comstock himself and might be taken as symbolic of the fears that many Americans had about the growing evil believed to be present in the rapidly growing cities where traditional morality was under attack. Comstock seems to have been somewhat of an innocent, basically unaware of many forms of sexuality, and in a sense, he was like the mythical little Dutch boy, putting his finger in the dike to keep the floods of sexuality at bay.

OSCAR WILDE

Sometimes the late-nineteenth-century response to sex, popularly called Victorianism, was like the three monkeys who saw no evil, spoke no evil, and heard no evil. This did not mean that evil did not exist, and neither did Victorianism mean that sexuality and sensuality were not present. In fact, there were a number of events that brought the unmentionable to public attention, despite all efforts of a Comstock to prevent this. One such event in the English-speaking world was the charges of homosexuality leveled against Oscar Wilde (1854–1900). His subsequent trial received widespread public attention throughout the Western world. Wilde, poet, novelist, playwright, and founder of an aesthetic movement of "art for art's sake," had long hair, wore flamboyant clothes, and like Andy Warhol nearly a century later, was a superb self-publicist. His attention-getting behavior led to his being satirized in the Gilbert and Sullivan operetta *Patience* in 1881 and in such magazines as *Punch*. It also made him a popular lecturer. In 1891, he became sexually involved with Lord Alfred Douglas, the handsome son of the marquess of Queensberry, now remembered for his boxing rules. Believing that Wilde had drawn his son into homosexuality, the marquess brought matters to a head in February 1895 by leaving his card for Wilde at the Albemarle Club, to which both men belonged, with the notation: "To Oscar Wilde, Posing as a Somdomite! [sic]" This misspelled missive led Wilde to sue Queensberry for criminal libel.

Three trials followed. The first trial was the one initiated by Wilde, but when the defendants gathered enough evidence to support the marquess's allegation that Wilde indeed engaged in homosexual relationships, Wilde withdrew his suit. On the strength of the evidence gathered against him,

Wilde himself was put on trial, and when the jury in the second trial could not agree on a verdict, a third trial was held. He was found guilty of several acts of gross indecency and sentenced to two years at hard labor.[62] It was in part his attempt to be clever that led to his conviction, because he had managed to deny everything until he blundered into saying that he never kissed a certain boy because he was much too ugly.

RICHARD BURTON

Though Wilde was a victim of the prejudices of his time, the publicity given the case opened up a whole new world of sex for many. They wanted more information, and there were a number of individuals out there willing to make it available, although it was much safer to do so by looking at non-Western cultures. One missionary for new attitudes toward sexual activities was Richard Burton (1821–90), the explorer and polymath. In his notes to his translation of the *Thousand Nights and a Night*, he summarized Western attitudes toward sexuality by recounting the story of the newly married husband who came into the bedchamber on his wedding night to find his bride chloroformed and a note stuck on her pillow that read: "Mamma says you're to do what you like."[63] This was an attitude that Burton felt he had to challenge.

Burton's father was a retired lieutenant colonel in the British army, and his mother came from a well-to-do Herefordshire family. Early in his life, Burton's parents moved to Tours in France, because his father thought the hunting was good, the prices were cheap, and educational facilities were available for his children. Burton matriculated at Trinity College, Oxford, but did not graduate and instead became a cadet in the Indian army. He found his niche when he was assigned to do a survey of the Sind area in India. This allowed him to plunge into Indian life, live among the natives, learn new languages, write learned books about his experiences, and have all sorts of adventures. As part of his Sind survey, he compiled information on homosexual brothels in Karachi, and his report to his commanding officer in India, Sir Charles Napier, contained such detail that when it was later circulated secretly among some of his fellow officers, his army career was ruined. It was not so much that he personally was suspected of homosexuality but that he demonstrated such poor judgment in discussing such a subject so dispassionately. His dismissal was recommended, and though this did not happen, he left India for a life of adventure and publication.

He took a pilgrimage to Mecca in disguise with the assumed name of Al-Haz and later published an account about it. He went exploring for the source of the Nile with John Hanning Speke, he visited the Mormons in Salt Lake City, he served as British consul to various Near Eastern cities, and he lectured and published widely. Burton's exploits made him almost a household name, but he was more anthropologist than explorer. He had mastered twenty-five languages plus numerous dialects; wrote a grammar of the Jataki dialect in India; compiled dictionaries in Hagar, Dahomey, and Brazil; and made transliterations of proverbs in ten different African argots. He translated works from Sanskrit, Portuguese, Neapolitan Italian, and Latin as well as Arabic and Persian. He also produced some forty-three volumes on his travels and explorations, which usually included information about the sexual and marital customs of the peoples he had seen.[64]

Burton wrote on prostitution, homosexuality, pederasty, castration, and infibulation—much of it buried in monographic reports. In 1863, he had founded the Anthropological Society of London in the hope that its journal would provide a scholarly means for publishing more such studies. Although he did publish his notes on a hermaphrodite there, the subject so frightened the editors that Burton had to find other means of publishing material on sexually related topics. His research, such as his essay on homosexuality that was based on his long-repressed study of Sind, appeared as a supplement to his translation of *The Book of the Thousand Nights and a Night*, originally published in ten volumes in 1885.[65]

Publication of this work was part of Burton's effort to translate and publish erotica, an effort in which he was engaged from 1876 to his death in 1890. With his friends Foster Fitzgerald Arbuthnot and H. S. Ashbee, Burton conceived the idea of a pseudonymous publishing house with a fictitious headquarters. This led to the formation of the Kama Shastra Society of London and Benares (India), although the Benares listing was meaningless, because the actual printing of the books took place in Stoke Newington in England.

Burton had amassed a large library of works on the art of love and sexual practice and was able to write about sex in a language that did not seem obscene to any but the most innocent readers. In addition to his translations of the *Nights*, he translated and published the *Kama Sutra*, the Hindu classic that offers advice on the many ways of sexually satisfying women and a description of the various positions possible in sexual intercourse. This was followed by the publication of the *Ananga Ranga*, which gives explicit recommendations for enhancing marriage as well as advice on how to seduce a new partner. Among other things, he also published *The Perfumed Garden of the Cheikh Nefzaoui*, containing bawdy stories and much advice on how to have better sex, and an English translation (from Latin) of *Priapei or The Sportive Epigrams of Divers Poets on Priapus*.[66]

OTHER POPULARIZERS

Burton was not alone in trying to educate the public about sexuality. Particularly influential in this respect was the German gynecologist Hermann Ploss (1819–85), who was proclaimed by one of his contemporaries to have founded "a new branch of science called anthropological and ethnographical gynecology." His two-volume work on women in nature and culture, *Das Weib*, contained up-to-date discussions of women's anatomy and physiology, together with legends, myths, rituals, and beliefs that were influential in shaping women's lives. He collected a vast amount of data concerning every aspect of the female known at that time, plowing through anthropological, philosophical, and psychological data as well as through research into physiology and aesthetics. In a sense, he might be considered a pioneer in women's studies.[67] In a subsequent revised posthumous edition edited by Max Bartels, the compilation was considerably expanded, aided by Ploss's notes. Eventually illustrations were added. Though a feminist commentator has complained that *Das Weib* was not a natural history of females but a history of woman as a "sex object,"[68] it still is regarded as a major effort to deal with some of the sexual issues that the emerging sexologists of the nineteenth century felt were important.

Perhaps the most influential of the researchers using historical and anthropological data was the German dermatologist Iwan Bloch (1872–1922). Though much of early sex research literature had included historical case studies, Bloch broke new ground in advocating the establishment of *sexualwissenschaft* ("sexual science"), which was to include not only biological and psychological data but cultural, social, and historical information as well. One of his first efforts in this regard was *Beiträge zur Aetologie der Psychopathia Sexualis*, published in two volumes in 1902 and 1903. Bloch held that perversions were found in every culture and in every historical period. One of the richest sources, he found, was religious writings. Religion for him was a museum of sexual beliefs and institutions in which everything ranging from sacred prostitution to fetishism to phallic cults to exhibitionism and learned discussion on such topics as sadism, masochism, and homosexuality could be found. From his data, Bloch concluded that every sensory organ could function as an erotogenic zone and thereby form the basis for a perverse sexual impulse.[69] The wonder was not that there were people with perverse sexual instincts but that more of us did not exhibit such behavior.

In this work, he criticized Krafft-Ebing's notion of congenital psychopathia sexualis (he later changed his mind, at least on homosexuality) and held that "aberrations" were due to the need for varied sexual stimuli as well as the influence exerted on the sexual instinct by "accidental external conditions." These three words—*accidental external conditions*—made up the descriptor he used to explain perversion arising in childhood, while the *need for varied stimuli* explained those arising in later years of adulthood.[70] Bloch examined childhood sexuality in some detail and noted the great frequency of copulatory attempts and other forms of sex play among children of primitive people. He emphasized that the elders in these groups did not look down on such childhood activities as abnormal or indecent.[71]

His study *The Sexual Life of Our Time and Its Relations to Modern Culture* in its German original is perhaps the best overall survey of sexual knowledge at that time.[72] Bloch became convinced by his earlier studies that the key to current problems of human sexuality was historical understanding. To this end, Bloch undertook a major study of prostitution, a topic, he stated, that enabled him to study the role and status of women as well as many aberrant forms of sexual behavior.* In spite of its limitations, Bloch's study of prostitution was the most comprehensive study of the subject up to his time. Unfortunately, he never completed it and thus never could draw the conclusions he felt were so important.[73]

Similarly, he argued that the problem of venereal disease, particularly syphilis, was emblematic of the problems of sexuality in society and that once the medical problems posed by syphilis could be overcome, humanity could look forward to a brighter future.[74] It is important to note that the few works of his that were translated into English are usually heavily abridged and are not indicative of the breadth of his scholarship.

Hirschfeld, after reading Bloch, felt for the first time that it was possible for sexology to be a real science and started *Zeitschrift für Sexualwissenschaft*, the first journal devoted to sexual science. Rohleder, also impressed by Bloch's concept of sexual science created a special category of "sexual science" in *Reichsmedizinalanzeiger*, the medical journal he edited.[75] Freud credited Bloch with having replaced the pathological approach to the study of sexual inversion with an anthropological one.[76]

Bloch was certainly more involved in the scientific and scholarly circles than Burton, but both were significant in challenging traditional ideas about sex. Also busy at work in gathering ethnographic-historical data was Friedrich S. Krauss, on whom Bloch in part relied and who influenced both Freud

---

* Bloch was important both in making definitions and in gathering data. Sometimes it seems as if he is a vacuum, sucking up information from historical, medical, and legal records. He is said to have possessed a personal library of eighty thousand volumes, but readers of Bloch should carefully check his citations, perhaps because he tended to cite things from memory. Many of his notes do not hold up.

and Hirschfeld. In 1904, Krauss founded the journal *Anthropophyteia* to publish ongoing research on the history and ethnology of sexual morality. Among other things, he did a study of homosexuality in Japan.[77] An ethnographic study of homosexuality was also done by F. Karsch-Haack, this time in various primitive cultures.[78]

Resources for finding data also increased. Symbolic of this new effort to hunt down sexual information is the bibliographical collection of banned books by Pisanus Fraxi (Ashbee's pseudonym), which first appeared in 1877. He described many of the underground erotic classics available in the British Museum and elsewhere and was, it is believed, the author of an extraordinary eleven-volume anonymous sexual memoir, *My Secret Life*, in which he described affairs with more than twelve hundred women.[79]

Novelists and others were also challenging the prudery of the time, not always successfully. George Eliot (Mary Ann Evans's pen name) was reprimanded in the *Saturday Review* on February 26, 1859, for discussing pregnancy at length in *Adam Bede*. Gustave Flaubert was legally prosecuted for publishing *Madame Bovary*, as was Charles Baudelaire for *Fleurs de Mal*. And even Alfred, Lord Tennyson was rebuked in 1855 for his emphasis on adultery, fornication, and suicide in his "monodrama" "Maude."[80]

THE JEWISH INFLUENCE

Within the medical-scientific community, there was a growing number of individuals who were willing to emphasize the pleasures of sexuality, including Hirschfeld and Ellis (who will be discussed more fully in chapter 3). Even those not quite willing to go as far as Hirschfeld and Ellis, however, found it important to pay attention to human sexuality, if only to help their patients. The largest number of such medical professionals at the turn of the century were in German-speaking areas, particularly Austria and Germany. Interestingly, the generation of physicians following Krafft-Ebing in medical discussions of sexuality were mainly those with Jewish backgrounds, perhaps because this segment of the medical community was somewhat freer of Christian ideology about certain forms of sex.[81]

The real or alleged influence of Jewish physicians in sex research is somewhat controversial for a variety of reasons. One of the difficulties is in determining who is and who is not Jewish; such a subject also brings up painful memories of the Nazi era. With the exception of Enoch Heinrich Kisch (1841–1918), a gynecologist who wrote on female sexuality,[82] few of the physicians with Jewish backgrounds involved in sex research were practicing Jews. Eulenburg, though of Jewish descent, had been baptized a Christian when he was seven.[83] Dessoir, who emphasized the psychological origins of sexuality as well as the ability of men and women to reshape their own souls and world,[84] had a Jewish father but had been raised a Protestant.[85] In fact, most of those physicians of Jewish background, even the politically more conservative ones such as Moll, were very secularized, and only the Nazis would classify them as Jewish.[86]

This raises the question of whether this "new" view of sex was part of a growing secularism or whether it was restricted to persons of Jewish background. The evidence favors the first rather than second explanation, because Orthodox Jewish views of sex were as hostile to nonheterosexual activity as Christian views. Moreover, Eulenburg, if asked, probably would have replied that it was his scientific objectivity rather than his Jewish origins that led to his attempts to understand human sexuality. In his book on sadomasochism, Eulenburg wrote, "It is of course impossible for me as a doctor to throw morally critical stones at the living as well as the dead."[87] On the other hand, there was some consciousness of a difference in Jewish attitudes toward sex that was expressed by some of those who wrote about it at a slightly later date. The Jewish sexologist Max Marcuse (1877–1963), for example, wrote, "Christians tend to give the sexual life of man a stigma of baseness which is not at all the case in the Jewish community."[88] Moreover, some of the Jewish physicians might well have had some hostility to Christian conservatism about sex. Fritz Wittle, for example, a Jewish physician who was a member of a psychoanalytic group following Freud, admitted early in the twentieth century that it was "his extreme delight to hurl the importance of sex into the teeth of society."[89]

On balance, it was probably not so much Jewish attitudes about sex that accounted for the disproportionately large number of Jewish physicians in the sex field as the fact that unlike in most areas of medical practice there was little hostility to Jewish physicians entering this somewhat stigmatized field, because it had mostly been ignored by other physicians. Though by law (originally signed on July 3, 1869) Jews could compete officially for any occupation they chose in Germany,[90] there is ample evidence to indicate that they had difficulty getting high civil service and university positions. They could, however, become physicians, and a disproportionate number of Jews who entered the professions did so to study medicine. Although there were difficulties in entering certain medical specialties, research and practice in the areas of sexual behavior were not among them. Obviously, there also was increasingly widespread public interest in various aspects of human sexuality, as demonstrated by the explosion of literature on the topic. In Vienna, much of the study of sexuality came to be associated with Freud and psychoanalysis, and most of Freud's early followers were Jewish. In other

German-speaking areas, the sexological movement was much less psychoanalytical, but the Jewish presence was still very strong.

The Jewish presence in the German sexological movement was emphasized by the Nazis, who classified sexology as a Jewish science. Sexologists
such as Hirschfeld were among the first persons they attacked when they
took power. Unfortunately, much of the information and resources that had
been collected by these early Jewish investigators were either lost or deliberately destroyed: Books were burned and sexologists fled, died under house
arrest (as did Moll), or were sent to concentration camps. It has only been in
the last twenty or so years that the early German Jewish contribution to
modern sex research has begun to be better understood.

Two of the three giants of modern sex research in the first part of the
twentieth century—Hirschfeld of Germany and Freud of Vienna—were Jewish, and though Freud's psychoanalytic movement found a haven in the
United States, the work of Hirschfeld was generally ignored and dismissed.
The other major figure, Ellis, was British, and it was Ellis, even more than
Freud or Hirschfeld, who carried the message of the new kind of sexology to
the English-speaking world. Increasingly, the study of sex in the twentieth
century became broader, moving from its early focus on psychopathology to
general sexual behavior. It is to this topic, as it centered around the lives and
works of Hirschfeld, Ellis, and Freud, that the next chapter is devoted.

3

HIRSCHFELD, ELLIS, AND FREUD

*T*hree men dominated sexology during the early years of the twentieth century: Magnus Hirschfeld (1868–1935), Havelock Ellis (1859–1939), and Sigmund Freud (1856–1939). Hirschfeld and Ellis could be called empirical data gatherers, while Freud was a system maker who, on the basis of his system, developed a new therapy for those afflicted with sexual and other problems. Though each man knew of the work of the others and had contact with the others, Freud increasingly distanced himself from not only Hirschfeld and Ellis but other sex researchers to devote his energy to developing his own model.

For a time, at least in the United States, Freudian ideas about sexuality were the dominant ones. One of the reasons for this is that Freud, through his treatment modalities, provided a way for those interested in sexology to earn a living. Hirschfeld was independently wealthy, and though he practiced medicine and treated patients, his research was supported by his own funds. Ellis, while also a physician, supported himself almost entirely by his own writings, many of which were outside of the sex field. Freud, on the other hand, earned his living as a practicing physician and as such was far more interested in treatment than the other two. Thus while the data that Hirschfeld and Ellis collected were invaluable, these men did not necessarily provide treatment modalities that a practicing physician could use to help patients. Inevitably, Freud, the new system maker, became the model for much of the medical community, particularly in America, where the developing field of psychiatry came to dominate not only the treatment of patients but the writing and research about human sexuality until well past the middle of the twentieth century.

The psychiatrist or psychoanalyst saw patients who sought help with their problems, and the professional then usually published the case histories, including analysis and treatment. The result was almost a circular process: Because the Freudians dominated the publications dealing with sexual problems, they received the patients with such problems. While the urologist and gynecologist could deal with some aspects of sexuality, the psychiatrist and psychoanalyst had a theoretical construct into which all aspects of sexuality could be included. Moreover, Freudian theories inevitably spread to many other elements in the intellectual community, further adding to the dominance of his ideas. Though succeeding generations of psychoanalysts and disciples of Freud added to or slightly modified Freudian theories, the system itself was challenged only as a new generation of empirical data gatherers appeared, mainly in the universities, a setting previously not receptive to sex researchers except in the biological sciences. The new generation held professorships, which provided them with the financial security that Ellis never achieved through his writings and that Hirschfeld was born to. It also enabled them to challenge the medical dominance of sex research.

MAGNUS HIRSCHFELD

The most neglected, at least in America, of the three men discussed in this chapter is Hirschfeld. Undoubtedly influenced by his own homosexuality and transvestism, he did not pretend, at least at first, to be the dispassionate reporter of the varieties of human sexuality as Krafft-Ebing claimed to be. Instead he seemed, particularly in his early years, to have had an almost missionary zeal to bring the "truth" about sexuality to everyone. Though Hirschfeld started out as a political propagandist for homosexuality, he eventually became a significant researcher into human sexuality. A major reason for his comparative neglect, however, is that many of his contemporaries never forgot the fact that he had been a strong advocate for homosexuality and that sometimes, in his zeal, he tended to go to excess. In his later life, he was also a radical in politics, believing that only through changing the system could long-delayed changes in laws about sexuality take place. There were other reasons as well.

Hirschfeld's writings, for example, were often poorly organized and early on were not so well thought out, although he tended to improve with age. He turned out a variety of books and articles, some of which were outstanding and some of which seem to have been hurried into production to meet deadlines or fill space. His own lifestyle also worked against him, and he was ready to give battle anytime he felt homosexual rights were being threatened. Sometimes he seemed to lack common sense. A good illustration of this last is his participation in the Harden-Eulenburg trial, an action brought by the enemies of Kaiser Wilhelm II and the imperial court.

HARDEN-EULENBURG-VON MOLTKE AFFAIR

Critics of the policies of Kaiser Wilhelm were afraid to attack the kaiser openly and thus sought to attribute to a cliquish group of his advisers, some of whom were believed to be homosexual, those policy actions that they opposed. A small group of opponents came to believe that an attack on the alleged homosexuality of his advisers might force the kaiser to dismiss them, which would then result in a change of policy. The seed for such an attack came from the kaiser's support of his friend Friedrich Krupp (1854–1902), who at age thirty-three had inherited control of the Krupp industrial empire. Though married, Krupp lived much of the time on the island of Capri off the coast of Naples, away from his wife. There he allegedly brought young fishermen, mule drivers, and others, some of whom were legally minors, to engage in sexual relations with him. Though homosexual activities in themselves were not against Italian law, corruption of minors was, and Krupp, after being declared persona non grata, was expelled from Italy for his alleged involvement with minors. In the ensuing scandal, Krupp died, probably by committing suicide, but the kaiser tried to quell the public uproar and defend the house of Krupp by giving his friend a state funeral.[1]

Just how much influence the kaiser's enemies had in encouraging the Italians to bring charges is unclear, but his opponents saw the kaiser's efforts to minimize the scandal as a chance to claim that his court was riddled with homosexuality. Matters came to a head when Maximilian Harden, publisher of a Berlin periodical, *Die Zukunft*, and an opponent of imperial policies, charged that the kaiser was surrounded by a group of catamites who were perverting Germany policy. When this failed to bring a response, Harden mentioned two individuals by name: Prince Philip Fürst zu Eulenburg, former ambassador to Austria-Hungary, and Count Kuno von Moltke, military commander of Berlin. In October 1907, von Moltke launched a libel suit against Harden, but Harden produced extensive data about the alleged homoerotic tendencies of von Moltke, and Hirschfeld testified as an "expert witness" that von Moltke was a homosexual on the basis of such evidence. Harden was acquitted, but von Moltke appealed. In the second trial, Harden was convicted and sentenced to four months in jail, and much of the evidence produced in the first trail against von Moltke was demonstrated to be fraudulent. Eulenburg, who was initially charged with perjury for denying his homosexuality, never was brought to trial.

Although Hirschfeld may have thought that he was only performing a professional service by testifying, his testimony played into the hands of those who wished to label homosexuals in high places as a peril to the fatherland, and neither the original conviction nor the eventual acquittal helped the cause of homosexuality. It also threw doubt on Hirschfeld's expertise, and more important, made him anathema to the kaiser and his court, whose support was essential if homosexuality was to made legal.

Some of his critics also opposed Hirschfeld on more professional grounds. Sexologists like Moll, though initially somewhat supportive of Hirschfeld's ideas, ultimately ended up in open opposition both to his theories and to the nature of his research. The disagreements between Moll and Hirschfeld in their later years, however, were more than scientific or scholarly disputes. Although the disagreements did have a professional basis, the men increasingly seemed to have been motivated by personal hostilities and rivalries. Their conflict forced many in the sexology field to choose sides.

Freud also had initially praised Hirschfeld, and in fact, Hirschfeld had joined with Karl Abraham in founding the Berlin Psychoanalytical Society. In 1911, at the Weimar Congress of the Psychoanalytical Association, Freud treated Hirschfeld as an honored guest and described him as the Berlin authority on homosexuality.[2] When Hirschfeld, however, left the society shortly after the Weimar Congress, Freud put him down, calling his "departure no great loss" and Hirschfeld "a flabby, unappetizing fellow, incapable of learning anything."[3] This not atypical Freudian putdown of his critics or "deserters" meant that many psychoanalysts, particularly those active in the United States, later ignored Hirschfeld's work.

HIRSCHFELD'S EARLY YEARS

Hirschfeld was the son of Hermann Hirschfeld, a well-known physician and philanthropist in the seaside spa of Kolberg in Pomerania, and Frederika Mann, a member of a prominent Jewish family from Pomerania. He, like his two brothers, decided to follow his father's footsteps and started his medical education at Strasbourg. He soon left there for Berlin, then moved to Munich. Hirschfeld also studied at Heidelberg and finally returned to Berlin to complete his studies. His dissertation was on the effects of influenza on the nervous system. He then visited the United States and returned by way of Morocco, Algiers, and Italy.

In 1894, he opened an office in his hometown as a general practitioner and obstetrician, but two years later he moved to Berlin where he became a specialist in hydropathy. It was in Berlin that he launched his career as an investigator of sex. His first entry into the field was a thirty-four-page pamphlet titled *Sappho und Socrates, Wie erklärt sich die Liebe der Mannër und Frauen zu Personen des eigenen Geschlechts?* (Sappho and Socrates, How Can One Explain the Love of Men and Women for Individuals of Their Own Sex?). Hirschfeld wrote the pamphlet shortly after learning of the suicide of one of his patients, a young homosexual officer who shot himself through the head on the eve of his marriage. Just before doing so, he had mailed Hirschfeld a letter in which he announced that he killed himself because he felt so tortured by the double life he was forced to lead. He urged Hirschfeld to tell others his tragic story in the hope that they could better understand the difficulties under which homosexuals lived. Writing under the name of Th. Ramien, Hirschfeld argued that homosexuality was part of human sexuality, that both its causes and its manifestations should be the object of scientific investigation, and that the penal laws against homosexuality should be changed in the interest of society.[4]

The pamphlet opened with a quote from Friedrich Nietzsche—"what is natural cannot be immoral"—and aroused more interest than might have been expected because of the publicity generated by the trial of Oscar Wilde in England. Hirschfeld, relying heavily on the work of Moll,[5] and to a lesser extent on Krafft-Ebing, stated that all of the sciences had demonstrated that homosexuals composed a third sex. He then went much further than Krafft-Ebing, however, and declared that homosexuality was simply a variety of human sexuality. The key to his theory lay in embryology (as did that of Ulrichs), although he was not fully aware of what Ulrichs had written. He attempted a 10-point scale to classify people based on his basic three principles of development; actually there were six principles, because he felt females and males went through the same three phases but with slight differences.

HIRSCHFELD'S THEORY

Most people, according to Hirschfeld, were originally bisexual, but in the course of their natural development, they lost their desire for members of the same sex. These people were the heterosexuals who loved members of the opposite sex. The second category of individuals was made up of the psychohermaphrodites—men and women whose sexual organs had developed normally but whose feeling centers for one or the other sex were imperfect, and as a result, these people could love individuals of both sexes. The third category consisted of those individuals whose sexual organs developed normally but in whom the desire for same-sex individuals in the feeling center failed to recede. The results were men who loved men and women who loved women. Hirschfeld continued to modify his theory of the causes of homosexuality over the next four decades but never really came to a satisfactory formulation, probably because none of what he said could really be proven.

The pamphlet *Sappho und Socrates* represents the strength and weaknesses of much of Hirschfeld's later work. He dismissed outright those people who disagreed with him, was sloppy in his historical data (he had Sappho killing herself because of unrequited love for a woman), and was quick to claim earlier historical figures as homosexuals or lesbians without much evidence. He also assumed that his was the only correct explanation for homosexuality, a claim that was quickly challenged by other homosexuals. In fact almost immediately after the appearance of Hirschfeld's pamphlet, another pamphlet was brought out by the same publisher, Max Spohr. The anonymous author of this pamphlet held that homosexuality was not an inborn condition but rather was acquired through an individual's passage through life. The problem, however, was not with the individual who developed into a homosexual but that society punished the homosexual, when it should really accept him or her.[6]

Many of the ideas in this second pamphlet were similar to those expressed by Benedict Friedländer and Adolf Brand, who opposed Hirschfeld's notion of a third sex, and may well have been written by them. Friedländer later argued that theories such as Hirschfeld's made all homosexuals effeminate (or in the case of lesbians, masculine), while they regarded homosexuality as an idealized aspect of male bonding such as had existed in ancient Greece. For them, homosexual love was spiritual and not a physical or animal desire; in other words, sexual intercourse was to have no place in such relationship. Friedländer and Brand did note that sexual intercourse could (and did) take place, but in circumstances in which the idealized love between two such individuals led to an intimate sharing of bodily fluids, the act was different from animal love.[7] This view was later more fully developed by Hans Blüher, who divided homosexuals into three types: the heroic male, the effeminate invert, and the suppressed homosexual.[8] Though Friedländer and like thinkers recognized that society was organized primarily around the family and the state—a heterosexual base—they also believed it had a secondary base in male bonding, which involved homoerotic feelings; this was the major role for the heroic male.

HOMOSEXUALITY AND POLITICS

This split over possible theories for the existence of homosexuality went beyond hypotheses that neither side could prove or disprove; it was a political split as well—and Hirschfeld was very much a politician. On his twenty-ninth birthday, May 14, 1897, Hirschfeld founded the Wissenschaftlich-Humanitäre Komitee (Scientific Humanitarian Committee) to give new life to the struggle started earlier by Ulrichs and others for the repeal of the antihomosexual provision, by then section 175 of the imperial penal code as adopted in 1871. The imperial law imposed a maximum of two years' imprisonment for "lewd and unnatural conduct" between males. As part of their campaign, the committee members circulated petitions to be signed by supporters of the legal change, and many people prominent in public life signed their names. For a time, the cause was adopted by some of the political parties. August Bebel, the leader of the German Social Democratic Party, spoke on the floor of the Reichstag in favor of the petition. As a result of his efforts, the petition was put on the agenda, although it was not officially discussed until 1905, at which time it was quietly removed.[9] While awaiting such discussion, Hirschfeld and the committee persuaded district attorneys in several of the larger German cities to refrain from prosecution if consensual sex was involved.

The failure of the committee to achieve its political goals tended to accentuate the split between the followers of Hirschfeld on the one hand and Friedlander and Brand on the other. The unity of the group was further damaged over the Harden–Eulenburg–von Moltke trials, although Hirschfeld continued to push for reform all of his life.

More important in the long run than the political activities of the committee were its scholarly activities on behalf of homosexuality, particularly the publication, starting in 1899, of *Jahrbuch für Sexuelle Zwischenstufen* (Yearbook for Sexual Intermediates), the title of which reflected Hirschfeld's concepts about a third, or intermediate, sex. Hirschfeld edited the twenty-three volumes that appeared (under slightly varied titles) between 1899 and 1923. Many of those published during World War I, when paper rationing was severe, were little more than newsletters, and several issues were combined into one. The series was briefly revived in stronger form after the war, only to fold with the monetary collapse of Germany in the early 1920s. The journal was a mixture of scholarly articles, reprints of classical articles such as Kertbeny's earlier pamphlets, propaganda pieces, political essays, biographical studies, and special pleading. Though in its early issues it had significant contributors such as Krafft-Ebing, the journal generally was ignored by official science and scholarship in Germany. For anyone studying the history of sex, however, the series remains invaluable.

HIRSCHFELD THE RESEARCHER

Hirschfeld also began to carry out sex surveys, and his 1903 report that 2.2 percent of all those surveyed were homosexual led Moll to break with

Hirschfeld over what he felt were exaggerated statistics. Though Hirschfeld's claim seems a remarkably accurate one in light of current knowledge, it brought nothing but hostility to Hirschfeld, and it served as one more excuse for his enemies to attack him.[10]

The rejection of the petition for reformation of sex laws, and particularly the defection of various groups from Hirschfeld's committee, forced a rethinking on the part of Hirschfeld. Though he had claimed that science had demonstrated that homosexuality was not pathological, few had agreed with him, and his writings on the topic had been far more polemical than objective. For a time, he felt that science was not on his side, because of the various disagreements among both his followers and his opponents. It was in this setting that he turned to Iwan Bloch, who had been emphasizing *sexual-wissenschaft*, or sexual science. It was through this new kind of sexual science that Hirschfeld sought to move the discussion of sex from the political arena to the scholarly, scientific disciplinary one, and in the process, he hoped to provide solutions to sexual problems.

Hirschfeld quickly rededicated himself to finding a scientific basis for his beliefs, and he embraced Bloch's new view of *sexualwissenschaft*. One of Hirschfeld's first acts was to begin publishing in 1908 a new journal devoted to sexology as a science: *Zeitschrift für Sexualwissenschaft*. The very first issue of this journal contained an article by Freud titled "Hysterical Fantasy and Its Relation to Bisexuality," and subsequent issues presented original works by Alfred Adler, Karl Abraham, and Wilhelm Stekel, among others. Hirschfeld, at that time, made a significant effort to include the Austrian psychoanalytical movement as part of the legitimate study of sexological science. Hirschfeld also traveled to Italy to solicit personally articles from Paolo Mantegazza and Lombroso, an indication that he hoped that sexology could become a new international science.

Hirschfeld also encouraged controversy. Helene Stöcker, an early Berlin feminist, contributed an article on the differences between the love lives of women and men; this was a response to a chauvinistic article on more or less the same topic by Wilhelm Sternberg. Stöcker was somewhat upset at the diversity of views in the journal and took Hirschfeld to task for publishing the Sternberg article, which she said was contrary to Hirschfeld's own views on the topic.

Journal articles came from a variety of disciplines, and during the first year, articles dealt with historical, philological, pedagogical, biological, medical, and ethnological aspects. Serving with Hirschfeld as coeditors of the journal were the Viennese ethnologist Friedrich Salomon Krauss and the Leipzig physician Hermann Rohleder, both instrumental in broadening the concept of sex research.

HIRSCHFELD, ELLIS, AND FREUD

Unfortunately, the plans were far more ambitious than the finances. After only a year of publication, the journal was merged with *Sexual Probleme*, a more popular and less scholarly journal. The resulting amalgam appeared under the title *Zeitschrfit für Sexualwissenschaft und Sexual Politik* and later under still different titles, as other attempts were made to revive it.[11] Undaunted by the failure of the journal, Hirschfeld, for his part, continued to apply what he believed was his newfound scientific objectivity to his research. His first work to qualify as a major contribution was *Die Transvestiten* (1910), a term he coined. This ignored classic (it was translated into English only in 1991) challenged the view that all cross-dressers were homosexual, since Hirschfeld found many such individuals were heterosexual. After examining possible correlations of cross-dressing with homosexuality, fetishism, and masochism, he said that while all might have some bearing, the transvestite was different. The difference between the homosexual man and the transvestite man (he also included some women) was not in behavior but in the focus of pleasure. Transvestites differed from fetishists, because fetishists tended to attach the object of the fetish to a beloved person, while transvestites focused on themselves and their clothing. While there was some masochistic tendencies, since male heterosexual cross-dressers tended to seek out masculine women, he felt that this was not a major causal factor.[12] His observations and data on cross-dressing, if not his theory, were not matched until the last decade of the twentieth century.

Hirschfeld then published *Die Homosexualität des Mannes und des Weibes* (1914), in which he repeated his ideas, summarized above, with slight modification. His use of the term *homosexuality* consolidated it into the community.[13] What is most valuable about Hirschfeld is the amount of data he compiled about homosexuality, transvestism, and other forms of sexual activity. He held that a variety of sexual behaviors were normal, and he was more interested in describing this variety than condemning it.

Hirschfeld was not content to rely solely on his practice or on case studies passed on to him by others; he set out to seek information from a large variety of informants. Shortly after 1900, he developed what he called a psychobiological questionnaire, which contained some 130 questions and which he administered to more than ten thousand men and women. On the basis of this, he wrote what he called his first sexobiological book, *Naturgesetze der Liebe*, which marked a breakthrough in his research: He was no longer just interested in the "pathological" but in sex behavior in general. This study was strongly influenced by the German evolutionary biologist Ernst Haeckel (1834–1919), who laid stress on the fundamental biogenetic law that ontogeny recapitulates phylogeny, or that the organism in its development demonstrates, to a great extent, the morphological changes that occurred during the evolution of the species. Haeckel, who developed a theory called monism, held that the material basis of true-life phenomena—nourishment and reproduction—was due to an intricate chemical interaction and said that an "erotic chemotropism" was the very source of love.

Hirschfeld adopted this belief and asserted that some kind of internal secretions, what we now call hormones, were the principal source of the feelings of love and sexual attraction. He held that the testes secreted a chemical substance that he called "gandrin" and the ovaries, something he called "gynecin," although such substances had not yet been isolated.[14] Part of the difficulty with Hirschfeld is that, as his biographer, Charlotte Wolff, said, "he tapped at the door of modern science but could not get it to open."[15] He wanted to find a biological explanation for all kinds of sexual behavior, and when the science of the time could not give them, he hypothesized such explanations. He often was on the right path, but sometimes his ideas were based on fallacies. In his defense, he occasionally seemed to realize there were difficulties with his concepts, such as the existence of a third sex, because he admitted that physically normal people could be homosexuals or bisexuals, but he still insisted on a third sex.

Part of the Hirschfeld's difficulty derived from his use of the monist theory, which lay at the core of both his and Moll's theories. Monism tended to deprive the idea of sexuality of its traditional, limited meaning. As Lawrence Birken pointed out, evolutionary theory posed a dilemma for sexologists, because by relating all forms of desire to each other, it gave them a generic unity that subverted difference. Yet, at the same time, it emphasized the possibility of controlling these desires by relating them to each other in a developmental hierarchy. In sexology, women and children became sexualized even as they continued as sexless denizens of the innocent world outside competitive society. Sexuality was simultaneously universal and the function of the adult male alone. The problem was the necessity of upholding the idea of difference with a theory that emphasized differentiation from a common sameness. As long as the accent was on the differentiation, difference could be sustained; but there was a gradual tendency to shift the emphasis to the common sameness underlying the apparent difference. In sexology, this saw a shift from the idea of an adult male sexuality to one of a universally defined sexuality.[16] Birken held that "the phylogenetic origins of sexuality in primeval undifferentiated desire undercut any attempt to distinguish the sexual from the nonsexual. In this context, social energies might appear as nothing more than a rarefied form of sexual energy. In other words, the social appeared as a higher stage of the sexual, arising out of but in opposition to primeval desire."[17]

It was Freud who broke through this difficulty, replacing the difference between social and sexual desire with a concept by which social and sexual love become differentiated unequally from each other. In replacing difference with differentiation, Freud was able to explain precisely how relations within the family were connected to relations outside the family in a hierarchical order. The Oedipus complex of Freud, in a sense, was a defense against an even more unbearable idea, the dissolution of the hierarchical order of the sexes.

Hirschfeld, however, never visualized any of this debate as a problem and so never dealt with it. He remained interested in organization, continually trying to extend the network of sexologists and inform the public. The Humanitarian Committee, in spite of disaffections of some, continued to agitate for change, but Hirschfeld also wanted more data and information. In 1913, he was instrumental in the founding of the Ärtzliche Gesellschaft für Sexualwissenschaft und Eugenik (Medical Society for Sexual Science and Eugenics), which was conceived of as the beginning of a worldwide movement for sexual reform. This led to a revival of the *Zeitschrift für Sexualwissenschaft* under the editorship of Eulenburg and Bloch.

World War I proved to be a major setback to the German sexological movement. Hirschfeld, who had long been a pacifist, initially threw himself into the German war effort, and in his tendency to exaggerate, which always plagued him, he became the supreme German patriot. In his absolute certainty that Germany was in the right, he made statements that never should have been made. When his passions began to cool, he quickly abandoned his early enthusiasm for the war, ending up involved in the movement to oust the kaiser and establish a democratic government in Germany. For the rest of his life, he was strongly left wing—probably, at least for a time, a Communist—and this influenced the reception of his ideas about sex.

During the war he began publishing *Sexual Pathology*, which he regarded as an update to Krafft-Ebing. Though it has a large number of excellent observations, it is weakened either by poor theory or by unfortunate remarks. Still, one of his most important contributions was to challenge Krafft-Ebing's baneful beliefs in the effects of masturbation. Amplifying on the studies of the German sexologist Rohleder, who had reported that 90 percent of all people younger than twenty had masturbated,[18] Hirschfeld found that if anything this was an understatement and reported that in his estimation 96 percent had done so. He wrote that the harmfulness and consequences of masturbation have been greatly exaggerated. "In most cases the exaggerated fear of the harmful consequences of masturbation is far more harmful to health than the act itself. A certain lassitude and inability to concentrate may, of course, be induced by excessive masturbation, but will pass very quickly of itself if the subject's mode of life is natural and normal."[19]

Hirschfeld, however, could not quite overcome the nineteenth-century fears of masturbation, and believed, following others, that there was such a thing as hyperactive self-gratification and recommended sterilization of men and clitorectomy in women to prevent it.[20]

Overall, his *Sexual Pathology* is a considerable advance over what had gone before, and when he wrote of chromosomal abnormalities or hormonal abnormalities, he was reporting on the cutting edge of the known research, but when he advanced his beliefs about psychoendocrinism—the interaction of organic and psychological factors—he was going beyond what the science of the day could confirm. Science had not advanced far enough to give the kind of answers that he tried to give, and unfortunately, he did not always distinguish between what he believed and what the existing evidence could demonstrate. When data were lacking, he fell back on theories that, in the end, turned out not to be valid. In many areas, however, he was quite cautious. For example, though he rejected Freud's theory about sexuality per se, he agreed with Freud about the sexual origin of many neuroses and with his emphasis on it in hysteria and obsessional ideas. He did not agree with Freud on psychic influences and complexes or with infantile experience.[21]

In 1919, Hirschfeld finally realized a long-term dream with the foundation of his Institute of Sexual Science in Berlin; there he could consolidate and extend his data and house his library of more than twenty thousand volumes and thirty-five thousand pictures that supported his research. Using his psychobiological questionnaire, he continued his wide-scale study of sexual habits. He established a marriage counseling service, gave out advice on contraception and sex problems, and continued his prolific writing (he himself claimed 187 works).[22] Increasingly, Hirschfeld and his colleagues branched out into studies of female sexuality, marriage, contraceptives, and prostitution, becoming less concerned with sexual variance and more concerned with general sexual problems.

INTERNATIONAL SEXOLOGY CONGRESSES

Still, Hirschfeld increasingly felt the key to understanding sexuality was endocrinology. In 1921, his dreams of having an international sexological movement came to fruition with the International Conference of Sexual Reform Based on Sexual Science, which took place in Berlin from September 15 to 20. The theme of the conference was the importance of internal secretions for human sexuality, but the papers from the thirty-six speakers were wide ranging and did not always hold to the theme. Included in the audience were physicians from Germany, Finland, the Soviet Union, Czechoslovakia, Hungary, Italy, Sweden, Denmark, Norway, Holland, France, the United States, Argentina, China, and Japan as well as the emerging independent Baltic states. Four of the participants were women. The fact that Hirschfeld was host and keynote speaker solidified his standing in the sexological community, and out of this meeting eventually came the World League for Sexual Reform. The economic and other postwar difficulties in Germany made it difficult to follow through on this until 1928, when J. Leunbach of Copenhagen organized the second meetings, this time known as the Congress of the World League for Sexual Reform. Coming from United States to the congress was Harry Benjamin, Margaret Sanger, and William Robinson, all of whom continued to be active in international sexological meetings; in a sense, these Americans were involved in the transitional development of one wing of the sexological movement.

Again Hirschfeld gave the opening lecture, and the league became formally organized with three co-presidents: Auguste Forel of Switzerland, Ellis of the United Kingdom, and Hirschfeld. The chief points of the league policy was an advocacy of sexual education, sexual equality of men and women, reform of marriage and divorce laws, encouragement of contraception and birth control, reformation of the laws on abortion, protection of the unmarried mother and the illegitimate child, prevention of sexually transmitted diseases, removal of the economic factors that led women into prostitution, promotion of a rational attitude toward sexually "abnormal" persons, and reformation of the laws regarding sexual offenses. In short, the platform combined Hirschfeld's ideas of sexual reform and research but the emphasis was on reform.

The Third International Congress, held in London, again saw Hirschfeld give the keynote address. He was followed by one hundred other speakers including such nonsexologists as the philosopher Bertrand Russell, the dramatist John van Druten, and the writer Desmond MacCarthy. Many of the talks were devoted to contraception and were marked by an effort by those in attendance to try to distance themselves somewhat from the eugenics movement.

The Fourth International Congress, held in Vienna in 1930, was again keynoted by Hirschfeld, still one of the three co-presidents (Norman Haire of the United Kingdom and Leunbach were the others). The conference had fewer participants than the earlier ones, and many of the scheduled speakers, such as Ellis and Benjamin, could not attend, although their lectures were printed in the book of congress papers.

The Fifth Congress had originally been planned for Moscow and then for Paris but ended up, in 1932, in Brno, Czechoslovakia. This congress was important because, unlike the others, it took place in an academic atmosphere at a university. It also had the sponsorship of the Czech president, Jan

Masaryk. Many of the participants broadcast their speeches back to their own countries over the radio. In spite of such a public relations coup, the congress was the last to be held, partly due to the rapid decline of the German sexological movement but also due to the strong differences between Haire and Leunbach. Leunbach wanted the league to join with the revolutionary workers' movement, but Haire was determined to keep all revolutionary activity out of the league and to concentrate on educational projects. The ultimate result was the dissolution of the league.[23]

HIRSCHFELD, THE NAZIS, AND HIS DEATH

The Brno congress marked the end of one phase of European sex research in other ways as well. Within a year of its conclusion, Nazi hoodlums, encouraged by the newly legitimatized Nazi government, on May 6, 1933, broke into Hirschfeld's institute in Berlin. They destroyed the greater part of his collection and data and removed books from the library and publicly burned them. Hirschfeld was traveling at the time and learned of the destruction in Paris, where, in a cinema, he saw with his own eyes the destruction of much of his life's work. He tried to start again in France, but he died in Nice on May 14, 1935—his sixty-seventh birthday.

This portrait of Hirschfeld hints at the disputes in the sexological field of the time, but there is not enough space to amplify all the differences. Distinctions were strong and Moll in particular was quick to counter almost every move that Hirschfeld made. When Hirschfeld organized his Society of Physicians for Sexual Science and Eugenics, Moll offset it with the formation of his International Society for Sexual Research. When Hirschfeld organized the first international conference on sexology in Berlin, Moll planned another one. Economic conditions in Germany postponed Moll's conference until 1926, when the International Congress of Sex Research was held in Berlin, to which Hirschfeld was pointedly not invited.[24] Moll claimed that if he had invited Hirschfeld, many of the others would not have attended, because, he stated, that Hirschfeld was seen as an apologist for homosexuality. Most of those who attended the Moll conference did not know until after they arrived that Hirschfeld had been left out. Moll and Hirschfeld were on opposite sides of almost every issue. Hirschfeld himself, however, frequently appealed to Moll to lay aside their personal differences and asked Moll to join him in the search for the scientific truth, which Hirschfeld felt could be found. Hirschfeld could never understand why Moll was so opposed to him. He knew they disagreed, but felt that ultimately science would give the answers.

Hirschfeld was erratic, was sometimes extravagant in his claims, and did not always think through his actions; he was left wing, very close to the Communist Party, and homosexual. Moll was cautious and conservative, a German nationalist, heterosexual, and very much concerned that homosexual researchers such as Hirschfeld would "infect" sex research. Moll challenged Hirschfeld's belief that he could diagnose homosexuality in children, was critical of his theory of sexual intermediaries, and could not abide Hirschfeld's belief that homosexuals had special virtues and were more democratic and more altruistic than heterosexuals.[25] Moll was particularly irked by Hirschfeld's claim that the science of sexology had been founded by Bloch and instead insisted that Krafft-Ebing deserved to have the credit.

Haire, who had studied briefly under Hirschfeld in Berlin and who served later as co-president of the World Congress, was fond of Hirschfeld. Still he found that Hirschfeld was often hard to take. He wrote, "As the rest of us, he had his imperfections. He was not always tactful. He didn't always stop to think how his actions might be interpreted by persons of ill will. He could be very selfish and exigent in small matters. His appearance was, I think, unprepossessing."[26]

Hirschfeld's ultimate importance to sex research is not so much his theory, although he had important insights, but his data collection and his organizing ability. Though many of the sources for his data were destroyed by the Nazis, he had published significant amounts of his data, and later generations of researchers have found his cases to be invaluable. So-called research on sex had started out mainly as learned opinions and theories, based on historical and cultural data and a few clients, but the growth of case studies and the ability to compare backgrounds of different peoples (as both Moll and Hirschfeld did) opened up new horizons in sex research and led to challenging and modifying traditional ideas. For the most part, however, the state of knowledge in the biological nature of humans was not yet advanced enough to go beyond this. The fact that some researchers, such as Hirschfeld, were openly challenging societal attitudes and threatening to undermine traditional ideas was a major factor in the opposition to Hirschfeld as expressed by Moll and others.

HAVELOCK ELLIS

More successful and much less dogmatic and antagonistic was the English sexologist Havelock Ellis, whose *Studies in the Psychology of Sex* (1896–1928) popularized the concept of the individual and cultural relativism in sex. In a sense, Ellis was a naturalist, observing and collecting information about human sexuality instead of judging it. Always cautious, he avoided unitary theories. Faced with the question of whether homosexuality was inborn or acquired, physical or psychic, he felt there was perhaps some truth in all views. Although he tended to believe that sexual differences were inborn and nonpathological, he was willing to grant that perhaps there was a higher number of neurotics among deviants than among other groups. This, however, he also qualified, by stating that the neurosis may be due to societal rather than biological factors.

Essentially, Ellis's work was a plea for tolerance and for accepting the idea that deviations from the norm were harmless and occasionally perhaps even valuable. He, like Hirschfeld, was a sex reformer who urged society to recognize and accept sexual manifestations in infants and to realize that sexual experimentation was part of adolescence. Ellis held that it was important to repeal bans on contraception as well as laws prohibiting sexual activity between consenting adults in private.[27]

We know a great deal about Ellis from his autobiography, including the fact that he had a fascination with the act of urination, which he called urolagnia. This fascination he believed resulted from observing his nurse urinate in a public park in his presence. She was pushing his perambulator when she suddenly stopped, a pause followed, and then he "heard a mysterious sound as of a stream of water descending to the earth."[28] This was in the days when there were no public toilets and women wore long skirts and no underpants; park walks were made of crushed stone to enable them to answer the call of nature.

His sea captain father, Edward P. Ellis, was absent during much of Havelock's youth, and he was raised primarily by his mother, Susannah Wheatley Ellis, who at the age of seventeen had undergone conversion to the evangelical principles of John Wesley. She made great efforts to protect him and her other children from the ever-present dangers of evil, packing them off regularly to religious services. Ellis's means of escape from his rigidly circumscribed life was through reading in the library of his grandparents, who allowed him to choose what he wanted to read. He seems to have been somewhat feminine as a boy with what were called "overfine sensibilities," and he was the subject of considerable bullying by his classmates.

At sixteen, he accompanied his father on part of a long sea voyage that ended for him in Australia. He had been seasick for much of the voyage, and when he went ashore in Sydney, he was diagnosed as too weak to continue on with his father to Calcutta. Instead, Ellis stayed behind, supporting himself as an assistant schoolmaster, a job at which he was not particularly successful. Away from his family for the first time, he found himself much troubled by sexual impulses that he did not understand. To come to terms with them, Ellis thought it necessary to "explore the dangerous ocean of

sex."[29] While browsing in a Sydney bookshop, he came across an anony-
mously written book, *The Elements of Social Science; Or Physical, Sexual and
Natural Religion*, which greatly influenced him.

The author was the physician George Drysdale (1825–1904), who was a
pioneer in pointing out the joys of sex in a manner that could avoid the Victorian censors. Drysdale, however, true to the standards of his day, held that
immoderate amounts of sexual activity were dangerous, and he expressed
horror at the variant forms of sexuality. Still, he emphasized that sexual
intercourse could and should be a delightful thing. He held that the main
obstacle to enjoying sex was the ever-present possibility of children, and it
was this fear that caused men to turn to prostitutes and other forms of sexuality and women to become sexually inhibited. A neo-Malthusian, he also
believed that overpopulation was a cause of poverty. All of these problems
could be overcome by the use of contraceptives.[30]

Drysdale was very un-Victorian in believing that sex itself was to be a central issue in the future. Love, for him, was a necessity rather than a luxury,
and moral virtues were, in the last analysis, simply matters of mental
hygiene. Though Ellis later was critical of much of the book, at the time it
influenced him. In the introduction to his series on the psychology of sex,
he wrote:

> The origin of these *Studies*, dates from many years back. As a youth I
> was faced, as others are, by the problem of sex. Living partly in an Aus-
> tralian city where the ways of life were plainly seen, partly in the soli-
> tude of the bush, I was free to contemplate and to meditate many
> things. A resolve slowly grew up within me: one main part of my life-
> work should be to make clear the problems of sex.[31]

After his return to England, Ellis continued to teach, but proved less successful at each job. He remained troubled by sex, and in his soul searching
he vowed to devote himself to a life of science. He set out to become a
physician, the only profession through which he felt he could safely devote
himself to the study of such a forbidden topic. In this, he was also strongly
influenced first by the writings of and then by friendship with James Hinton,
who, like Ellis, had been troubled about what to do with his life and had followed medicine. Hinton had also begun to study sexual behavior.

Helped financially in part by an inheritance from his mother and a loan
from Hinton's sister (women were continually coming to Ellis's rescue), Ellis
began his medical studies. To help support himself, he turned to writing, an
occupation that gave him a livelihood for the rest of his life. Though he
completed medical school, he never really practiced medicine, and he used

his training primarily to give him authoritative status for his sex writings. In December 1891, Ellis married Edith Lees, and though the marriage was consummated, sex thereafter did not play a significant role in it. His wife had strong lesbian tendencies, and neither Ellis nor she wanted any children. After a few years of marriage, the couple agreed not to engage in sexual intercourse, although both embarked on extramarital affairs. For a time, Ellis was impotent. His mistress, Françoise Delisle, with whom he finally found himself potent, wrote an account of her relationship with him in which she told about some of the sexual difficulties he had.[32]

As his success as a writer and editor grew, he began his first efforts toward a study of sex with the publication of *Man and Woman* (1894), in which he set out to investigate to what degree sexual differences were artificial or biological. He concluded that men were designed to make history and women to make children, a relationship that he said maintained a "perfect equipoise."[33] This was because civilization was the creation of mentally and physically "abnormal" persons, and as more men than women fit that description, civilization drew its inspiration from men. This conservative view of the importance of women was strongly criticized by the eugenicist and statistician Karl Pearson. Others criticized the book for, among other things, its lucid description of menstruation.

Nonetheless, his study on male-female relationships aroused enough attention that he gained the courage to proceed further with his study of sex, although he continued to write on a variety of other topics. Ellis was most impressed by the way in which sexual topics had been treated by writers such as Nietzsche, Émile Zola, and Giacomo Casanova. Increasingly in his general literary efforts, he began making gentle attacks on the hypocrisy of modern morals, a theme that appeared throughout his sexual studies as well. Homosexuality seemed to him to be a key to understanding the nature of human sexuality, and as he began investigating the topic, he realized that several of his acquaintances were homosexuals, most notably John Addington Symonds and Edward Carpenter.

Ellis researched his topics through literature and history, supplemented by a rather haphazard collection of individual sex autobiographies, what might be called case studies. Many of his friends not only wrote their own sexual autobiographies for him but solicited their friends and sexual contacts to do so as well. He asked his wife about her homosexual friends, and though this upset her, she complied with his request to solicit biographical information from them.

Others before him had gathered information in the same way, particularly his friend Symonds (1840–93), best known for his studies of Renaissance Italy. Symonds, a homosexual who was married and the father of four children, had written two anonymous books on homosexuality: One was a long essay on homosexuality in ancient Greece called A *Problem in Greek Ethics* (1883), and the other, a historical overview titled A *Problem in Modern, Ethics* (1891). He had done much research on the topic.[34] When Ellis realized the extent of Symonds's research, he proposed they collaborate on a study of homosexuality.[35] Symonds had used the term *inversion* to describe homosexuality in his published studies, as did Ellis for a while; however, Ellis ultimately discarded it for the term *homosexuality*.

HOMOSEXUALITY

The collaboration was an uneasy one, because the two men disagreed with each other, perhaps because Symonds, who was far more emotionally involved in the subject than Ellis, held such strong views. Symonds denied both that masturbation was a cause of homosexuality and that homosexuality might be inborn. Instead, he had romantic notions about male bonding similar to those of Friedländer, Brand, and Blüher, discussed earlier in this chapter. While Ellis agreed with Symonds about the harmlessness of masturbation, he was doubtful about the romantic notions of male bonding and instead held that heredity played some part. Symonds also did not believe that homosexuality was either a morbid or perverted condition, and although Ellis could accept this, he held that inverts were often neurotic.

Neither man's method was particularly scientific. Ellis himself rarely conducted in-depth interviews of the subjects of his case histories, unless they were close, personal friends. Instead, much of his material was secondhand written accounts given to him by the individuals themselves or by intermediaries, and Ellis did no further investigation. Moreover, a disproportionately large number of the cases he used had been gathered by Symonds, who was not always the most dispassionate observer or chronicler. To further complicate the data, Ellis, to disguise his subject's identities, sometimes deleted details and changed facts. The case history of his own wife, for example, is listed under the title "Miss H.," and it neglects to mention crucial factors in her life.

Though the agreement made between Ellis and Symonds stipulated that Ellis write the introduction and the case studies, and Symonds, the historical background, Symonds died before the project was completed. Ellis, however, gained permission to use much of Symonds's material, and it formed a significant part of the book.[36] The book appeared in 1896 in German, *Das Konträre Geschlechtsgefühl* (to avoid problems of British censorship) under the names of both authors, and it was scheduled for an English version in 1897. Before the English edition was distributed, however, Symonds's family, to the surprise of Ellis, sought to remove his name from the publication, since they did not want any public association of the family name with homosexuality. Because the book had already been printed, the family was allowed to purchase all the copies of the first English edition and have them destroyed. They also secured an agreement with Ellis that if any further editions were published Symonds's name would be removed from the title page.

While Ellis accepted these conditions, he found he had trouble getting a publisher for his revised work in England. His book was finally accepted by Watford University Press. This new English edition, published under the title *Sexual Inversion*, soon ran into censorship problems, not so much for its content but because of its publisher. This was because Watford University Press was also the publisher for the Legitimation League, a group that advocated divorce by mutual consent as well as the removal of any stigma of illegitimacy on children born out of wedlock. The London police believed that such advocacy was an indication of a conspiracy by anarchists to undermine British morals, and they kept careful watch on a bookstore run by the league. An undercover agent saw Ellis's book on the shelves of the store, and the police decided to use what they felt was an indecent book to stamp out the league, the publisher, and the anarchists who were encouraging divorce. In spite of considerable support by literary figures such as George Bernard Shaw, Frank Harris, and Edward Carpenter, the bookseller pleaded guilty and was fined £100. Ellis, anxious to rid himself of his controversial publisher, took back his rights, transferred publication to the United States, and only later released the book in England.

This discussion of Ellis's early publication difficulties emphasizes what might be called much of the theme of Ellis's life and work: a desire to compromise, to avoid association with dogmatists, and to avoid unnecessary controversy. Unlike Hirschfeld, who pushed his way through obstacles at full speed, Ellis tried to win people over to his view. When he had to choose between stating that homosexuality was inborn or acquired, he said there was truth in both views. Ultimately, however, he tried to distinguish between homosexuality, which involved any physical or sexual relation between two people of the same sex, and inversion, which had definite congenital origins. He held that inborn homosexuality was inescapable and, therefore, socially acceptable, whereas acquired homosexuality might be open to cure and thus not acceptable. After admitting this, he concentrated on congenital inversion and so avoided the issue of dealing with acquired homosexuality.

He concluded his study on homosexuality by discussing what the attitude of society should be toward the congenital sexual invert. Typically, he believed that the correct answer lay in avoiding two extremes.

On the one hand, it cannot be expected to tolerate the invert who flouts his perversion in its face and assumes that, because he would rather take his pleasure with a soldier or a policeman than with their sisters, he is of finer clay than the vulgar herd. On the other, it might well refrain from crushing with undiscerning ignorance beneath a burden of shame the subject of an abnormality which, as we have seen, has not been found incapable of fine uses. Inversion is an aberration from the usual course of nature. But the clash of contending elements which must often mark the history of such a deviation results now and again—by no means infrequently—in nobler activities than those yielded by the vast majority who are born to consume the fruits of the earth. It bears, for the most part, its penalty in the structure of its own organism. We are bound to protect the helpless members of society against the invert. If we go far-ther, and seek to destroy the invert himself before he has sinned against society, we exceed the warrant of reason, and in so doing we may, per-haps, destroy also those children of the spirit which possess sometimes a greater worth than the children of the flesh.[37]

Though Ellis used the terms *aberration* and *deviation* in this passage, he, for the most part, struggled to avoid any language of pathology. Though he often used the word *abnormal*, he did so in the purely statistical sense of "not average." He also used the term *anomaly*, sometimes *sport* (dalliance), and at other times *variation*. To emphasize the biological nature of homosexuality, he reported that inversion seemed to reappear in different members of the same families and implied that it might well be inherited. He also, unlike the author of any study that had yet appeared, attempted to emphasize the achievement of homosexuals. Here, using materials gathered by Symonds, he identified a large number of historical individuals, including Renaissance humanists whom Symonds had studied in detail, such as the Frenchman Marc-Antony Muret (1526–25), the "Prince of the Humanists" Desiderius Erasmus (1466–1536), the artist and poet Michelagniolo Buonarroti (Michelangelo) (1475–1564), the polymath Leonardo da Vinci (1452–1519), and a whole series of British men of letters from Christopher Marlowe to Francis Bacon to Oscar Wilde. Included in his list were also scientists such as Alexander, Baron von Humboldt (1769–1859).

Ellis insisted that homosexuality was not a disease but that sexual ambiva-lence, what is now called bisexuality, persisted in some degree in almost everyone. He held that an awareness of our shared hermaphroditic constitu-tion helped render homosexuality comprehensible. Like Hirschfeld, he went beyond embryology and argued that internal secretions might eventually furnish the key, but, ever cautious, he felt scientists did not yet have the final answer.[38] He rejected the possibility of a cure for homosexuality and disagreed with Freudian analyses of psychic mechanisms, such as the Oedipal complex, as being factors in homosexuality, but he agreed that there was a tendency for homosexuality to arise in heterosexual persons who are placed under conditions in which the exercise of normal sexuality is impossible.[39] He simply concluded that the doctrine of acquired inversion was difficult to document, since it might only be a case of retarded differentiation.

Ellis paid more attention to female homosexuality, lesbianism, than others had before him, but he confined his discussion to a single chapter. Interestingly, in his discussion of male homosexuality, he did not assert that congenital homosexuals were always effeminate, but he did note that female homosexuals had more virile temperaments than other women.[40] He also felt lesbianism was specially fostered by those conditions that kept women in constant association, not only by day but often by night, without the company of men. Though he emphasized that homosexuality was as common in women as in men, its pronounced forms were less frequently met with in women than in men. He also seemed to assume that dildos were in common use among lesbians, and he deemphasized the importance of the clitoris.[41] Ellis stated that homosexual

> passion in women finds more or less complete expression in kissing, sleeping together, and close embraces, as in what is sometimes called "lying spoons," when one woman lies on her side with her back turned to her friend and embraces her from behind, fitting her thighs into the bend of her companions's legs, so that her mons veneris is in close contact with the other's buttocks, and slight movement then produces mild erethism. One may also lie on the other's body, or there may be mutual masturbation.[42]

He emphasized, however, that mutual contact and friction of the sexual parts seems to be comparatively rare.

Much of this work is similar to what Hirschfeld said but without Hirschfeld's excesses and with much more caution. His homosexual study is much less detailed than his later volumes and marks an uncertainty in Ellis about his real object. This study does set the pattern for his later writing, in trying to avoid some of the major controversies of his time, all the time arguing for tolerance and understanding.

MODESTY AND OTHER STUDIES

It is in the second volume, later to be listed as volume one in the revised complete series, that Ellis hit his stride. This volume is devoted to the study

of modesty, sexual periodicity, and autoeroticism, and Ellis followed the
same pattern as he did in his study on homosexuality, turning to history,
anthropology, literature, and biology for answers. Again he tried to remove
stigma surviving in the use of earlier terms by seeking more neutral ones.
Perhaps the key to the volume is his study of masturbation, to which most of
the volume is devoted. Ellis believed that if he could eliminate the fear and
anxiety that pervaded much of the scientific writing on sexuality when it
came to masturbation, he could set the study of human sexuality on a new
course.

To describe the phenomena that he grouped under masturbation, Ellis
invented the term *autoeroticism*, which he defined as

> the phenomenon of spontaneous sexual emotion generated in the
> absence of an external stimulus proceeding, directly or indirectly, from
> another person. In a wide sense, which cannot be wholly ignored here,
> auto-eroticism may be said to include those transformations of repressed
> sexual activity which are a factor of some morbid conditions as well as
> of the normal manifestation of art and poetry, and, indeed, more or
> less color the whole of life.[43]

In defining masturbation as autoeroticism, he managed to bring together
a number of different psychosexual phenomena, including erotic dreams,
daytime fantasies, narcissism, hysteria, and masturbation, and held that the
most typical form is "occurrence of the sexual orgasm during sleep."[44] In the
words of Paul Robinson, thus "by means of an inference that might be called
innocence by association, masturbation was transformed from a malignant
vice into a benign inevitability."[45] Ellis, however, did not dismiss the possibility that masturbation might result in some slight nervous disorders, and
he felt that in its extreme form it marked the divorce of the physical and
psychological dimensions of sexual expression. Still, he removed it from the
category of illness.

What is interesting in the developments of sex research in this period is
seeing how each of the researchers interacted with the others' findings.
Freud, for example, was working with hysteria and its association with sexuality, and so was Ellis; each was familiar with the other's work. The researchers
of the time—Hirschfeld, Ellis, Freud, and others—also were extremely
pleased when others cited them favorably.[46] Moreover, each borrowed from
the others.

One of the keys to Ellis's physiological concepts was the theory of tumescence and detumescence, which he had adopted from Moll. For Ellis, however, tumescence described the "accumulation" of sexual energy during

arousal and detumescence, the "discharge" of that energy at the moment of climax.[47] The idea is similar to Freud's libido theory, because Ellis recognized that the detumescence could discharge its force in other than sexual ways.

What made Ellis different from the other sexologists of his time was his much greater emphasis on love and courtship. He himself held that his studies differed from those of earlier investigators primarily in the attention he devoted to normal sexuality as distinct from abnormal. Moreover, he consistently tried to relate that which might be abnormal to the normal. For example, in his analysis of sadism and masochism, Ellis argued that the principal element in both was the association of love with pain, something he also observed in animals.[48] For this reason, he sought to eliminate the two categories, combining them into sadomasochism, using the term *algolagnia*. He ended up arguing that the whole of sexual psychology, including its various deviations, derived from the exigencies of courtship.[49] Similarly, he ultimately included other forms of sexual variation—fetishism, exhibitionism, bestiality, transvestism (which he called eonism), urolagnia, coprolagnia, and others— into the same scheme. He regarded most of these anomalies as congenital, and he could relate each to some aspect of normal sexual life. Coprolagnia, for example, was related to the attraction of the female buttocks; exhibitionism, to the pride of the male in his genitals; bestiality, to the primitive belief that animals were really disguised men, and so on. Even so-called normal men and women had unusual fixations. In *Erotic Symbolism*, he argued that all sexual deviations involved an imitation of both the actions and the emotions of normal sexual intercourse.[50]

Ellis emphasized that women were sexual creatures and demonstrated that the Victorian belief that women lacked sexual emotions was mainly a nineteenth-century idea without empirical foundation. He argued that women had sexual desire no less intense than men and that women's capacity for sexual enjoyment was comparable with men's.[51] He concluded, however, that the sexual impulse in women differed from that in men, because in women (1) the impulse was more passive, (2) the impulse was more complex and less spontaneous, (3) the impulse grew in strength after sexual relationships had been established, (4) the threshold of excess was less easily reached, (5) the sexual sphere was larger and more diffused, (6) there was more periodicity, and (7) there was greater variation—both among women and within a single woman.[52] Much of this was due to the fact that sexual excitement in men was wholly contained in a single event, penile erection. Though clitoral erection occurred in women, Ellis felt that behind the clitoris was the much more extensive mechanism of vagina and womb, both of which demanded satisfaction. He ridiculed Freud's idea that female sex was exclusively vaginal and held that it could have been advanced only by someone who lacked any direct knowledge of woman's sexual experience. Whereas all erotic sensitivity in the male was concentrated in the penis, it spread in women to several non-genital areas such as the breast. Ellis concluded that women were far more sexual beings than men. It was this diffuse nature of women's sexuality that made courtship necessary and essential.

In general, Ellis's goal was to study and identify various forms of sexual expression without stigmatizing them. When he studied cross-dressing, he found that Hirschfeld's term, *transvestism*, was both too narrow and too stigmatizing and urged the use of the term *eonism*, after the Chevalier d'Éon, although this term never caught on. Following Hirschfeld, however, he recognized transvestites were different from homosexuals, and that many could be heterosexual. He concluded, however, that the ultimate causal explanation might lie in an imbalance of the endocrine system, something he, as did Hirschfeld, more strongly supported as he grew older.[53]

ELLIS'S AND HIRSCHFELD'S STRENGTHS AND WEAKNESSES

Ellis essentially was empirical, pragmatic, and tolerant. Though he was trained as a physician, his methods were really more closely tied in with the humanities than with the sciences. Historical data furnished a significant part of his database, as did cross-cultural data. At heart, he was a sex reformer, and he strongly supported the struggle to gain greater contraceptive information, marriage reform, and rights for women and sexual minorities.

As indicated at the beginning of this chapter, one of the difficulties both Ellis and Hirschfeld posed to the medical community was that because they put so many forms of sexual behavior within the normal range of possibilities for all humans, they offered little for the physician. Ellis and Hirschfeld could enable the physician to make a better diagnosis and give explanations for sexual variations, but they provided no role for treatment. Physicians, even liberal ones, were reluctant to tell their troubled patients that their sexual behaviors were more or less part of being human.

It might well be that one of the barriers to the acceptance of Ellis and Hirschfeld by the medical community was their willingness to go beyond the medical audience, although Hirschfeld did sometimes write specifically for physicians. Hirschfeld also maintained a medical office and retained more of a belief in a medical model than Ellis, at least for treatment purposes. Ellis was a physician who never practiced medicine, and he wrote for the general public, despite the necessary legal claims to the contrary in those geographic areas in which censorship would allow sex books to be distributed only to professionals.

This appeal to the public at large was Ellis's and Hirschfeld's strength and weakness. Neither developed the kind of disciples that were so common among major medical figures of their time, and neither, since they were not associated with a university, developed a group of student followers. Hirschfeld certainly was more conscious of the need for continuity than Ellis, but he was unable to put into practice his ambitious plans. Ellis felt that his writings would be enough.

It was Freud who most effectively maintained the medical model, and this gave him an advantage in the professional community. He also carefully cultivated disciples who believed in his work and in his treatment modalities. He claimed that those who adopted his viewpoint could cure their patients, and although Ellis could be easily read and was widely available (Hirschfeld less so), it was Freudian ideas that dominated the treatment world of the first part of the twentieth century. Ellis, Hirschfeld, and others might well have been the permission givers, but it was Freud who pushed the theory and the treatment. This is something the organized medical community could accept, even if the treatment ultimately lay in the hands of a specialist rather than a generalist.

SIGMUND FREUD

Unlike Ellis and Hirschfeld, Freud did not set out to be either a sex reformer or a sex researcher, although he ended up as both. Born in Moravia, Freud moved with his father to Vienna in 1859, where he spent much of the rest of his life. A brilliant student in grammar school, he went on to study at the University of Vienna, where he became particularly interested in physiology. For a brief time, following his graduation in 1881, he continued his research work in physiology, but when he became romantically involved with his cousin, Martha Bernays, Freud decided to turn to the practice of medicine to support himself and his hoped-for family. He then interned at the Vienna General Hospital, spending part of his time in the psychiatric division. In 1885, the university awarded him a traveling fellowship that enabled him to study in Paris under the famous neurologist Charcot, who had demonstrated the value of hypnosis in treating patients. In 1886, Freud began practicing in Vienna as a specialist in nervous diseases and a few months later married Bernays.

In the 1880s, specialists in nervous diseases who had private practices generally treated patients who were neurotic and only occasionally some who were psychotic. The psychiatrist looked on patients with nervous diseases as suffering from hereditary degeneration or from lesions in the central nervous system. This meant that the practitioner was more or less helpless, because treatment, such as the then-popular brain surgery, often resulted in the death of the patient. Freud, impressed by Charcot, made tentative steps toward using hypnosis as a treatment technique and originally was pushed further in this direction by the experiences of Josef Breuer (1841–1925). Breuer held that neurotic symptoms were physical expressions of repressed emotions and that such symptoms would vanish if the painful experience were recalled and the emotion belated expressed. The result of this initial collaboration was *Studies on Hysteria*,[54] which took the term *unconscious* from German Romantic literature and philosophy and tied it in with *repression, conversion*, and *abreaction*, a term employed by psychoanalysts for the process of releasing a repressed emotion by reliving the original experience in the imagination.[55]

PSYCHOANALYSIS AND SEXUALITY

There is a vast literature on the development of Freud's thought, much too vast to cover in a brief overview. The focus here is on psychoanalysis and sexuality, and not on Freudian concepts in general. Freud's sexual theories gradually emerged and changed over time. Early in his practice, for example, he adopted the belief that those individuals suffering from neurasthenia (nervous anxiety) were masturbators. For a time, he even believed that nocturnal emissions were as damaging as masturbation. Gradually, he changed his mind,[56] influenced by the work of Moll, Ellis, and others who demonstrated that masturbation was part of the experience of childhood sexuality. His most important work in this respect was his 1908 paper titled "'Civilized' Sexual Morality and Modern Nervousness," in which he criticized Beard's theory of neurasthenia (discussed in chapter 1) as well as downplayed the exaggerated dangers of masturbation. He further emphasized that the sexual instinct in humans served the purpose of procreation and that pleasure and pleasurable gratification were natural aspects of child development.[57] About the same time, Freud had also abandoned his ideas about the influence of childhood seductions on the development of later behaviors such as homosexuality.[58]

Increasingly, in fact, sexuality emerged as the key to much of Freud's thinking, and he went so far as to claim that every neurosis had a specific sexual cause. Sexuality, in the words of Frank Sulloway, became for Freud the indispensable organic foundation for the scientific explanation of mental disease.[59]

The transition in his theoretical approach can be dated from 1899, when he began exploring the stages of sexual development. He held that the autoerotic was the first stage of development and that it preceded alloeroticism (homoeroticism and heteroeroticism) but that it survived as an independent tendency.[60]

Much of the change of ideas resulted from his own intense self-analysis. At the close of this experience (about 1902), he emerged with the conviction that he had discovered three great truths: that dreams are the disguised fulfillment of unconscious, mainly of infantile wishes; that all human beings have an Oedipus complex in which they wish to kill the parent of the same sex and possess the parent of the opposite sex; and that children have sexual feelings. He later added two ideas to these emerging principles of psychoanalytic thought, namely the division of the human mind into superego, ego, and id and the concept of the death instinct *(thanatos)*.

Freud was no advocate of sexual freedom, although he eventually believed that sexual energies had to be directed, not repressed. This led him to postulate that variant sexual behavior came from sexual drives that were misdirected in their aim or object. The cause of this misdirection lay in the nervous system and the mind through which the instinctual drive operated. Though Freud paid comparatively little attention to most forms of variant sexual behavior, his followers seized on his concepts to emphasize, far more than Freud himself did, the environmental and accidental causes of variant impulses. Later behaviorists, who stressed learning and conditioning of animals and humans, carried these environmental and accidental determinations to the extreme, but the practical result of both Freudianism and learning psychologies was to suggest that everyone had the potential to channel his or her drives toward any form of gratification and with any object. The major effect of such a conclusion was to undermine the assumption that certain forms of sex were against nature, for nature itself—the instinctual drive—was visualized as being able to express itself in many ways.

Freud came to endorse the belief that all human beings are bisexual, and he emphasized the importance of this in human psychosexual development. He also extended a more general phylogenetic paradigm of sex to encompass numerous other "component impulses" of the human sexual instinct, such as sadism, masochism, and coprophilia (love of excrement). As a result, Bloch described Freud as having "gone further than any other writer in biologico-physiological derivation of sexual perversions."[61]

He also emphasized the role of the environment, accepted the ubiquity of perverse sexual drives, and enthusiastically welcomed the efforts of anthropologists and cultural historians in the collection of data on various forms of sexual behavior. It was not only the social and cultural environment that influenced sexual behavior but also the individual's environment as he or she grew up. Freud noted how pathogenic fantasies developed out of the germs of infantile sexual fantasies. His hypothesis of the effect of infant or childhood repressed or unconscious erotic experiences proved to be a major breakthrough in explaining certain forms of sexual behavior. This was because it clarified why a particular form of fetish, for example, a shoe fetish, might develop in some individuals, whereas a rubber fetish or a leather fetish would develop in others. The influence of chance events in childhood on possible fetishism had already been put forth by Binet and developed by others, but the theory had been criticized by Moll because it remained unclear to him why one kind of behavior developed and not another or why similar experiences did not effect everyone in the same way. Freud's concept of unconscious repression and conversions offered a logical way out of this impasse.

Freud was an advocate of nature and of nurture, relying on both to explain psychosexual pathology. He acknowledged the possibility of heredity predisposition but also recognized environmental determinism. His approach allowed him to adopt and exploit the best theoretical ideas advanced by the two opposing camps.[62]

Exclusive homosexual orientation, for example, was conceived by Freud as a complex blend of the biological (bisexual potential) and psychical factors. He strongly opposed a strict distinction between inborn and acquired characteristics, since, in practice, he found a mingling and blending of both.[63] In addition to the idea of inborn homosexuality, Freud considered a number of other potential causal factors, including the Oedipal complex, fear of castration, regression to a primary autoerotic state, obstacles put in the path of ordinary sexual satisfaction, hatred of a father or a sibling, and fixation on the notion that women have a penis. He also tried to distinguish among manifest, latent, aim-inhibited, and sublimated homosexuality.[64]

Later, he stated that psychoanalysis had established two factors in the etiology of male homosexuality. The first factor was the fixation of erotic needs on the mother. The second was that even the most normal individual is capable of making a same-sex object choice and if he had done so at some time in his life, he either still adheres to it in his unconscious or else protects himself against it by vigorous counterattitudes.[65] In short, in some individuals, inversion is possible simply because human beings have the capacity to be attracted to objects of the same sex under certain psychical or social circumstances.

Freud rejected the idea of a third sex. although he recognized the possibility of biological or genetic factors. The problem, he felt, was to explain why homosexuality prevailed over the heterosexual potential of human bisexuality.[66] Freud was opposed to any effort at separating homosexuals from the rest of humanity as making up a special class, since all human beings are capable of a homosexual object choice. At least twice in his career, Freud emphasized that homosexual persons were not sick. The first time was in 1904, when he was interviewed by the editor of the Vienna newspaper *Die Zeit*, to whom he stated that the homosexual did not belong before the court of justice and that he or she was not sick.[67] Later, in 1935, he wrote to a mother whose son was homosexual, indicating to her that he could not eradicate homosexuality in a person and that it was a great injustice to prosecute such behavior in individuals. He advised her to read Ellis on the topic.[68]

Freud's belief in the underlying biological forces involved in forming sex-specific behavior was a major factor in his concept of latent homosexuality. This led him to equate femininity with passivity, gentleness, and timidity and to imply that females who were ambitious, athletic, aggressive, or in other ways masculine were showing latent homosexuality and losing their femininity. Such assumptions have been heavily criticized by some modern feminists, most notably and uncharitably by Kate Millet,[69] but the difficulty with criticizing Freud is that he said different things at different times in his career.

MEGA THEORY

As his theoretical thinking matured, Freud moved from specific to grand theory. Ultimately, he concluded in *Civilization and Its Discontent* that the problems in sexual development were attributed to the development of civilization. Expanding on Krafft-Ebing he came to hold that the deepest roots of sexual repression that advanced along with civilization, lay in the human effort to overcome an earlier animal existence.[70] Freud seemed to be subscribing to a kind of organismic theory of history, in which there was only a limited amount of vital energy in every culture and the losses suffered through sex could not be replaced. One aspect of this line of thinking was the need to control sexuality and especially female sexuality. Freud believed that the female held great feelings of inferiority over the lack of male genitalia, while the male, conscious of his possession of such an important body part, feared that women wanted to castrate him. In brief, the narcissistic rejection of the female by the male is liberally mingled with fear and disdain.[71] This aspect, however, was further developed by some of Freud's followers, who put the need for female subjugation in the extreme.[72]

SEXOLOGY

Freud, however, cannot be blamed for the excesses of his disciples, and the theories he developed seemed able to answer many of the questions facing society. They proved particularly attractive to those involved in the study of literature, anthropology, and some of the other social sciences. Like Marx and Engels, or for that matter John Locke, Freud posited an original state of nature in which the sexual assumed great importance. He suggested that civilization developed as it learned to control the sexual. This was a radically different formulation from those advanced by Ellis and Hirschfeld, who were less ambitious in their explanations. It was the grandiose nature of the Freudian answer—which, of course, was not subject to any kind of empirical verification—that appealed to so many. Although he had strongly grounded his ideas in biology, he gave free range to his imagination when he left it.

During much of his early career, he depended heavily on many of the early sexologists, but over time he tended not to cite them, ignoring or denigrating Hirschfeld, Moll, and other sexologists, including many on whom he had relied in his early writings. What he did was to seize on selected and historically transient evidence and generalize it into universal law. He shifted from proximate causal to ultimate causal theory in his efforts to attain a synthetic, psychobiological solution to the problem of the mind. In the process, he began to ignore later research findings of some of his sexologist colleagues and went on his own to base his whole theory of psychoanalysis on sexuality, his "indispensable premise." Freud himself made important contributions to sex research, and his concept of the unconscious was a major breakthrough. His emphasis on both biological and psychological factors is also important, although some of his disciples tended to ignore the biological. Perhaps his greatest significance is that he made an understanding of sexuality a key to the understanding of human nature. His writings, far more than those of any of his contemporaries, broke down the barriers against the discussion of sex, thus encouraging others to look seriously at human sexuality.

As he aged, Freud grew more insistent that he was correct, bringing along a coterie of individual disciples who generally accepted what he said without looking at what was being said outside of the psychoanalytic arena. The rise of the Nazis dispersed these disciples to other countries, where they promoted the ideas of Freud, not those of the other contemporary German sexologists. In the English-speaking world, and particularly in the United States, much of the sexological research preceding Freud and contemporary with Freud was neglected or overlooked by Freudian-oriented psychoanalysts, who dominated the therapeutic and scholarly writing on sex.

With the death of Freud and Hirschfeld and particularly with the rise of Nazism, German-speaking areas lost their preeminence in sex research. Ellis in England had been increasingly isolated by his ill health, and no one of his stature rose to carry on his tradition. Instead, the leadership in sex research passed to the United States, and it is the beginning of such research in the United States that forms the topic of the next chapter.

4

THE AMERICAN
EXPERIENCE

*S*o far in this book, there has been an emphasis on homosexuality as a major factor in sex research. Krafft-Ebing, Hirschfeld, and Ellis all were concerned with variant sexuality, and the first works of each of them dealt with it, although Krafft-Ebing did so in less detail than the others. Freud was not so concerned with homosexuality per se, but many of his psychoanalytic followers devoted considerable attention to it. One reason for this attention, as indicated earlier, was a growing public awareness of the existence of individuals who loved members of their own sex more than those of the opposite. Another reason for the concern with homosexuality was the rapid growth of cities and the challenges this posed to traditional patterns and assumptions of living. Some of this concern by sex researchers with homosexuality, as in the case of Hirschfeld, was probably also to get a get a better understanding of themselves.

MEDICAL RESEARCHERS

American medical observers were conscious of some of the research taking place in Europe, but for the most part, they did not do any major research on sexual topics in the nineteenth century, although toward the end of the century, they did contribute case histories and discussions of various "deviant" sexual practices, including homosexual behavior.[1] Probably the best informed American medical writer on the subject was Lydston of Chicago (discussed in chapter 2). He recognized the complexity of the phenomenon of "perversity" and the difficulty in determining cause. He believed it was possible that a person might be born with a tendency toward perversion because of a physical deformity of either the brain (idiots, for example) or the genitalia (such as hermaphrodites) or because of congenitally misdirected impulses. He remained a believer in the dangers of masturbation, holding out the possibility that homosexuality was a result of "overt stimulation of sexual sensibility and the receptive sexual centers, incidental to sexual excesses and masturbation."[2]

Lydston, however, was an exception. Generally, in fact, most American physicians who did look at sex did so to emphasize its dangers, as Beard did with his concept of neurasthenia (see chapter 1). One reason for this is that American physicians in the last part of the nineteenth century were striving to professionalize themselves and to upgrade their standards to match those in Western Europe. In such a setting, the investigation of sexual activity was not seen as respectable. Instead, the respectable physician strove to be as much a guardian of traditional morals as he or she was a medical practitioner. One result was the enforcement of a more or less official prudery among the members of the increasingly powerful American Medical Association (AMA). Symbolic of this was the reaction to Denslow Lewis's paper "Hygiene of the Sexual Act," which was presented at the 1899 meeting of the AMA. Although Lewis was allowed to read his paper, the famous Johns Hopkins University gynecologist Howard Kelly objected to such a presentation on the grounds that the "discussion of the subject is attended with filth and we besmirch ourselves by discussing it in public." Later, the editor of the *Journal of the American Medical Association* refused to publish the paper because he was opposed to publishing "this class of literature."[3]

One way would-be physician researchers into sex could cope with such demands for respectability was to make negative judgments about what they observed, thus enforcing standards of morality, which in a sense is what Lydston did. Another way was to express shock and surprise that various unusual sexual phenomena could exist. Thus the following comment by the St. Louis physician C. H. Hughes, a specialist in nervous and mental diseases who apparently was aware of some of the European sex research at the time. In 1893, he published a brief description titled "An Organization of Colored Erotopaths," in which he wrote, "I am credibly informed that there is, in the city of Washington D.C., an annual convocation of negro men called the drag dance, which is an orgie of lascivious debauchery beyond pen power of description. I am likewise informed that a similar organization is lately suppressed by the police of New York City."[4] Hughes, in short, not only managed to express surprise but, by emphasizing that it was only "colored" who were involved, managed to shift the stigma of such conduct to what he clearly regarded as an inferior race.

Another example of the reluctance of the American medical establishment to deal with sexuality in any serious way is emphasized by the anonymous American reviewer in 1902 of Ellis's study on sexual inversion. While complimenting Ellis for pursuing his studies of sex on a scientific level, the reviewer concluded that Ellis was too inclined to fill his books "with the pornographic imaginings of perverted minds rather than cold facts, and the data which are collected are seemingly of little value. Whether any practical results can come from such labor is doubtful."[5]

Earlier, William Noyes had reviewed the first German edition of the same work by protesting

> against the appearance of such a work as this in a library (series) intended primarily for popular reading. Even Krafft Ebing [sic], although writing solely for the profession, has been severely and justly criticized for unnecessary emphasis and importance he has given this subject by his articles on the perversions of the sexual sense, and nothing but harm can follow if popular literature is to suffer a similar deluge.[6]

Noyes claimed that publicity given various kinds of nonprocreative sexual activity only allowed perverts to recognize their condition, and since understanding failed to lead to amelioration, it was best if it had not been mentioned at all.

Still, discussion of variant forms of sexuality appeared occasionally in the medical journals, usually of isolated cases among a physician's patient population, and were reported in the hope that the gathering of as many facts as possible might lead to scientific conclusions. By and large, the American medical establishment believed "degeneration" was congenital and probably caused by a hereditary physical weakness of some sort. Case reports highlight the peculiarities of the brain and genitals, both before and after death. Undoubtedly, this kind of reporting served to confirm the tendency to think of the individuals involved as being perverts who were isolated cases as well as physically handicapped.

There were exceptions. For example, Randolph Winslow reported on an outbreak of gonorrhea spread by rectal coition at a boy's corrective institution in 1886 Baltimore. He found anal intercourse to be common in such institutions and emphasized the difficulty in eliminating such activities. Younger boys, he said, sold their favors to older ones for economic gain, and when questioned about it, the boys justified their conduct by the existence of prostitution in the outside world.[7] Though Winslow and others knew that

European investigators had reported a tendency among homosexuals to group together, observers generally ignored the real life around them and rarely went beyond the facts of a particular case before them.

The psychologist G. Alder Blumer, who had read the Continental literature, reported a case of "perverted sexual instinct," but he classified his patient as insane and possibly epileptic. This description was frequently applied in the American medical literature to those departing from the accepted norms in sexual behavior. He added that his homosexual patients told him they were "able to recognize each other," but he did not explain how this came about.[8] Lydston again was the exception, writing,

> There is in every community of any size a colony of sexual perverts; they are usually known to each other, are likely to congregate together. At times they operate in accordance with some definite and concerted plan in quest of [a] subject wherewith to gratify their abnormal sexual impulses. . . . The physician rarely has his attention called to these things, and when evidence of their existence is before him, he is apt to receive it with skepticism.[9]

It was not only physicians who were reluctant to accept the existence of variant sexual behavior but other professionals as well. Such behavior to them was morally wrong. A member of the Philadelphia Bar, for example, told J. Richardson Parke that he would refuse to defend anyone accused of inversion on the grounds of social and professional decency. Parke himself had no such qualms and held that sexual inversion had "been perceptibly stimulated in our larger cities, and in our native born population, particularly, by the ever-growing desire to escape having children."[10]

John Burnham argued that American physicians of this period refused to recognize the social aspects of behavior,[11] and until they did, he noted that homosexuality as well as other variant forms of sexual behavior would remain merely a series of case studies. Allan M'Lane Hamilton was one of the leading psychiatrists at the turn of the century, and he pointed out, as had his Continental counterparts fifty years earlier, that the lack of such studies severely handicapped the court in dealing with cases that came to its attention. The result was that the court dealt with these individuals by classifying them as mentally unsound and grouped them with the insane.

> The attitude of the law so far is very harsh regarding punishment of offenders of this kind when detected, when they happen to be distinctly responsible, and it rarely recognizes any extenuating circumstances, and while possibly this restriction is best for society, there is no

doubt that in cases where a congenital taint exists, some degree of protection should be afforded the possessors of mental weaknesses who are apt to be the prey of persons of their own sex.[12]

There were just some things that Americans did not want to think about. For example, in 1889, A. B. Holder, who had lived and practiced in Montana, writing in a medical journal, described the custom of *berdache* among the Crow Indians where the men adopt the role and clothes of women and even took another man as husband.[13] Yet twenty years later a Zuni Indian healer, a *berdache*, could be taken by an anthropologist to Washington, D.C., and introduced as a woman without either the anthropologist or Washington society aware of the biological sex of the individual.[14] Obviously Americans had difficulty in believing such people could exist and have an important role in any society.

CIVILIZED MORALITY

Official attitudes of American society were characterized by what has been called *civilized morality*, defined by Mark Connelly as the prescriptive system of moral and cultural values, sexual and economic roles, religious sanctions, hygienic rules, and idealized behavioral patterns that emerged in the Jacksonian period and influenced American middle-class life up to the beginning of World.War I. In the ideal world of civilized morality, all moral values were absolute and timeless, masculine and feminine roles were sharply defined and demarcated, and sexuality was seen as a potentially destructive force.[15]

Believers in this civilized morality held that the code's viability depended on an

> unremitting effort to root out all opportunities for moral lapse. This understanding fueled the purity crusade of the late nineteenth century, particularly the shenanigans of Anthony Comstock and like-minded individuals who were determined to protect American society from salacious books, prostitutes, poker, and other forms of mental or physical licentiousness. In a sense, the purity crusade was an attempt to force the reality of social conditions into line with the dictates of civilized morality.[16]

Two fundamental components of this civilized morality were the conspiracy of silence[17] and the double standard.[18] Institutionalizing the enforcement of this morality was the American Purity Alliance, which had been formed in 1895 through a merger of local and state purity groups.[19] Inevitably, such attitudes gave Americans doing sex research a different perspective from the one researchers had on the Continent. To put it simply, Americans were more prudish in their public discourse. Though they read Ellis, and Freud's works were gradually reaching an American audience, attitudes toward public discussion were changing only slowly. Most Americans publicly writing about sex in the pre–World War I era remained prisoners of the assumptions of civilized morality; that is, they viewed public discussion of sexual matters as not quite respectable. Still, to come to grips with what they believed to be the evils brought on by sexual activity, American investigators thought it necessary to study and understand such topics. It was with this kind of justification that the serious analysis of sexuality began in the United States.

WOMEN AND PROSTITUTION

Basic to much of early-twentieth-century American sex research was prostitution, because so many evils seemed related to it: exploitation of children, pornography, disease, and crime.[20] Prostitution was called the "great sin of great cities,"[21] but as William E. H. Lecky wisely noted in his *History of European Morals* (1869), it was the prostitute, the very symbol of degradation and sinfulness, who ultimately proved to be "the most efficient guardian of virtue."[22] By this he meant that only by tolerating the kind of double standard that prostitution implied could the good woman—the nonsexual woman—be preserved in her chastity.

But women in the last part of the nineteenth century were beginning to challenge some of the traditional male assumptions. Their weapon was their very weakness, the assumption that somehow they were less sexual than men (unless some evil force so aroused them that they fell from gentility) and, therefore, purer; as such, they were considered the designated guardians of family virtue. Chastity was the mark of gentility, and though chastity or its lack might be difficult to prove, the assumptions of chastity were not. Once women were married, however, it was motherhood that counted.

Women, or at least proper women, were taught to think of themselves as a special class, and having become conscious of their unique sexual identity, they could no longer accept uncritically the role definitions drawn up for them. Motherhood came to be elevated into a mystique, which Freud made into a pseudoscientific basis of existence. Even though the Victorian conception of women as wan, ethereal, and spiritualized creatures bore little relation to the real world in which women operated machines, worked the fields, hand washed clothing, and toiled over great kitchen stoves, this view was endorsed by both American science and religion. Even fashion conspired to the same end, for the bustles, hoops, corsets, and trailing skirts in which women were encased throughout much of the nineteenth century can be seen, in retrospect, to have been designed to prevent them from entering the world of men.[23]

At one level, prostitution preserved this official female innocence, but at another level, it threatened to destroy the innocence because of the dangers of disease. It was this issue that caused so many people to condemn prostitution. The use of prostitutes as sex partners was fairly widespread in the last part of the nineteenth century, and patrons came from all classes. A good illustration is the case of the Reverend William Berrian, pastor of New York City's Trinity Church, probably the wealthiest parish in the United States, if not in the world, at that time. In a sermon preached in 1857, he virtuously proclaimed that during a ministry of more than fifty years, he had been in a house of ill fame not more than ten times.[24] If such a guardian of morals could somehow visit a prostitute, one has to conclude that large numbers did so, at least at some time in their lives. In fact, so widespread was prostitution during the last part of the nineteenth century, that Ellis thought prostitution to be a necessary product of civilization and urban life.[25]

SEXUALLY TRANSMITTED DISEASES

The trouble with prostitution went beyond the moral because of its close association with disease. Though Ricord had plotted the three stages of syphilis shortly after the middle of the nineteenth century, the full impact of the disease began to be realized only at the end of the century. By then numerous pathological conditions were recognized as being the result of syphilis, including the appearance of gummas on the skin, on the bones, and in vital organs; the weakening of blood vessels; and the destruction of various parts of the nervous system. Though the knowledge of the sequelae of the disease was itself important, it was the realization that the disease could be passed on to innocent wives and children that proved a major challenge to civilized morality.[26] The difficulty was compounded by the fact that, until well into the twentieth century, physicians could offer little therapeutic hope for those afflicted with syphilis. Most physicians treated syphilis with mercury, either orally, in vapor baths, or topically. Mercury was used as much for magical reasons as for any evidence of cure, because the justification for its original use had been based on ancient theories of humor and health, which by the end of the nineteenth century had been discarded in other areas of medicine. The mercury treatment was widely used, and the element was given in such doses that the therapy itself was often lethal. In fact, many of the symptoms then attributed to syphilis are now believed to have resulted from mercury intoxication.[27]

The other major venereal disease being diagnosed at that time was gonorrhea. Treatment for it depended in large part on oral medications, and it was believed that when the medications were taken in sufficient quantity, they had an antiseptic effect as they were excreted through the urethra. Ricord, however, had emphasized how little the physician could help when he noted: "Gonorrhea begins and God alone knows when it will end."[28]

Inevitably, venereal infections were high, although estimates of the numbers range widely. Perhaps the most accurate determination of the prevalence of the venereal diseases among young males in the United States is one from 1909 based on U.S. Army physicals; the army found a rate of one out of every five. Some medical groups set higher rates of infection based on their own practices. One committee of New York physicians estimated that as many as 80 percent of the men in the city had been infected at some time in their lives with gonorrhea and from 5 to 18 percent had suffered from syphilis. A Boston physician during the same period found that more than 33 percent of a sample of male hospital patients had gonorrheal infections.[29] Prostitution was inevitably blamed, and fears grew, especially as it was realized that gonorrhea could lead to sterility in women and blindness in infants and that syphilis could cross the placental barrier. The military was also concerned because gonorrhea in particular was temporarily so disabling in males, and it made them unfit for battle for ten days to two weeks, and occasionally even longer. Some of the estimates of the rate of infection are just guesses.

In 1904, Fritz Schaudinn and Erich Hoffmann identified the causative agent of syphilis. Soon after, August Wasserman, Albert Neisser, and Carl Bruck developed diagnostic tests that relied on a complement-fixation reaction; these tests allowed physicians to make accurate diagnoses. Theoretically, gonorrhea could be diagnosed much easier by examining under the microscope a slide smeared with the exudate, but few physicians had the technical facilities to do this until well after the first decade of the twentieth century.

With the isolation of the syphilis spirochete, there was an immediate search for a cure. In 1910, Paul Ehrlich and Sahachiro Hata announced the discovery of Salvarsan, an arsenic compound, which was the first effective treatment for syphilis. It was the 606th compound Ehrlich and Hata had tested on syphilitic rabbits, and their success marked the beginning of modern chemotherapeutics.[30] Although Salvarsan was hailed as a miracle drug, it had serious side effects, and many individuals succumbed to the high toxicity of the treatment, since it was a race to see whether the spirochete or the patient would be killed first, and many times the patient lost.

Ehrlich continued to experiment with other substances, and in 1912, he found what he called Neosalvarsan, or No. 914, a less-toxic, but less-effective, drug than the original. This finally gave the physician a treatment modality over syphilis and emphasized further the importance of medicine to the study of sexuality. This medical control was strengthened by the discovery of the sulfanilamides by Gerhard Domagk in 1921. The sulfanilamides' ability to kill bacteria, including those causing gonorrhea, was recognized in 1936. Penicillin, which had been discovered in 1929 by Alexander Fleming, was finally produced in quantity during World War II, and by 1945 it had been shown to be effective against both gonorrhea and syphilis.

PREVENTION OF SEXUALLY TRANSMITTED DISEASES

Because the new treatments did not really develop until after World War II, and Ehrlich's earlier treatment was very costly and time-consuming, many medical professionals looked to prevention as an answer. In their search for answers they looked not at the male but at the female, both as a cause of sexually transmitted diseases and as an innocent victim. It was the bad women, the prostitutes, who were the cause, and the innocent wives and children who were the victims. Inevitably, many came to believe that the best way to deal with the diseases associated with prostitution was to regulate the practice and require medical inspection and isolation of infected prostitutes. This, however, implied direct government entry and supervision of what large numbers of people believed to be immoral conduct. When this Continental practice was introduced into certain port areas of Great Britain in the last part of the nineteenth century, it ran into a storm of opposition, in part from evangelical religious figures but also from women, the very preservers of public morals. The initial intent of this opposition was to eliminate officially tolerated, medically inspected prostitutes, but in the long run, it led to a growing movement for abolition of prostitution altogether, particularly in the United States,[31] and a full-fledged assault on the double standard.

SEXUAL ALTERNATIVES

So far in this chapter, reticence of Americans, especially the medical professionals, about sexuality has been emphasized. This emphasis is somewhat misleading, because not all Americans were as reticent as the medical professionals in discussing sex issues. Many had quite different attitudes, and there were a number of fringe movements that set out to challenge the hypocrisy surrounding sex. One such movement was the free-love movement, which had as its goal the abolition of traditional marriage and the establishment of a new and better kind of relationship, based on passionate attraction. In one free-love community, couples signified their union by tying strings of the same color on their fingers; when the passion fizzled, they simply removed the strings. Many of the free-love communities were founded by religious prophets, a fact that allowed the adherents to claim that both God and morality were on their side.

In Oneida, New York, John Humphrey Noyes and his Christian Perfectionists promoted his theories of "complex marriage" and "marital continence" as well as the use of coitus reservatus in intercourse. This involves putting a finger at the base of the penis to create pressure on the urethra, which prevents ejaculation, or simply ceasing motion before ejaculation, or using both techniques. Though coitus reservatus is not 100 percent effective as a contraceptive, it certainly cuts down the probabilities. The Mormons instituted polygamy to keep straying husbands within the confines of marriage, and Sylvester Graham tried to establish chastity in "Grahamite boarding houses" to protect men from "venereal indulgence."[32]

Victoria Woodhull, one of the woman leaders of the free-love movement, taught that sex was not only central to human existence but essential to preserve one's health and vital strength. "Show me a man or a woman who is a picture of physical strength and health and I will show you a person who has healthy sexual relations," she claimed. While never condoning overt promiscuity, she remained hostile to marriage as an institution, because it inevitably led to sexual starvation and slavery for women and too often produced unloved and unwanted children. The way to combat such evils in her mind was to give people the freedom to love, "love wrought of mutual consent based upon desire."[33]

MANY FORMS OF REFORM

Prostitution and sexual problems were just two of the many social ills that seemingly plagued Americans at the end of the nineteenth century. Alcoholism was another. Pornography was still another. Lack of rights for women was another. Slavery, an earlier problem, had been eliminated by the Civil War and the Reconstruction that followed it. Inevitably, following the abolition of the evil of slavery, solutions to other social ills were also seen to be the responsibility of government. Emphasizing this turn to government was the fact that traditional solutions of community pressure or a religious call to moral reform no longer seemed to work in the growing urban setting. Organized groups were formed to agitate for a variety of government interventions, including the prohibition movement, which dealt with alcoholism, and abolitionism, which dealt with prostitution. Giving the right to vote to women was supposed to raise the moral level of America. What ties the various social movements together in American history is the demand for the state to intervene in the country's social and economic life. The result was the Progressive movement, a movement that embodied conflicting impulses, ranging from social justice to efficiency and from the power of education to change behavior to a belief in the need for greater coercion to force the recalcitrant to conform.

ANTHONY COMSTOCK

An early forerunner of this felt need for coercion had been the so-called Comstock Law passed by Congress in 1873, signed into law by President U. S. Grant, and titled "An Act for the Suppression of Trade in, and Circulation of Obscene Literature and Articles of Immoral Use." The articles of immoral use were specifically aimed at contraceptives, emphasizing just how much the fear of sex played in the various reform movements. Comstock, the founder of the New York Society for the Suppression of Vice, supervised the enforcement of the law from his position as an unpaid U.S. postal inspector in New York City (see also chapter 2). He almost singlehandedly prosecuted those who wrote, published, and sold literature or art that he considered obscene.[34]

Comstock was motivated by his belief in the absolutes of civilized morality, which necessitated that sexuality be restored to the private sphere and that any public expression of sexuality was by definition obscene. He also believed that lust itself was dangerous, and he and his allies attacked not only sexual literature sold for profit but also any dissenting medical or philosophical opinion that supported the belief that sexuality had other than reproductive purpose.[35] The result was censorship of the mails and a curtailment of the public discussion of sexuality. There was open war between those who thought that sex was best regulated by restricting its public discussion and those who held that public education and information were the keys.

Comstock's success depended on public support, and for a time a sort of self-censorship was imposed by the publishers themselves. Increasingly, however, many of his allies broke away from his campaign of entrapment and intimidation. Undeterred, he expanded his attack to include suffragists, whom he felt threatened traditional family morals. Ultimately, it was Comstock's inability to compromise and his willingness to extend his net ever wider that weakened his movement and led to him becoming an almost ludi-crous figure.

PRINCE A. MORROW

Replacing Comstockery as a major factor in determining American views of sexuality was the growing public health movement. Many of those in this emerging coalition were concerned with sexually transmitted diseases, and they were handicapped by the activities of Comstock in carrying out their felt need for public education. Giving leadership to this new focus was the New York dermatologist Prince A. Morrow (1846–1913).

Morrow had spent a year in Europe after completing his medical training, and while there, he had made contact with some of the leading figures in the study of sexually transmitted diseases. On his return to the United States, he translated Jean-Alfred Fournier's discussion of syphilis and marriage into English and wrote his own manual about syphilis, which was geared to both students and practitioners.[36] Morrow's interest was reinforced in 1899 when he attended the first of two international conferences, both held in Brussels, to consider the public health aspects of sexually transmitted diseases. At this first conference, those in attendance, mainly physicians and public health officials, concluded that sexually transmitted diseases were more prevalent than generally believed and that medical inspection of prostitutes, the stan-dard European practice, was not effective.[37] After the conference, Morrow made a moving appeal to the New York Medical Society emphasizing that among the major sufferers of sexually transmitted diseases were innocent women and children and unwitting youngsters who had no idea of the dan-gers to which they exposed themselves.

Morrow helped organize the second international conference, which was held in 1902. There, delegates heard about the success of a French educa-tional campaign to warn youth against the dangers of venereal disease and to urge infected persons to seek treatment. Morrow returned to the United States committed to founding an American group to educate professionals and the public about the dangers of the venereal peril. The first step was to write a full exposition of the disastrous implications of introducing venereal disease into marriage.

In 1904 he published *Social Disease and Marriage*,[38] which attracted a wide medical audience as well as considerable attention from lay people. Morrow attacked the double standard of sex morals and believed and preached that the male should be held as guilty as the prostitute he patron-ized. He argued that sexually transmitted diseases should be reportable, and urged women, as the principal sufferers of the silence about sexually transmitted diseases, to take a leading part in the struggle against disease and sexual vice. Morrow also thought that sex education was the best answer to the problem.

One result of Morrow's efforts was the establishment of what were called social hygiene societies, the first of which appeared in Chicago in 1904. These groups quickly took root and formed the basis for a national campaign to wipe out the ignorance and prejudice that allowed venereal disease to infect the nation. In 1905, Prince himself established the Society of Sanitary and Moral Prophylaxis (soon renamed the American Society for Sanitary and Moral Prophylaxis), which took as its mission the education of the public about sex and sexually transmitted diseases. Morrow believed that although sex instruction should be given at home at an early age, the majority of parents were not qualified to give it. Thus, he contended, the duty fell to teachers and sex education should be an integral part of the course of study in all teacher training schools. This idea appealed both to the American medical profession and to the leaders of the American Purity Alliance, who in 1908 announced that sex education would become one of their objectives.

The main purpose of sex education was to emphasize the importance of sexual purity for both sexes and to eliminate the false impression often held by young men that sexual indulgence was essential to health and that chastity was incompatible with full vigor. Though it was recognized that sexual activity could have nonprocreative purposes and even pleasurable purposes, sexual intercourse was to be reserved for marriage, in which only husbands and wives were to enjoy such pleasures. These pleasurable aspects of sex did not enter into the public discussion.[39]

The purity movement gained further impetus in 1911, when the two major purity organizations—the American Purity Alliance and the American Vigilance Committee (founded by Jane Addams, Grace Dodge, David Starr Jordan, and others in 1906)—elected the same officers. This was the first step to the organizations' consolidation as the American Vigilance Association, which was dedicated not only to fighting prostitution but to educating the young about the dangers of immorality. In 1910, many of the medically oriented groups that were supportive of Morrow's ideas had met and elected him the president of a new national organization, the American Federation for Sex Hygiene. In 1913, after Morrow's death, the two forces, medical and purity, merged formally and symbolically to form the American Social Hygiene Association.[40]

To be successful required the education not only of the young but also of the general public. This was a more difficult task than many of the reformers envisioned. In 1906, when Edward Bok, the editor of the *Ladies' Home Journal*, published a series of articles on venereal disease, he lost some seventy-five thousand subscribers.[41] This reaction took place even though the advocates of sex education in that period carefully emphasized the themes of chastity and abstinence, reinforcing this by portraying the horrors of disease. Attitudes began to change only after World War I, when the influence of Ellis and Freud began to be felt, but the American public did not significantly change its view of either sex education or sex in general until after World War II.[42]

CONTRACEPTION

If public discussion of sexually transmitted diseases aroused such public reaction, the discussions in the United States of contraceptives was even more controversial. Generally, medical professionals in the last part of the nineteenth century and first part of the twentieth century had turned their backs on both contraception and abortion, regarding them as something that poorly trained or irregular medical practitioners got involved with. Physicians in the increasingly powerful American Medical Association were not only active in campaigns to outlaw abortion but, in keeping with their view of their moral guardianship, refused to deal with contraceptives. Some of these actions derived from the prevailing medical belief that women had a special sphere and were designed for the maternal role, thus any attempt to avoid this role was dangerous to their health. Though individual physicians might ignore these publicly stated assumptions, contraceptives and information about them went underground in America in the last part of the nineteenth century, just as effective birth control devices were reaching the market. The newly invented latex rubber condoms were being sold in the United States as early as 1860 as effective prophylactics; they were available in barber shops and other places where men congregated. Though pushed as a prophylactic against venereal infection, its contraceptive value was also recognized.[43]

More widely available to women were various devices that were advertised to deal with prolapsed uterus or "female complaint"; in one way or another, they formed a barrier over the cervix. Many of these pessaries were patented by the U.S. Patent Office, even during the period when Comstock was most active, although their alternative use was never specified in patent applications. The most effective of these pessaries in terms of contraception was that developed by W. P. J. Mensinga (C. Haase). Most of the early pessaries had used a hard rubber ring, but after the vulcanization of rubber, it became the favored material. What made Mensinga's diaphragm, dating from the early 1870s and made of latex, more effective than earlier models was the incorporation of a flat watch spring in the rim to keep it in place. Later, a coiled spring was used.[44]

In the 1860s, Edward Bliss Foote had introduced in the United States something he called "The Womb Veil," a device very similar to the Mensinga diaphragm. Unfortunately, the pamphlets recommending its use were seized and destroyed under the Comstock Law of 1873, but clear descriptions of it exist in earlier editions of his book.[45] The Mensinga diaphragm was popularized by Mensinga's student Aletta Jacobs, who, in 1882, opened the first contraceptive clinic in the world in the Netherlands and taught women how to use the Mensinga diaphragm. It was the device used by Planned Parenthood clinics in the 1920s, and it was probably the birth control device most widely used by women up through World War II.

A number of effective spermicides had also appeared on the European market by the end of the nineteenth century. Usually regarded as the first person to become active in manufacturing and selling spermicidal suppositories is the English chemist W. J. Rendell, who in about 1880 first put his quinine and cacao pessaries on the market. Cacao, or cocoa butter—a yellowish, hard, and brittle vegetable fat that is obtained from the seeds of the plant *Theobroma cacao*—contains about 30 percent oleic acid, 40 percent stearic acid, and various other fatty acids. It is a fairly effective material for a suppository because of its low melting point, and it probably worked as a contraceptive by blocking the cervix with an oily film. On the other hand, the quinine served as a spermicide, because it is a general protoplasmic poison; unfortunately, many individuals have a toxic reaction to it. Other chemical spermicides also began to appear on the market, and some news of them reached the United States, more by word of mouth than by any references in the medical literature.[46]

HOSTILITY

So strong was the official American attitude of hostility to sex that it even carried over into the treatment of soldiers in World War I. The international moral order visualized by President Woodrow Wilson led to the portrayal of the American soldier as a knight crusading for democracy who kept himself pure for his lady fair by abstaining from alcohol and sex. Such emphasis on ascetic dedication might have lessened the fears and anxieties of the mothers and wives left behind, but a most unfortunate side effect was that the government planners believed their own propaganda, and as a result did not make any plans to deal with venereal disease.[47] Wilson himself lent his authority to antivenereal fervor: "The federal government has pledged its word that as far as care and vigilance can accomplish the result, the men committed to its charge will be returned to homes and communities that so generously gave them with no scars except those won in honorable conflict."[48]

Though military-controlled prostitution had been part of the American tradition in early wars, no such activity could be contemplated under the new moral tradition so associated with the Progressive Era. The result was almost a total failure in planning on how to deal with millions of young men at the height of their sexual urges. Pulled off the farms and out of the small towns and tossed together in mass camps without women and without the moral constrains of their families, communities, and churches, these men were victims of sexually transmitted diseases, which had become endemic. The War Department chose to emphasize abstinence to the troops, although as the last step of its six-point program, it had suggested the distribution of prophylactic packages when other solutions had failed. Inevitably, the American policy proved a total failure in Europe, and the major successes against venereal diseases came from those commanders who ignored official policy, which was finally changed by orders of General Pershing in 1918.[49] One result was a vast number of Americans who suffered the sequelae of third-stage syphilis and were confined to Veterans' Hospitals in the 1930s and beyond.

CLELIA MOSHER

In spite of such official attitudes, there was a vast outpouring of information, or rather misinformation, about sex in the popular literature and in the professional journals. There was also the beginnings of serious sex research, much of it by women, that went beyond the simple reporting of individual case studies so common in medical writing. Several of the researchers might be called closet researchers, as the results of their early work was not published until much later. For example, the work of Clelia Mosher (1863–1940) did not reach print until 1974, more than eighty years after she had originally started it. Her story might be regarded as symbolic of the ambivalence about sex that most Americans held at the turn of the twentieth century.

Mosher's research grew out of an 1892 questionnaire that she designed for married women. Mosher used the survey to help her prepare a lecture on the marital relation, which she gave to the Mothers' Club, the members of whom were mostly faculty wives at the University of Wisconsin, where she herself was a graduate student. Her motivation for the research was to give better sexual advice to young women who came to her for counseling before they married. Of course, many of her survey questions dealt with sex; she was trying to determine the extent of women's knowledge of sexual physiology before they married, whether they habitually shared the same bed with their husbands, whether they had a venereal orgasm, and what in their minds was the true purpose of sexual intercourse. A total of forty-seven women filled out Mosher's questionnaire over the years. After the first batch of interviews, it seems that others were conducted only sporadically from among her patients. Mosher became a physician and practiced in Palo Alto from about 1900 on; she joined the faculty at Stanford in 1910.

Mosher found that the majority of women had known little about sex before they were married, and one women said that until she was eighteen she did not even know where babies came from. These women learned about sex in marriage, and most shared the same bed with their husbands. Thirty-five of the forty-four who answered the question said they felt a desire for sexual intercourse, and one wrote that sex was not only agreeable to her but delightful. Thirty-four of the women regularly experienced orgasm during sex. One said when she did not have an orgasm it was depressing and revolting, another said the absence of orgasm was "bad, even disastrous, nerve wracking." The most detailed and personal responses were elicited by a series of questions on the "true purpose of intercourse," and though nine believed that intercourse was a necessity for men, thirteen claimed it was a necessity for both sexes. The other fifteen who answered this question did not believe it was a necessity to either sex. While reproduction might be the primary purpose of sex, twenty-four women believed that the pleasure exchanged was a worthy purpose in itself. At least thirty of these women used some sort of contraceptive—withdrawal, douching, condoms, and cocoa butter—and two used a rubber cap over the cervix. Some felt that intercourse once a month was probably enough, but most thought it should be more often. Several of the postmenopausal women still enjoyed and desired intercourse.

Mosher's last interview was dated 1920, and all but one of her respondents had been born before 1890. Thirty-three of the women were born before 1870, and of that number, thirteen were born before the Civil War. Though the sample is fairly representative geographically of late-nineteenth-century America, socially and economically it was not. Of those whose education is known, 81 percent had attended college or normal (teacher training) school, and the remainder had at least attended secondary school. Most were married to college graduates.[50]

Though Mosher undoubtedly used her data to advise her women patients and students, few others knew about it. This makes the data interesting to us but not particularly helpful to her contemporaries who sought information about sexuality. For her time, Mosher was a feminist, and early in her career, she did publish a paper on normal menstruation in which she demonstrated four factors contributing to disabilities then believed to accompany menstruation: constricting clothing, inactivity, chronic constipation, and the general expectation that discomfort was inevitable. Each of these, she noted, was reversible and not physiological. She earlier had challenged the widespread myth of the time that women breathed costally (using only the upper chest) because of the physiological requirements of pregnancy, while men breathed diaphragmatically (using the diaphragm). Mosher found that while most women she examined did breathe costally, she believed that physiology was not the cause but constrictive clothing and a sedentary life; when the restrictive clothing was not present and women exercised normally, they breathed diaphragmatically. These data served as her master's thesis at Stanford.[51]

ROBERT LATOU DICKINSON

Another, and more significant, closet researcher was the physician Robert Latou Dickinson (1861–1950), who conducted studies between 1890 and 1920 and published them in the 1930s. Dickinson, an innovative specialist in obstetrics and gynecology, began practicing in 1882. He introduced the use of electric cautery in the treatment of cervicitis, was among the first to use aseptic ligatures for tying the umbilical cord, and gained widespread prominence for his innovative teaching methods for medical students—particularly for his use of rubber and sculpted models to teach female anatomy and to show fetal growth from fertilization to birth.[52] Dickinson became active in the campaign for better sex education and especially birth control, but both of these public roles were assumed when he was in his sixties and his professional reputation was secure. He has to be regarded as the most prominent American physician associated with the campaign for birth control in the immediate post–World War I period as well as the most significant sex researcher during the first three decades of the twentieth century. In 1923, he founded the Committee on Maternal Health, which began compiling data on contraception.[53] He tried to persuade Margaret Sanger to allow accredited physicians to have more control in her New York clinic and do more active research in contraceptives, but she did not take his advice. He broke new ground in the medical profession in 1920 with his presidential address to the American Gynecological Society, in which he urged his fellow physicians to do more work in the fields of contraception, infertility, artificial impregnation, and voluntary sterilization.[54]

Dickinson's most significant contributions were to the study of female sexuality, and though he had written scholarly articles on some of his data during the course of his active career, most of his data were published late in his life. Dickinson strongly believed that the key to effective medical practice was a good patient history, and during the years of his most active practice, 1890–1920, patients were not received by him until they had filled out a four-page questionnaire that accounted for general and family history. As he examined the patient, he also made at least five drawings: one each of the uterus, cervix, and vulva and two of the pelvic difficulty for which the woman had sought his assistance. The maximum number of drawings for any one patient was sixty-one. Later, as photography became easier to do, he supplemented the drawings with photographs. His standard practice was to read the patient's answers to the questionnaire and use them as a basis to question her further. He found that his patients often confided in him information about sexual problems, which he recorded and treated, if possible.

By 1923, Dickinson had compiled data on more than five thousand cases, which he turned over to the Committee on Maternal Health, after they had agreed to help him publish it. Lura Beam, a writer with a background in education and applied psychology, was called in to make a preliminary review of the data, and she proposed publication of three books, two of which were actually published. The original fifty-two hundred case histories were divided into two groups: one consisting of four thousand married women and the other of twelve hundred single women. These data served as the source material for A *Thousand Marriages*, in which he described the analysis of one thousand cases; data from other researchers were used as a control group. In *The Single Woman*, Dickinson analyzed three hundred fifty of his case histories, comparing them with information from other groups of women used as two control groups.[55] A third volume that he wanted to do on lesbianism was never finished, although he contributed his data to other studies.

One of the advantages of Dickinson's data was that he often saw his patients at different points in their lives and could plot the changes in their attitudes on sexual issues. For example, he illustrated with twenty cases how "passion and frigidity" could appear and disappear. At age thirty-four one of his patients became disgusted with coitus, but her husband did not. Later, the husband lost interest, but after six years without an orgasm, the woman again became orgasmic and the husband and wife found it difficult to remain apart from one another.[56]

All kinds of data are available from the Dickinson studies, including information on frequency of intercourse (two to three times a week in his married sample). Some 11 percent, however, had intercourse once a year or less. He has information on length of intromission before ejaculation and the attitudes of brides. Dickinson also collected data on more treatment-oriented issues, such as frigidity, dyspareunia, minor menstrual disturbances, sexually transmitted diseases, and fertility, as well as on such psychological issues as anxiety and fear. One in twelve of his patients in the married sample had venereal disease, usually gonorrhea, emphasizing just how widespread the disease was among his upper-middle-class patients.

In comparing single women with married women, he found that relatively more wives than single women reported masturbating. Twenty-eight of his sample of single women had been involved in same-sex relations, but he found no evidence of maleness of feeling in his subjects (i.e., lesbians were believed to have masculine qualities), some of whom he examined in the period before 1900. He reported that seventeen of those who had lesbian experiences later married and had ordinary fertility.

Dickinson's work has often been overlooked by later generations, but he proved to be innovative in his research in almost every way and in many areas. Very much interested in the physiology of intercourse, a subject that had been first studied by Felix Roubaud in the 1870s,[57] Dickinson preceded William Masters and Virginia Johnson in trying to observe what happened in the vagina during intercourse. To do so he used a glass tube resembling an erect penis in size and shape while women masturbated to orgasm; in the process, he proved once and for all that women did have orgasms involving physiological changes.[58] One of Dickinson's major contributions was a definitive summary of human sex anatomy, based on his work and on that of others.[59] From his discussion and references, it is possible to follow the growing knowledge about sex. He indicates, for example, that it was Ernst P. Boas of Mount Sinai Hospital in New York City who plotted the pulse rate of a couple during sexual intercourse in 1928 and 1929, demonstrating the rise at intromission, gradual increase, and then drop after orgasm.[60]

Along with W. F. Robie and LeMon Clark, Dickinson was responsible for the introduction of the electrical vibrator, or massager, into American gynecological practice. This device produced intense erotic stimulation and even orgasm in some women who previously had been unable to reach a climax. Dickinson and collaborators theorized that once a woman had achieved orgasm, even with a vibrator applied to her genitals, she was more likely to proceed to orgasm during coitus or through digital masturbation.

MAX J. EXNER

That either the Mosher or the Dickinson studies could have been published in the United States before World War I is extremely doubtful. Evidence of this is Max J. Exner's prewar study of the sexual activities of 948 college men, which was based on questionnaires. Exner, a physician associated with the Young Men's Christian Association (YMCA), was a leading sex educator of the time and a strong upholder of the importance of chastity. He believed that sex education should be designed to curb "morbid curiosities" and erase sexuality from consciousness.[61] Although he later modified some of his ideas, Exner was one of the leaders in the struggle to keep the American army pure during World War I. Even before the United States entered that war, he had conducted, in 1916, inspections of troops involved in the war against Pancho Villa and reported that the character of the camps was "sensualizing," with the coarse elements prevailing. In his mind, venereal disease merely represented the physical repercussions of a far more dangerous moral decay. Exner's solution was to urge the military authorities to raise the moral environment of the camps by removing the temptations; they were to cut down on the sale of liquor and eliminate prostitution.[62]

Thus even though Exner's study is the pioneer attempt to secure statistical data on the sexual behavior of American college men, it is badly flawed because of the researcher's biases. Exner asked a small group of men who were primarily from the Northeast and another more national group based on mail questionnaires to respond to thirteen questions, which were not particularly well designed.[63]

He analyzed the data by simple tabulation, and since the population was homogeneously college male, the data should have been significant, but they are not now so regarded, partly because of the content of the questions.[64] Exner tended to give the impression in most of his questions that sex outside of marriage was wrong. Still, in his sample, 518 men responded that they had engaged in sexual practices of one kind or another, 62.5 percent having practiced "self-abuse," 17 percent having practiced both "self-abuse" and intercourse, and 2 percent having practiced "various perverted practices," which remained unnamed.

The goal of Exner's research was to document the need for effective sex education, which he felt had to begin early, in the elementary schools. He found that most individuals had bad sources of sex information, having been told about sex by boy associates (544 of the 676 who answered this question), girl associates (33), or hired men or older men (22). The rest had received information about sex from their parents or reliable sex educators. He argued that the facts were learned too late in these men's development and that only by creating "an inspiring atmosphere with reference to sex, through a true and full interpretation of its meaning" in a classroom could boys learn the basic importance of chastity.

The importance of the Exner study is more symbolic than significant. He did his work publicly, demonstrated that individuals would respond to questions about sex, and made an early attempt to quantify replies. Interestingly, in spite of his assumptions and limitations, Exner turned out in the end to be a major force in bringing about organized sex research in the United States.

KATHARINE BEMENT DAVIS AND JOHN D. ROCKEFELLER JR.

The story of how organized sex research came about is somewhat complicated, and the key was the involvement of John D. Rockefeller Jr. In 1910,

Judge Thomas C. O. O'Sullivan of New York City charged a grand jury chaired by Rockefeller with determining whether white slavery rings existed in the city. Rockefeller later stated that he never worked harder in his life and added that he "was on the job morning, noon, and night."[65] Scheduled to sit for a month, the committee sat for six, and though it found no evidence of a formal organization of white slavers, it did find informal associations of brothel owners and even prostitutes. It also stated that recruitment to prostitution took place on a national level, that prostitution was linked to criminal elements, and that there were mutually beneficial ties between those engaged in trafficking in prostitution and the police.[66] One of the recommendations of the commission was the establishment of another commission to study the laws relating to the methods of dealing with prostitution in the leading cities of the United States and Europe.

When the mayor of New York City refused to set up such a commission, Rockefeller decided to do so himself, and in 1911, he established the Bureau of Social Hygiene, with an advisory board of Paul N. Warburg, Starr J. Murphy, and Katharine Bement Davis (1860–1935). Rockefeller served as chair, and Warburg and Murphy were chosen because they were friends of Rockefeller.

Davis was a different matter. She had been chosen because she was warden of the new Reformatory for Women at Bedford Hills in New York, many of whose inhabitants had been prostitutes. Davis had been appointed warden shortly after receiving her Ph.D. from the University of Chicago in 1900, and in her new position, she emphasized education and rehabilitation. Concerned that employment opportunities for poor and working-class women were too often restricted to domestic service, Davis believed part of their rehabilitation was to enable them to get better jobs. Because the state budget often proved to be erratic as well as insufficient, Davis put her female charges to work renovating the grounds and running the physical plant, to save money that could be used to better educate them. Inmates mixed concrete, laid foundations, graded lawns, cut ice from the river in winter, and even slaughtered pigs.[67] Davis believed that with respectable employment and adequate learning, the women could return to society without further danger of law breaking. She likened her reformatory to an educational institution and often referred to the prison's former inmates as graduates.[68]

To better manage her reformatory, she had developed a program that tried to identify different types of offenders and to separate the potentially reformable from the more hardened or irreclaimable. This program led her to recommend that an offender be studied by experts after being convicted but before being sentenced, a subject on which she wrote a pamphlet, a sort of forerunner to the presentencing report now commonly used in most courts.

In 1911, Davis's pamphlet came to the attention of Rockefeller, who, after conferring with Davis, bought land adjacent to the Bedford Hills Reformatory and set up the Laboratory of Social Hygiene under her direction in 1912.[69] Her success and work there led Mayor John P. Mitchell to appoint her commissioner of corrections for all of New York City's prisons in 1914, a first for a woman. Among other things, she lobbied for creation of a parole commission, the use of probation and parole, and indeterminate sentences. When the Parole Commission was created in December 1915 she was appointed its first chairman. Though her administration was often the center of controversy, she proved extremely successful. She ultimately lost out, however, when Mitchell was turned out of office, and her career as a city official came to an end in 1917, after which she became the salaried secretary of the Bureau of Social Hygiene.

Like his father, Rockefeller was convinced that the solution to any problem required the gathering of data and then the establishment of a plan of action. What was needed first were the data, and the plan for gathering these data was provided by Davis. Data were collected at the new laboratory from examination and testing of women, particularly prostitutes, and she also examined prostitution as an institution. Beginning in 1912, and for the next six years, the bureau undertook the most significant studies of prostitution undertaken up to that time. The bureau hired George J. Kneeland, who had directed the Chicago Vice Commission, to undertake a study of prostitution in New York City.[70] This was followed by studies of prostitution in Europe by Abraham Flexner, a study of European police systems by Raymond B. Fosdick (including how they dealt with prostitution), and a study of prostitution in the United States by Howard Woolston.[71]

Davis was involved in all of these, although in some more than in others. She wrote a chapter in Kneeland's book about the prostitutes committed to her reformatory, and she delayed publication of the Woolston book for several years while she made revisions. She actively maintained contacts in Europe to assist with the European studies.

Usually, Rockefeller acted behind the scenes, but on the issue of prostitution and vice, he was very public, not only because of his grand jury experience but also because he served as a member of a Blue Ribbon Committee sponsoring Eugène Brieux's play *Damaged Goods*, the theme of which centered on the effects of venereal disease. The success of the play was the final death blow to Comstockery in the United States, and Comstock died shortly after, a figure more of derision than respectability. Rockefeller attended the opening performance of the play and emphasized that the key to understanding the "evils" springing from prostitution was frank public discussion.[72] Rockefeller also wrote the introduction to Kneeland's book.

Under Davis's impetus, the Bureau of Social Hygiene gradually became interested in other areas of sexual behavior, and Rockefeller became less noticeably involved, perhaps because it soon became evident that there was no easy solution to the problems of prostitution. Early in the spring of 1920, Exner, who was then the director of the Department of Educational Activities at the American Social Hygiene Association, tried to persuade Davis to serve as coordinator of a proposed study of the sex life of women. Davis agreed on the desirability of such a study but believed she needed to consult women physicians about the feasibility. Her correspondents agreed that such a study was needed, but they opposed the involvement of the YMCA and YWCA in the project, an involvement that had been suggested by Exner. As a result, Davis agreed to form a committee to study the sex lives of five thousand women, subject to the approval of Rockefeller.[73] Rockefeller not only agreed to this but made an initial appropriation of two thousand dollars toward the project at a bureau meeting[74] and gave additional financial support in succeeding years.

This marked a departure from the earlier activities of the Bureau of Social Hygiene, but one that was a logical extension of its endeavors and one that threw the bureau more clearly into the field of sex research. Davis eventually extended her study to include a section on lesbian women,[75] a fact that is specifically mentioned in the minutes of the bureau and that indicates the basic change in attitude among the advisory members toward sex. Prostitution and other sex-related problems were increasingly being seen as related, and to deal with such problems successfully, it was necessary to better understand sexuality itself.

The data in the Davis report were drawn from responses to questionnaires—eight pages for married women and twelve pages (two of which were taken up with definitions) for the unmarried women. Names of the women were drawn primarily from the alumnae registers of the leading women's colleges. To obtain the required number of subjects, nearly equally divided into single and married women, a preliminary letter asking for cooperation was sent to twenty thousand women, and the final questionnaire was sent only to those who responded as willing to answer the questions.

The answers from the twenty-two hundred respondents were tabulated, and to interpret them, Davis undertook explanatory studies into the following topics: contraceptives, frequency of intercourse, happiness of the married women in terms of both general factors and sex, backgrounds of the unmarried sample, autoerotic practices among both married and unmarried women, periodicity of sex desire among both married and unmarried women, and prevalence of lesbianism in both married and unmarried women. About 50 percent of all women in the study reported experiencing "intense emotional relations with women," but the number giving these feelings overt sexual expression was less than two hundred. This is a higher percentage than Alfred Kinsey later found, although the difference might well be one either of class or of definition. Davis remained, however, very much a woman of her time in many ways. She found it "probably not surprising" to find that a large number of her subjects had "engaged in various erotic practices," although she did not bother to explain what these might be.[76] She did, however, discuss masturbation as an erotic practice and not as self-abuse. Some 64.8 percent of her unmarried college women admitted to masturbating at some time, although only 40.1 percent of her married sample did.[77]

Among Davis's married sample, only 71 women had sexual intercourse before marriage. The vast majority of these women used some form of contraceptive after marriage (730), and most of those who did not use one themselves approved such use in principle. Some 9.3 percent of the married women had at least one induced abortion, and one had eight such abortions.[78] Davis also found that of the women she classed as highly erotic, a much higher percentage had received sex instruction from what she called responsible sources before they were fourteen years old. She discovered that women who desired intercourse more frequently than they engaged in it with their husbands were more likely to be unhappy than were the wives who had husbands who agreed on the frequency of sex. The mode of greatest frequency was once or twice a week.[79]

Although she had not done the kind of study that Exner had suggested, Exner remained friendly with her and encouraged Davis to push the bureau even deeper into the kind of serious sex research that she was doing. Encouraging Exner was Earl F. Zinn, a young psychology graduate student from Clark University, whom Exner had hired in 1920 as coordinator of questionnaires on research projects at the American Social Hygiene Association. Zinn, after some preliminary work, made a series of proposals to Exner about sex research, which Exner took to Davis in 1920. After Davis made some modifications, she and the bureau were persuaded that a large-scale study of human sexuality should be undertaken. Conscious of the possible consequences of such a study to the social hygiene movement with which the bureau was associated, Davis and others felt more "scientific backing" for such a study was needed. It was therefore suggested to Rockefeller by his advisers that the study be conducted by an independent agency, specifically the National Research Council (NRC).[80]

Established in 1916 by the National Academy of Science to coordinate research funding during World War I, the National Research Council continued to function after the war as a conduit for research funds and projects. Since the Rockefellers supported projects of the NRC in other areas, it seemed natural to turn to the NRC to obtain the necessary respectability for the new kind of sex research that Zinn and Davis now proposed. What the two advocated was a systematic and comprehensive research into all aspects of human sexuality in its individual and social manifestations, the prime purpose being to evaluate conclusions now held and to increase our body of scientifically derived data.

Included in the proposal would be studies of physiological and psychological aspects of continence, masturbation, intercourse, and "aberrations" and investigations into prostitution, venereal disease, the family, marriage, divorce, family planning, and sex education.[81] Though Simon Flexner, one of the advisers to Rockefeller's father, originally opposed the proposal, he later, perhaps after some discussion with the younger Rockefeller, thought the project might well be worth attempting. Before going further, Flexner suggested that a group of scientists should be brought together for a conference and then action should be "based on the way the project appeared to them." As a result, in 1921, Rockefeller authorized the employment of Zinn to work under Davis's direction in an effort to secure the support of the National Research Council for such a conference. A budget of ten thousand dollars was allocated by the bureau for Zinn's expenses and proposed conference.[82]

Robert M. Yerkes, the resident salaried officer of the National Research Council, was supportive of the proposal but believed many of the council members would not be. Initially, the proposal was presented to the newly formed Division of Anthropology and Psychology but that division refused to deal with sex research, even if funds were available. Because there was no social science division, Yerkes then tried the Medical Sciences Division, where he also encountered an unwillingness to have anything to do with sex. Fortunately, there was a serendipitous change in the division chair, and the new chair was able to overcome the qualms of the committee; a special conference on sex research was held under the sponsorship of the division on October 28, 1921.[83]

The twelve invited participants at the conference (including Davis and Zinn) voted in favor of establishing the Committee for Research in the Problems of Sex (CRPS) within the Division of Medical Sciences and issued a statement to that effect. Included in the statement were perceived impediments to research into sex, including both the lack of data and the reticence and shame associated with such research. The committee reported, nonetheless, that it was convinced that through the use of methods employed in physiology, psychology, anthropology, and related sciences problems of sexual behavior could be subjected to scientific examination. The committee also listed a number of possible research topics, including sex and internal secretions, sex habits of primitive peoples, race and sex, variations in sexual impulse,

attitudes toward sex, physiological and psychological effects of masturbation, continence, a better understanding of sexual intercourse, and birth
rates in a variety of different groups.[84]

GILBERT V. HAMILTON

An initial budget of fifty thousand dollars was proposed,[85] and with this
action, the nature of sex research changed. Inevitably, Americans, with the
availability of research funding, quickly rose to preeminence in the field. An
early version of peer review of sex research, which was beginning to exist in
other fields, was also established in at least some aspects of sex research. The
results will be discussed in the next chapter, but everything did not necessarily progress smoothly. One of the first studies considered by the council, one
that had been solicited by Yerkes, was not approved for funding by the
National Research Council itself even though the Committee for Research
in the Problems of Sex had recommended it. This serves to again emphasize
the stigma. Involved in the turndown was Gilbert V. Hamilton, a physician
who had taken special training in psychology under Yerkes. His research was
ultimately funded by the Bureau of Social Hygiene, to which Davis had
taken it after it was rejected by the NRC.

Hamilton studied two hundred married persons—one hundred men and
one hundred women, including fifty-five married couples—most of whom
were either patients of New York City psychiatrists or friends of such patients.
Each subject was examined privately in Hamilton's consulting room and
were presented with questions on typed cards, 372 for women who had been
pregnant, 357 for those who had not been pregnant, and 334 for all men.
Though the questions were presented in typed form, the subjects were
encouraged to talk out their answers in a give-and-take conversation; these
were taken down word for word without comment. The interviews varied in
length from about 2 hours to more than 30. Among other findings, Hamilton reported the occurrence of multiple orgasm in some females.[86]

Hamilton's results, however, were judged so controversial by the Bureau of
Social Hygiene that it refused to be identified in any way as supporting what
he wanted to publish. Davis had retired from the bureau and from the CRPS
by this time, and Hamilton's failure to gain permission to list his granting
agency gives some indication of the influence she had wielded. Hamilton
had to agree that he would neither state nor imply that he had received
financial or moral support from the bureau. The only thing permitted to him
was to refer to a "group of scientific men" who had acted as advisers to his
project.[87]

Hamilton reported a wide range of sexual activity among white, married, college-educated men and women, far more than was generally assumed at the time. Interestingly, his findings were more or less ignored by the scientific community, in part because of the "undue" proportion (21 percent) of his subjects who sought psychiatric help before their participation in the study. Though the small size of the sample also limited its usefulness, the results more or less matched the later data collected by Kinsey for urban, white, college-educated males and females between the ages of thirty and thirty-five.[88] The findings differed somewhat from Davis's conclusions, though Hamilton's questionnaire was modeled on hers. Hamilton's subjects seem to reflect far more marital unhappiness than Davis's subjects. His subjects also demonstrate greater sexual variety, but this might be because the interview allowed more information to be gathered than Davis could acquire through a mailed questionnaire.

SUMMARY

Ellis, in surveying American sex research in 1931, held that the investigations of Davis, Hamilton, and Dickinson (he did not know about Mosher) represented breakthrough research in that they treated on a fairly large scale the sex activities and "sex relationships among fairly normal people, on a sufficiently large and systematic scale to be treated statistically."[89] He added that he had long ago realized that there was no rigid rule of normality and that in reality there was a wide range of variation, all of which legitimately should be admitted within the limits of normality. It was this wide range that the American investigators found. It was on just such studies that Kinsey and his collaborators built.

Although Dickinson once calculated that each interview compiled by Hamilton cost an average of three hundred and fifty dollars—far more than either he or Davis or, for that matter, Kinsey himself spent (the Kinsey interviews cost about two dollars each)—all these early studies emphasized that good sex research could be done and that it needed and deserved support. Though Dickinson's interviews were part of his patient load, he himself later received considerable financial support, some from Rockefeller sources, which enabled him finally to publish his studies. With these studies, the emphasis on purity and chastity—which had been such an important factor in what passed for sex research in the United States—was being replaced by a more realistic portrayal of what took place. It is worthy of comment that Exner, the exemplar of these traditional ideas, helped usher in the new age, but it is even more important to note that it was women such as Mosher and Davis who brought the new realism to American sex research.

5

ENDOCRINOLOGY
RESEARCH AND
CHANGING ATTITUDES

*W*hatever the original intent of Davis, Zinn, and Rockefeller had been in turning to the National Research Council to front serious research into human sexuality, once the formal scientific establishment was involved, the nature of sex research changed. It became university based, and the traditional physician was increasingly replaced by the Ph.D. trained in research methodology. As evidenced by the rejection of funding for Hamilton by the National Research Council, most scientists were not entirely comfortable with actual sex research conducted on humans. Because those at the NRC were laboratory scientists, they felt that the best science was done in a laboratory where conditions could be controlled. Inevitably, the Committee for Research in the Problems of Sex (CRPS) concentrated on what might be called "safer" subject areas (namely the kind of research that could be conducted in a laboratory), and consciously or unconsciously this allowed the committee to avoid the real "problem" areas (namely human sexuality, which was outlined in the original committee report).

Still the committee periodically gave lip service to the problem areas. For example, it sent an invitation to Ellis to visit the United States in the hope it could consult with him. Ellis, however, fell ill shortly after the receiving the invitation and never came. Zinn was sent to Europe to survey the research taking place there, and Davis coordinated his briefings. It is known he consulted with psychiatrists such as Freud, Eugen Bleuler, and Paul Schilder, but who else is not clear, although he made a report on his return.[1] The committee also consulted with various American experts on possible topics of interest to it. Clark Wissler, curator of anthropology at the American Museum of Natural History, was asked to consider the possibility of studies of sex in primitive cultures; Helen B. Wooley, director of the Bureau for Child Development at Teachers College, was asked whether the CRPS might give support to some of the bureau's studies of sex problems in children; and Dickinson met with the committee in 1925 to discuss the possibility of research in urological, gynecological, and obstetrical clinics. Little came out of any of these contacts, and Dickinson eventually was funded by other sources, some of which came from other wings of the Rockefeller Foundation. There was some discussion of a journal devoted to sex research, but this, too, was voted down.

These initial explorations and difficulties further emphasize the ambivalence that many in the scientific community had about being labeled sex researchers, even though their research might be supported by a committee set up to deal with sex problems. The answer is more complicated than simple fear of being labeled, however. Many of those interested in expanding the understanding of human sexuality argued that for findings to be meaningful they had to be accepted by their contemporaries in various specialty areas of science from which they came, something that might not happen if they were ghettoized as sex researchers.[2] The scientists involved wanted to publish in their own journals and to be judged by their peers. It was from their colleagues that they received their academic appointments, promotions, and recognition, and there was no department of human sexuality. The problem was that the study of sex was an interdisciplinary and interprofessional activity, and though the CRPS might coordinate grants, the researchers went off on their own. The committee was clear, however, about some of the minefields that were feared to exist, and even though it worked under the auspices of the medical division of the National Research Council, comparatively few grants were given to physicians. In addition, those physicians who did receive grants were essentially researchers and not practitioners, further weakening their influence on sex research. It also meant that the committee was not particularly motivated by a medical emphasis on diagnosis and treatment.

An old-boy network (few women were involved) of sex researchers quickly emerged, and the research most likely to be funded was that of interest to the committee members or their friends. After the retirement of Davis from the committee, most of the committee members either received significant grants of their own or gave such grants to their students or colleagues. Even Hamilton was well connected in this respect, since he had studied with Yerkes. At least one of the committee members became utterly self-serving, and Rockefeller Foundation officials were quite embarrassed by his special pleading.

FRANK R. LILLIE

In spite of such criticisms, what the CRPS concentrated on, it did well. It was also focused. Members of the committee had been requested to give suggestions of possible projects, and two such proposals dominated much of the early efforts of the committee. A key program was that put forth by Frank R. Lillie (1870–1947), one of the CRPS's original members, who prepared an outline on potential research areas in biology. The other was prepared by K. S. Lashley, a member of the psychology department at the University of Minnesota. At the request of Yerkes, chair of the committee for the first twenty-five years of its existence, Lashley drew up a program for research in sex neurology and psychobiology, areas of interest to Yerkes himself. Both proposals received important financial support from the committee, although it was Lillie's outline that served as the main guide.[3]

Lillie was one of the great entrepreneurs of early-twentieth-century biological sciences. He was Canadian; received his Ph.D. from the University of Chicago in 1894; returned to Chicago in 1900, where he spent the rest of his career; and was important in the development of several areas of biology. He was early associated with the development of the Marine Biological Laboratory at Woods Hole, Massachusetts, and from 1910 on was director of that lab as well as the biology department at Chicago. Lillie married into the wealthy Crane family of Chicago, and through his family contacts he was able to raise funds to help him establish many of his research programs. It was under Lillie that the Woods Hole Oceanographic Institution was formally incorporated. Among other things, he served as president of the National Academy of Sciences and during part of that time was also chair of the National Research Council.[4]

Lillie, however, was more than an entrepreneur, he was also a major researcher in his own right and expanded our knowledge of the factors that went into sex determination. Each new discovery about sex determination led to new questions. In 1902, the American investigator Clarence E. McClung found that the accessory chromosome, what is now called the X or Y chromosome, determined the sex of the fetus.[5] But complications arose with such a simple answer almost immediately. One such contradiction was resolved by Lillie in his study of freemartins. A freemartin is the female twin of a male calf who is born sterile, because it has been partially masculinized before birth by the male's testicular hormones, which are transferred to the female by the fusion of the two placentas. Hormones had not yet been isolated, but Lillie concluded that some internal secretions produced by the male's fetal testes entered into the female's embryonic circulation and inhibited the development of her gonads into ovaries, stimulated the development of mesonephric ducts and tubules into male-type sperm passageways, and inhibited the development of the müllerian ducts into the components of the female reproductive system.[6] To put it succinctly, Lillie demonstrated that in utero developments could effect the development of the visible sexual organs of individuals. It was the investigation of these internal secretions that occupied much of the CRPS's concern.

As of this writing, there is a debate in scientific circles over the growing inclination of the government granting agencies to prioritize areas of research instead of relying on investigator-initiated research.[7] Such prioritization has become the norm in many of the health-related agencies, such as the National Institutes of Health and National Institute of Mental Health, and it is becoming so in the National Science Foundation, which is being urged to do more applied research aimed at improving life and standards in the United States. Though there are arguments on both sides, because so many potential breakthroughs in science seem to have resulted from the serendipitous results of trying to answer basic questions rather than a deliberate attempt to deal with any so-called technical or social problem, the experience of the CRPS is perhaps the first major long-term example of the results of directed research. Sophie Aberle and George Corner, in their comments on the research efforts of the CRPS in sex and reproduction, claimed that more was learned about sex and reproduction in mammals in the first twenty-five years of CRPS than in any other similar quarter century in history.[8] This, however, is attributing too much to the CRPS, because although almost every American researcher in the field was sponsored by CRPS, there were also major breakthroughs in Europe, simply emphasizing that this period in time was one of those most ripe for contributing major additions to such knowledge. The general history of these developments has been outlined by others,[9] but it is important to summarize the breakthroughs here. Though it is difficult to separate these findings into categories, major discoveries concerned the female cycle, biochemistry, and reproduction, all of which overlapped. The result of this research not only led to a more complete understanding of the menstrual cycle and of fertility but finally led us to understand what happened at puberty. One of the more practical results was the development of oral contraceptives, a development that will be recounted in a later chapter.

ENDOCRINOLOGY

Both Ellis and Hirschfeld believed that what came to be called endocrinology would provide the answers to their questions. Some aspects of endocrinology, however, such as the effects of castration on animals, had long been known to science. Aristotle, for example, described the castration of birds, boars, stags, and cattle and gave precise directions for removing the ovaries from sows. He also described the effect of castration on the human male.[10]

The first modern report of the effects of castration was published in 1849 by the German zoologist by A. A. Berthold (1803–61). He had performed experiments on four young roosters. In two of them, he completely removed the testes, after removing the testes in the other two birds, he made an inch-long incision into the abdomen of each and inserted one testicle. He observed that the castrated roosters showed all the typical capon characteristics, behaving more like hens than roosters and having small combs, wattles, and heads. The roosters with the transplanted testes, however, behaved like other roosters. When he killed the birds, Berthold found that the testes he had inserted had become attached to the intestine, where they had acquired a good blood supply and were in good condition. He concluded from this that there must be some substance within the testes, other than sperm, which affects the temperament of the male chicken and which influences the development of the secondary sex characteristics. He also theorized that this substance was carried through the body by the bloodstream, because the blood vessels were the only channels connecting the transplanted testes to the rest of the body.[11]

What was true for the male gonads was also true for the female. In 1896, the Viennese gynecologist Emil Knauer (1867–1935) proved the existence of female internal sex secretions by transplanting the ovaries of fully mature animals into immature female animals who quickly displayed mature sexual characteristics.[12] These findings were more or less ignored until 1910, when Artur Biedl (1869–1933) published the first comprehensive study on glands and their secretions.[13]

Some ideas of the possibility of biochemical reactions in the body had been realized much earlier by researchers not particularly interested in sexuality. A pioneer in this was Thomas Addison (1793–1896), who had a patient who, for some reason, grew more and more emaciated until he finally died. Addison soon noticed other patients suffering from the same symptoms. He did an autopsy on the second patient and found that every organ was seemingly normal except for the adrenal glands, which lie above the kidneys. These had become withered and shrunken into tiny fibrous beads. Before publishing his findings, Addison gathered evidence over a period of five years on eleven such patients and concluded that the adrenals helped maintain normal function of the heart, digestive organs, blood, and skin.[14] Addison, however, did not know how the adrenals exercised their tremendous power and hesitated to guess.

At about the same time that Addison was writing, the French scientist Claude Bernard (1813–78) realized that the liver secreted a sugar-forming substance, which he called an "internal secretion." Later, he isolated the secretion, which he named glycogen, and went on to show that both the thyroid and adrenals similarly produced internal secretions.[15]

Up to this point, scientific caution had prevailed in unraveling the secrets of internal secretions. Changing this cautionary roll was Bernard's successor in experimental medicine at the Sorbonne, Charles-Édouard Brown-Séquard (1817–94). He jumped to the conclusion that the male sex glands secreted some substance that controlled not only sexual vigor but many other bodily functions. This is a valid assumption, but anxious about his own waning sexual power, he then reasoned that all he had to do was to extract this powerful secretion from animal testicles, inject it into himself, and thereby renew his sexual vigor.[16] This he proceeded to do, and though he reported much benefit, his advocacy of this method evoked skepticism and criticism, since others who tried it did not report the same results. Unfortunately, the experiments of Brown-Séquard led some of the more "respectable" scientists to avoid the study of internal secretions, because of its association with sexual rejuvenation, a topic that they felt might impair their academic integrity.

Brown-Séquard, however, was on the right track, and in 1891, George Redmayne Murray (1865–1939) demonstrated that it was possible to cure myxedema (a disease attributed to hypothyroidism) by injections of a glycerin extract of sheep's thyroid, although it was not clear what was contained in the extract.[17] It was not until 1902, however, that the science of endocrinology moved from speculation to reality. In that year, William Maddock Bayliss (1860–1924) and Ernest Henry Starling (1866–1927) demonstrated the existence of a substance that they called secretin in the duodenal secretions of the pancreas. They demonstrated the secretin activated the pancreas to secrete pancreatic juice.[18] With the discovery of glycogen and secretin, it was readily apparent that internal secretions were extremely important in regulating body functions.

Though others continued to work on internal secretions, it was the example of Brown-Séquard that caught the public's imagination. Adding to the public's imagination were the experiments of Eugen Steinach (1861–1944), who by 1912 had experimentally masculinized the mating behavior of female guinea pigs and feminized the mating behavior of males by castrating them at birth and transplanting heterotypic gonadal tissue into them.[19] By doing this, he demonstrated, for the first time, the prenatal hormonal control of the adult behavioral outcome of the male-female bipotentiality that exists in us all. Steinach, like Brown-Séquard, immediately sought to apply his findings to humans. He came to believe that by ligating the vas deferens (a method now used to bring about male sterilization) he could bring about sexual rejuvenation. The reason for this rejuvenation, he argued, was because the secretions associated with ejaculation would then flow back into the body.[20] Actually, as we later learned, this does not happen, since instead of flowing back it is secreted in the urine. Like Brown-Séquard, he initially reported success in his operations, a success apparently due to a placebo effect. His method was soon discredited.

QUACKERY AND ENDOCRINOLOGY

In the United States, however, in spite of the failures of Brown-Séquard and Steinach, a minor industry grew out of the promise to restore potency to the aging male. For example, the Packers Product Co., run by a Fred A. Leach out of Chicago, sold something called "Orchis Extract," imputed to be made from an extract of ram's testicles. The U.S. postal authorities put Leach out of business for fraudulent use of the mails, in part because of false advertising, because he implied he obtained his product from Armour & Co., one of the major American meat packing companies, and Armour wanted no association with him. He quickly returned to the market under a new name for the same product, this time the Organo Product Co., without any references to Armour. Postal officials, nonetheless, accused him of fraud in 1919, and their refusal to deliver mail to him again put him out of business.[21] Even Magnus Hirschfeld got into the act with a concoction labeled Titus Perlen to encourage sexual stimulation, an action that was used by his critics to discredit him.

Perhaps the ultimate salesman of the new potency products for males was John J. Brinkley (1885–1942), popularly known as "Goat Gland" Brinkley, who promised to restore virility to the weakened male by implanting goat testicles and to reinvigorate females by treating them with the royal jelly of the honey bee. Brinkley carried news of his operations and elixirs on the newly developed radio and owned his own radio station in Kansas through which he blanketed much of the Midwest. When the government revoked his medical license and moved to close his station, he ran for governor of Kansas in 1932 as a third party candidate. He received 244,607 votes, running third behind the winner, Alfred M. Landon (who received 278,581 votes), who ran for president in 1936 against Franklin D. Roosevelt. Brinkley then moved to Mexico and opened another radio station, which for a time was said to have reached one-third of the radio listeners in the United States and from which he pitched his secret remedies and his goat gland operation. He also sold advertising time on his station to other health hucksters. Eventually, the Mexican authorities closed the station because he was not properly registered in that country.[22]

Undoubtedly, the misleading claims of Steinach and Brown-Séquard, even Hirschfeld, made researchers in endocrinology cautious in their claims, and the examples of Brinkley and Leach worked as factors in the reluctance of many of the sex researchers of the 1920s and 1930s to have any identification with them. Still there were breakthroughs in understanding the importance of human internal secretions. One of the major breakthroughs, made at the University of Toronto, was the 1921 discovery by Frederick G. Banting (1891–1941), Charles H. Best (1899–1978), and John J. R. Macleod (1876–1935) of a way to extract insulin from the islets of Langerhans in the pancreas. This discovery provided a life-saving treatment for diabetics.[23] It was in such a setting, one of great potential and yet one that, as far as sex was concerned, could easily lead to exaggerated claims, that the Committee for Research in the Problems of Sex decided to give major support to the developing field of internal secretions.

FEMALE HORMONES

It was not in the study of males but in the study of females that the first major breakthroughs occurred. By 1920, after several false starts and assumptions, a theory of menstruation was beginning to develop. The essentials were that menstruation occurred because the lining of the uterus, after being prepared for implantation of the ovum, would degenerate if fertilization of the egg did not occur. This theory required that ovulation and the corpus luteum formation precede the premenstrual change. Still there were many unanswered questions, and the 1920s saw a number of researchers investigate problems that led not only to better understanding of the cycle but to isolation of female hormones. One of the first researchers to be supported by the CRPS was Edgar Allen (1892–1943) of Washington University in St. Louis (later he moved to the University of Missouri and ultimately to Yale University). In 1923, Allen and his colleague Edward A. Doisy (1893–96), who later won a Nobel Prize, injected a somewhat crude ovarian extract into immature female rats and found that within 48 hours the rats went into estrus (heat) just as if they possessed fully mature ovaries.[24]

They had been aided in their study by the earlier discovery of Charles R. Stockard (1879–1939) and George N. Papanicolaou (1883–1962) that in some laboratory rodents (guinea pig, rat, and mouse) the cycle of the ovarian function could be followed and ovulation detected by taking vaginal smears.[25] (Papanicolaou later developed the Pap smear test for cervical cancer.) The use of vaginal smears made it possible to test exactly what was happening in the female reproductive tract without killing or operating on the experimental animals. Allen and Doisy tried to develop more concentrated active estrus stimulating extracts from ovaries but were initially unable to do so.

Their problem was solved serendipitously by Selmar Ascheim (1878–1965) and Bernhard Zondek (1891–1966), who in their search for simple test of pregnancy found that if they injected a patient's urine into an immature laboratory mouse or rat, they could get an accurate yes or no answer about the patient's pregnancy. If the patient was not pregnant, the test animal showed no reaction; if the patient was pregnant, the animal displayed an estrous reaction, despite its immaturity.[26] This came to be called the Ascheim-Zondek test for the diagnosis of pregnancy. It also implied that with the onset of pregnancy there was such an increased production of hormones by the ovaries that substantial quantities of estrogenic substances were excreted.

When Allen and Doisy heard about the test, they realized there was a rich and easily handled source of hormones in urine from which they could develop a potent extract. In a sense, the scientific race for the isolation of a female hormone was on, as scientists in other countries—notably in France, the Netherlands, and Germany—were also active. Allen and Doisy's research was sponsored by the committee, while that of their main rival, Adolf Butenandt (b. 1903) of the University of Göttingen was sponsored by a German pharmaceutical firm. In 1929, both teams announced the isolation of a pure crystal female sex hormone, estrone, in 1929, although Doisy and Allen did so two months earlier than Butenandt.[27] By 1931, estrone was being commercially produced by Parke Davis in this country, and Schering-Kahlbaum in Germany.

Interestingly, when Butenandt (who shared the Nobel Prize for chemistry in 1939) isolated estrone and analyzed its structure, he found that it was a steroid, the first hormone to be classed in this molecular family. Steroids, from the Greek *steros* (meaning "solid"), are members of the alcohol family, but they differ from the common alcohols in that they are subject to crystalline solidification. All steroids have one thing in common: a group of seventeen carbon atoms arranged in four rings.

ISOLATION OF HORMONES

This discovery was important, because chemists had already learned to manipulate and alter steroid molecules, so there was a good possibility the substances could be duplicated by chemical synthesis. It also emphasized the possibility that since estrone was a steroid the secretions of the testes may be as well.

The result was a rapid spurt in isolating various internal secretions, now generally called hormones. In 1930, Guy F. Marrian (1904–81) of London isolated another substance from pregnant women's urine called estriol, which differed only slightly from estrone but was milder in its activity. The two substances are the excreted metabolites of estradiol, the active hormone produced in the ovaries. The estrogenic substances turned out to be responsible for the development of secondary sexual characteristics and for the cyclic changes in the vaginal epithelium and endothelium of the uterus.

Estrogens, however, are not the only female sex hormones. As early as 1903, Ludwig Fraenkel of Breslau had discovered that once the ovaries released an egg, a yellow substance formed within the ruptured egg sac. He had also found that if he removed this yellow body (corpus luteum) from a rabbit doe shortly after she mated with a buck the embryos would not develop. The American researchers George W. Corner and his then student assistant, Willard M. Allen, at the University of Rochester, were sponsored by the committee and succeeded in producing a corpus luteum extract in 1930. Corner later described it as a "few spoonfuls of a thick greasy semifluid material looking like a poor grade of cylinder oil."[28] Because it favored the process of gestation, this second female hormone was called progesterone, and by 1934, several scientists working independently isolated the pure hormone.

Fred C. Koch at the University of Chicago was another scientist supported by the committee. Koch had started by grinding up bulls' testicles and preparing extracts from them, which he injected into capon chickens to see how they influenced comb growth. His initial results were not very good. In 1929, he joined with T. F. Gallagher, another scientist supported by the committee, and they developed a many-staged extraction process that yielded a more active mixture than any Koch had seen before. Just 0.01 milligram of the new substance produced an upstanding red comb in five days.[29]

Again there was a concentrated effort to isolate the substance, and Butenandt managed to isolate a few grains of crystalline hormone in 1931 but from urine rather than testicles. Because it stimulated comb growth and because it was a steroid, he coined a Greek name for it: androsterone.[30] Though the substance was biologically active, it was found not to be the male hormone itself but a metabolically changed or degraded form. It was not until 1935 that a Dutch team, improving and modifying on Koch's method of extracting a substance from crushed bulls' testes, succeeded in extracting the true male hormone, which was called testosterone.[31] Within a year, the Yugoslavian chemist Leopold Ružička, working in Switzerland, managed to make a synthetic duplicate of natural testosterone.

The availability of injected hormones allowed a large number of investigators to carry out extensive studies with sex hormones in all sorts of animals from fish through amphibians to birds to mammals. The result was to demonstrate the idea first put forth by Lillie that the differentiation of the organs of the two sexes in utero was determined by hormones. Was there not then a second hormonal onset that was involved with developments at puberty and that changed girls into women and boys into men? Again, one of the researchers sponsored by the committee, Herbert M. Evans, found that when he injected female rats with fluids from the anterior part of the female glands, they produced abnormally large quantities of corpus luteum.[32] The substance was called a gonadotropin because it stimulated the ovaries and testes. Ultimately, one of these substances was identified as follicle-stimulating hormone (FSH), because it was at first thought to cause only the secretion of estrogen by the ovaries. This in turn led to the maturation of an egg cell in its follicle.[33] FSH was later found to stimulate the development of the ovaries and the production of sperm cells.

A second gonadotropin was also found by the same team: interstitial cell–stimulating hormone (ICS),[34] so called because it stimulated the secretion of testosterone by the interstitial cells of the testes, which in turn led to the growth of the penis and prostate and the development of secondary sexual characteristics. Later the name was changed to the luteinizing hormone (LH), because in females it triggered the formation of the corpus luteum and brought about ovulation.

Both gonadotropins need to be present to accomplish their effects on the sex organs. This is because there is a feedback action regulating the pituitary output of these hormones (hormones are also produced in the pituitary). This output depends on the level of testosterone, estrogen, or progesterone in the blood, and the pituitary either steps up or reduces the output of FSH or LH accordingly.

MENSTRUAL CYCLE

The result of such discoveries was a much better understanding of the whole physiology of reproduction, including the workings of the menstrual cycle, the period of female fertility, and the nature of conception.[35] One of the key breakthroughs was by G. W. Bartelemez and J. E. Markee, who explained the role of the blood vessels in the uterus during menstruation. The two researchers found that the endometrium is fed by arteries that come up into it from the underlying muscle. These arteries have two kinds of branches. One kind is unique: Such arteries are wound into coils and make their way rather tortuously toward the surface, where they break up into capillaries, which supply the inner one-third of the endometrium. The other kind is the standard artery: They run a rather straight and short course directly to supply the basal two-thirds of the endometrium.[36] As the endometrium thickens, additional coils appear in the coiled arteries, and this leads to a retarding of the circulation of the blood through them and their branches. Thus the immediate cause of menstruation was conjectured to be the injurious effect of the lack of blood supply to the endometrium brought about by the mechanical compression of the arteries.[37] This conjecture of mechanical causes of ischemia was discarded when the whole process was found to be hormonally mediated.

The menstrual cycle can be divided into phases. Some researchers have used the following five phases: (1) follicular, (2) ovulation, (3) luteal, (4) premenstrual, and (5) menstruation or pregnancy. Others have used four phases: (1) proliferative (or follicular), (2) ovulatory, (3) progestational (luteal), and (4) menstrual. Those writing for the general public tend to use five, because it seems to be more descriptive of what women go through, and some women are very conscious of their oncoming menses or the premenstrual phase. The premenstrual phase is not usually used in scientific writing, perhaps because much of the research on the menstrual cycle was conducted on nonhuman primates, particularly rhesus monkeys, who could not communicate with the researchers. Corner found that progesterone and estrogen had menstrual suppression properties and the discontinuance of progesterone brought on bleeding. He theorized that in the normal cycle the uterus would not bleed during the first half of the cycle (the follicular phase), because the ovaries are furnishing estrogen; it would not bleed during the second half of the cycle (the luteal phase), because the corpus luteum is furnishing progesterone. Demonstration of this hypothesis is essentially what made the oral contraceptive possible.

During the follicular phase, which lasts from six to fourteen days, several follicles (from which the phase gets its name) begin to ripen and mature within an ovary. This process is mediated by a hypothalamus decapeptide that was originally called gonadotropic-releasing hormone (GnRH) but is now usually called luteinizing hormone–releasing hormone (LH-RH) or luteinizing hormone–releasing factor (LH-RF). The release of LH-RH is in response to low levels of estrogen in the bloodstream. LH-RF in turn stimulates the pituitary to secrete FSH[38] and LH. FSH stimulates the follicle to produce and secrete estrogen into the bloodstream (the follicular phase). As the level of estrogen in the bloodstream rises, it signals the pituitary to stop releasing FSH. This then leads to ovulation, which involves the rupture of a mature follicle from the ovary. When the egg is released from the follicle, it begins its journey to the fallopian tube, where it may or may not be fertilized by a sperm. Some women are able to feel the rupturing of the egg *(middleschmerz)*, but most do not. The result is the luteal phase, since the empty follicle, now referred to as the corpus luteum (yellow body), secretes progesterone.[39] Progesterone leads to an increase in the growth of the endometrium (uterine lining) in preparation for the egg in case it is fertilized (premenstrual phase). After four to six days, the corpus luteum begins to disintegrate, leading to a decrease in the levels of both progesterone and estrogen and, finally, to menstruation, the disintegration and discharge of the uterine lining.

There are also variations in body basal temperature during the cycle, with the temperature rising slightly during the later part of the menstrual cycle. The temperature begins to rise one to two days after ovulation in response to rising levels of progesterone, and it is on this fluctuation in body temperature that some individuals used to chart their fertility cycle.[40] Other changes take place in cervical mucosa, which have also been used to plot safe periods for sexual intercourse.[41]

As estrogen levels drop, the cycle starts over again. If, however, pregnancy takes place, the placenta begins to produce human chorionic gonadotropin (hCG) at rapidly increasing levels for the first six weeks; this hormone delays the onset of menstruation and replaces LH.[42] The placenta also produces estrogen and seems to be the chief source of estrogen during pregnancy. Progesterone is also present in the placenta. The ability to detect hCG was the explanation for the success of the first pregnancy tests. Essentially, the key to the cycle was Corner's discovery that menstruation was not brought about by the positive action of the corpus luteum but by its decadence. The proof of this was a key to understanding the cycle of the human female, although our ideas on this continue to be refined.[43] Though not all of the researchers involved in such studies were supported by the CRPS, most of them were, and it enabled American science to move into a more or less dominant role in sex research. This move was undoubtedly assisted by the rise of the Nazis and the destruction of the German scientific enterprise, but the full effect of these discoveries depended on the effective and inexpensive production of the hormones, something that was not immediately possible.

GREGORY PINCUS

Generally, the research remained noncontroversial, but occasionally it received the kind of notoriety the committee tried to avoid. One of the researchers who received such notoriety was Gregory Pincus. Pincus picked up where another Rockefeller-supported scientist, Jacques Loeb (1859–1924), had left off. Loeb, who immigrated from Germany in 1891, taught at the University of Chicago and the University of California; and from 1910 on, he was affiliated with the Rockefeller Institute in New York City. In large part because Sinclair Lewis used Loeb as the model for the altruistic scientist in his novel *Arrowsmith* (1925),[44] Loeb in death became the symbol of the scientist as a lonely searcher after truth, immune to the temptations of the "Bitch God of Success." Earlier Loeb was also praised by Thorstein Veblen and H. L. Mencken for his willingness to stand against prevailing American orthodoxies, since he was not only a mechanist in science but a socialist in politics.[45]

As a mechanist in science, Loeb sought to treat living things as "chemical machines" whose behavior could be explained by the laws of physical science. To test his hypothesis about chemical machines, he immersed fertilized sea urchin eggs in saltwater with an osmotic pressure that had been raised, by the addition of sodium chloride, above that of seawater. When replaced in seawater, they underwent multicellular segmentation. Later, he subjected unfertilized eggs to the same process, and by 1899 had succeeded in raising larvae by this technique, the first notable triumph in achieving artificial pathogenesis.[46] The possibility of living things as chemical machines challenged the traditional assumptions of science, and in Donald Fleming's words was an "assault upon the crux of biological mysticism, the process of fertilization where so many ingenious theologians through the ages had seen their chance to slip a soul in while nobody was looking."[47]

Picking up on this was a young assistant professor at Harvard, Pincus, who was establishing himself as an authority on mammalian sexual physiology and was specializing in delicate work with rabbit eggs. His research was supported by the CRPS. Pincus received national publicity in 1934 when he announced that he had achieved in vitro fertilization of rabbit eggs (that is, inside a test tube). Unfortunately, in the aftermath, the *New York Times* portrayed him as a sinister character bent on hatching humans in bottles, similar to that of Professor Bokanovsky, created by Aldous Huxley in the novel *Brave New World* (1932), which was then being widely read.[48]

By 1936, Pincus had shown that rabbit ova could be artificially activated by a number of techniques, including exposure to salt solutions and to changes in temperature. He had also successfully transplanted many developing ova into female rabbits and shown that the ova had grown into embryos before the host does were sacrificed to check the progress of the experiment.[49] His "immaculate conceptions," as they were called, made good copy for journalists even though many of them were hostile to what he was doing. *Collier's* magazine, then one of the most widely sold magazines in the United States, ran an article titled "No Father to Guide Them," which was both antifeminist and anti-Semitic (Pincus was Jewish) and held that biologists were playing with nature. The article concluded that Pincus was threatening the American male, whose virility, already called into doubt by the Great Depression and the declining birth rate, was now threatened by a fate worse than simple castration, namely the very reason for his existence.[50]

Shortly after this, Pincus was denied tenure at Harvard, an action that served as a further warning to those engaged in sex-related research to keep a low profile. Harvard did give him a research grant that enabled him to go to Cambridge University in England for a year. But on his return, he was unable to get an academic position. Hudson Hoagland at Clark University (in Worcester, Massachusetts) came to his rescue by offering him a courtesy appointment as a visiting professor of zoology, but such a position had no salary and lacked faculty privileges. With financial support from Nathaniel, Lord Rothschild, Pincus was able to move to Worcester, and some of his research there was eventually supported by the CRPS, although not until 1940. His move to Massachusetts, however, did not isolate him from media attention, and periodically, they gave new publicity to the birth of the parthenogenetic rabbits.[51]

Still unable to get Pincus a regular academic appointment at Clark or elsewhere, Hoagland joined with Pincus in 1944 to set up the Worcester Foundation for Experimental Biology, a nonprofit corporation supported by local philanthropists and various foundations and agencies, including the CRPS. Pincus went on to conduct research about steroids. Much of the money eventually came from the G. D. Searle pharmaceutical company, since Pincus and his colleagues were interested in developing a process for the commercial production of cortisone, a secretion of the adrenal cortex.[52] By that time, however, Pincus was no longer supported by the CRPS.

In terms of practical application to sexual practices, perhaps the most significant result of the new endocrine discoveries was the development of more effective contraceptives. In the 1920s, Davis demonstrated that contraception was known and widely practiced by the majority of married women she studied. The problem was not a lack of effective contraceptives but dissemination of information about them and making them affordable to the average person. Physicians themselves were often reluctant to deal with contraceptives. When Sanger originally asked Dickinson to give his public support for her early campaign on contraception, he turned her down. He believed his career as a physician came first, and that his association with Sanger would end his influence among other medical men. Dickinson had even told Clarence James Gamble (1894–1966)—heir to the Ivory soap fortune as well as a pharmacologist interested in dissemination of contraceptive information—that first-class scientists neglected contraceptive research because of the belief that birth control was illegal and was associated with immorality.[53] Though Dickinson later changed his mind about working with Sanger (he was already an advocate of birth control and prescribed it for his clients), as did many other physicians, the reversal was brought about by changes in American society as much as it was by new scientific breakthroughs.

In fact, in spite of the studies by Davis and Dickinson, American researchers remained hesitant about studying sex in humans in any direct way through much of the 1930s. Animal studies, as has been emphasized, seemed safer, and though many of these had implications for human sexuality, such implications were usually never directly drawn. This was unfortunate, because in Europe studies in human sexuality reached an ebb. Germany, the center of most of the early-twentieth-century studies of sex, was under Nazi control and the Nazis, who were extremely prudish in discussions about sex, mounted campaigns to eliminate pornography, imprisoned homosexuals to cure them, sent off into exile many of the those engaged in sex research, and either destroyed or dispersed valuable collections of materials such as those of Hirschfeld.

Outside of Germany, many of the pioneer generation of physicians who had dominated sexology were dying off, and the younger generation of medical practitioners were not interested in studying sex. The major exceptions to this generalization were in the field of contraceptive research and in the area of sexually transmitted diseases, then called venereal diseases. Still there were difficulties even here. Many respectable scientists were unwilling to do research into contraceptives, and even the Rockefeller-funded CRPS was unwilling to investigate the topic. The Rockefeller Foundation remained interested in it, however, and when it decided to fund studies into the chemical composition of spermacides, the foundation could find no American scientist willing to do the work. Instead, they went to England, but after some preliminary finds had been made there, the English group refused to go further.

Human sex research in the eyes of much of the scientific community was simply too controversial to investigate directly. Following their own interests, these scientists concentrated on much narrower studies in nonhuman species. Even the topic of venereal disease was handicapped by an unwillingness to discuss the issue publicly. For example, when Thomas Parran Jr. was made head of the Venereal Disease Division of the U.S. Public Health Service in 1926, one of his goals was to tackle the problem of syphilis. He found it much more difficult to do so than he anticipated, because he was unable to discuss the topic publicly. As late as 1934, when he was scheduled to give a radio talk for the Columbia Broadcasting System (CBS) that was to review the major public health issues, he was told he could not use the words *syphilis* and *gonorrhea* on the air. It was not until 1936, after he was appointed U.S. surgeon general, that federal funds were finally appropriated for public clinics to carry out Wasserman testing for syphilis and to treat the disease.[54]

Sex education in the schools theoretically had been growing, but in most schools where it was taught, it was regarded as an adjunct to physical education (which supposedly included health education). The goal of such education was to emphasize abstinence, which was often done by focusing on the dangers of disease and infection. Though some anatomy and physiology was included, it was usually better taught in the girl's classes than the boys, since menarche and menstruation were deemed more important to them.

At the college level in the United States, however, particularly in the state universities and the private women's colleges, courses began to be offered in marriage and family, often in home economics departments. Such courses also included discussions of the anatomy and physiology of sex, although most of the students were female.[55] This movement for some sort of education in marriage and family continued to grow and was formalized in 1939 with the organization of the National Council on Family Relations and the establishment of the journal *Living*, which later changed its names to the *Journal of Marriage and the Family*.[56] One effect of the foundation of the NCFR was to encourage exploration into various aspects of dating and courtship as well as sex. It was in just such a course that Kinsey started his research.

MARRIAGE MANUALS AND THE SEXUALIZATION OF LOVE

The changing ideas about marriage and the place of sex can most easily be traced in the appearance of new types of advice manuals, which came to be popularly known as marriage manuals. They differed from earlier ones not only in accepting women as sexual creatures but in insisting that sexual satisfaction was part of their marital right. Many of these manuals were written by women, and even those that were not seemed to be aimed more at women than men, since the emphasis was often on having husbands (sex was still permissible only in marriage) learn the sexual needs of their wives. A major consequence of such manuals was the increasing sexualization of love. Using the writings of Ellis, Freud, and others as a base, the marriage manuals offered advice and help, of types not previously available, to the newly marrieds of the 1920s and 1930s.

A major reason for the appearance of these new manuals was the wide-spread popular belief in Western society that traditional marriage was failing. One source of fuel for this belief was the changing status and influence of women, which led to demands for greater equality in marriage. Though women had finally gained the right to vote at the end of World War I in the United States, the United Kingdom, and other countries, this had not led to anything like equality between the partners in marriage. Such a demand had already appeared in the last part of the nineteenth century, and it grew more influential as the twentieth century progressed.[57] One of the early symbolic representations of this unrest were the plays of the Norwegian dramatist Henrik Ibsen (1828–1906). The heroine in A Doll's House (1879) came to believe that her domestic happiness was specious and that her whole personality had been circumscribed and smothered by society's conventions. Her solution was to explore the world alone in search of selfhood, and though this seemed daring and attractive to many of the women of the time, few were willing to go quite that far. In his Ghosts (1881), Ibsen portrays the horrible hereditary effects that result from an unfit father who, in spite of his defects, continued to fulfill his procreative obligations to society.[58]

There was a growing plea (not yet a demand) to change laws and customs so that the woman would not feel forever chained to a partner in a repulsive relationship. Ibsen's ideas were not isolated ones but were carried over into feminine protest novels by such authors as Ellen Key (1849–1926), who urged that women must be faithful to their own personalities rather than conventional morality. Some of the same kind of message was conveyed in the works of Ellis and Edward Carpenter, and in fact, the whole European sex movement was strongly influenced by the need for reform in marriage, including the right of divorce. Divorce, in fact, was becoming much more common, and as early as 1908, Shaw, in his introductory remarks to his play Getting Married (1908), wittily proclaimed that divorce was a necessary condition for the maintenance of marriage.[59]

Obviously, there were literary counterparts to Ibsen, Key, and others who argued on the other side. Most notorious was Otto Weininger, whose Sex and Character (1903) is a misogynist tract emphasizing that women are heartless, shameless, and amoral.[60] August Strindberg (1849–1912) dramatized sex as a conflict between two well-armed antagonists. Though women, he wrote, were normally inferior combatants in most battles, in marriage they were stronger because they were more realistic and more unscrupulous. He said that though the world contained only one disagreeable woman the problem was that she was universal and each man drew the same woman for his companion. The American author Robert Herrick (1868–1938) added that the trouble with middle-class women was that they married to become queens and they wanted to rule and not to work.[61] Much of the psychoanalytic writing of the time, in fact, is devoted to attempting to explain female neuroses as due to repressed sex desire.

Various solutions were proposed. Among the popular American advocates of change were Judge Benjamin B. Lindsey and Wainwright Evans, who advocated what they called companionate marriage. This included the use of birth control, divorce by mutual consent for childless couples who were incompatible, education of the young in the art of love and successful marriage, and a reform in laws regulating alimony.[62]

Another way of dealing with the issues was to redefine marriage by emphasizing the sexual and affectional basis of intimacy and the importance of companionship in marriage. This change in emphasis is most reflected in the change in the nature of marriage manuals that began to appear in the United States. Interestingly, though Americans such as Sanger and Robinson wrote on sex and marriage, the two most widely read authors were Europeans: the British botanist Marie Stopes and the Dutch gynecologist Theodoor van de Velde.

MARIE STOPES

Marie Charlotte Carmichael Stopes (1880–1958) was one of the pathbreaking women who were so important in the development of a better understanding of human sexuality. Born to a well-to-do family in Scotland, she was tutored at home by her parents until she entered boarding school in London at age twelve. After graduating from University College in London in 1902, where she studied botany, geology, and physical geography, Stopes went on to receive a doctorate at the University of Munich. She then joined the science faculty at the University of Manchester, the first woman to do so. In 1911, she married, and leaving Manchester, she moved to London where she became a lecturer on paleobotany at the University of London from 1913 to 1920. During this time, Stopes published a textbook on ancient plants as well as a catalog of Cretaceous flora in the British Museum collection. During World War I, she engaged in research on coal with R. V. Wheeler. In short, Stopes was a first-class scientist and researcher.

Stopes's interest in sexuality came from the failure of her 1911 marriage to Reginald Ruggles Gates. In 1916 the marriage was annulled on the basis of nonconsummation, and in 1918 she married Humphrey Verdon Roe, an aircraft manufacturer. He was already interested in birth control, and shortly after their marriage they founded the Mothers' Clinic for Constructive Birth Control in London, the first of its kind in England.[63] After this, Stopes (who kept her maiden name) relinquished her lectureship at the University of

London and devoted herself to family planning and sex education for married people. She also wrote fairy tales under the name of Erica Fay. Among her other pseudonyms were Mark Arundel, Marie Carmichael, and G. N. Mortlake.

Strongly supportive of the concept of marriage, she also emphasized the rottenness and danger of the unhappy marriage, which she said came about when love was separate from sex. She believed that both were key to a happy marriage. Her first book on the subject, *Married Love*, was published in 1918, although she had originally drafted it in 1914 to crystallize her own ideas. It became an immediate success and was translated into numerous languages, causing a sensation at the time of its publication. Although the first edition dealt scarcely at all with birth control, she received so many requests for instruction on the subject that this was followed up by the short book *Wise Parenthood*, which was also an immediate success. Within nine years it had sold half a million copies in its original English edition. Stopes published a number of other books more or less dealing with same subject with titles such as *Radiant Motherhood* and *Enduring Passion*; many of these sound somewhat overromantic to later generations of readers.[64]

Her great achievement was to move the topic of birth control in much of the English-speaking world from the confines of the physician's office to public discussion. But equally significant was her emphasis on the importance of sex in the life of a woman. Stopes believed that both love and sex were essential parts of marriage. Because the sexual discontent so dangerous to marriage was something that she thought could be remedied, she took to writing marriage manuals on how to renew love through sex. Her general theme, as it was of several other contemporary writers on love and marriage,[65] was that in the past marriage had been held together by the economic dependence of the wife on the husband, by the sense of duty of the husband to his family responsibilities, and by law. She felt that in the new day and age there had to be a new basis for holding together a union of equals who chose to join their lives for companionship. This new force that was to bind love and marriage was sexual compatibility.

Where the acts of coitus are rightly performed, the pair can disagree, can hold opposite views about every conceivable subject under the sun without any ruffling or disturbance of the temper, without any angry scenes or desire to separate. They will but enjoy each other's differences. Contrariwise, I am sure that they can have ninety-nine per cent of all their other qualities and attributes in perfect harmony, and if the sex act is not properly performed; if they fail to adjust themselves to each other; if they are ignorant of the basic laws of union in marriage, all that harmony and suitability in other things will be of no avail, and they will rasp each other apart in sentiment, until they but endure each other for some extraneous motive, or they desire to part.[66]

Sanger, perhaps better known to American readers today than Stopes, was not quite so explicit, but Sanger also emphasized the importance of sex, holding that the bedrock of lasting happiness in marriage "lies in a proper physical adjustment of the two persons, and a proper physical management of their mutual experiences of [sexual] union."[67] The cornerstone of a proper physical management for Sanger was built on the use of effective contraceptives so the life of the woman would not be governed by frequent unwanted pregnancies.

It was believed and taught that once women's sexual needs were accorded an equal status to those of men, the norms of companionship and mutuality in marriage would inevitably extend into the domain of sex and become the seal of love. Hannah Stone and Abraham Stone, who also wrote marriage manuals, said that the sexual embrace should become the expression of mutual desire and passion. The joy of sex could only increase when it was mutual.[68] Isabel Hutton emphasized that "no matter how ideal the partnership in every other way, if there is want of sex life," the marriage cannot be a success.[69] Stopes had added a strong second to that: "In these modern days when friendship, mutual occupations, businesses, almost every phase of our civilized life, bring men and women together in innumerable ways, the only justification of marriage is the mutual need for and the mutual enjoyment in sex union."[70]

THEODOOR VAN DE VELDE

Some writers seemed to caution against quite so much emphasis on sex as a cure for all marriage problems. Theodoor Hendrik van de Velde (1873–1937), for example, whose marriage manual was the most popular one in United States, emphasized that marriage depended on a range of factors, including the emotional and social compatibility of the husband and wife, a sharing of interests, and an agreement on family.[71] Still, even he held that vigorous and harmonious sexual activity was the key to the temple of marriage: "It must be solidly and skillfully built, for it has to bear a main portion of the weight of the whole structure. But in many cases it is badly balanced and of poor material; so can we wonder that the whole edifice collapses."[72] In effect, almost all the manual writers, including van de Velde, agreed it was good sex that made marriage a success.

Van de Velde's first book, *Ideal Marriage*, was published in 1926 in both

Dutch and German. Though van de Velde anticipated he might be ostracized for publishing the kind of sex information he included in his book, this was not the case. The book went through forty-two printings in Germany before it was suppressed when Hitler came to power in 1933. The London edition went through forty-three printings, and while figures are not available for the American edition, it went through six printings in the first two years of publication. Random House, which later purchased the rights to publish it, reported that it sold half a million copies between 1945 and 1965, when it again was revised and republished.[73]

Van de Velde had first married in 1899 when he was twenty-six, but the marriage was not a happy one. After ten years, he left his wife for Martha Breitenstein-Hooglandt. This act, he said, forced him to give up his practice and live in sin until his wife finally divorced him in 1913, and he could marry Breitenstein-Hooglandt. The couple settled near Locarno, Switzerland, and it was there that van de Velde began writing his marriage manual based on his own experience as well as his friends' and patients' experiences.

Though there is considerable anatomical and physiological data in the first part of the book—some of it inaccurate even for the time—the key to the book is its discussion of sexual intercourse. For him, however, the only normal intercourse was between two people of the opposite sex. Having said this, he went on to emphasize the importance of foreplay, including kissing and touching the erotic parts of the body: the nipples, clitoris, and penis. Though he said an orgasm by cunnilingus or fellatio was pathological, such activity was perfectly permissible in foreplay. He saw no reason why, if they wanted to, a couple should not copulate during menstruation and that intercourse during pregnancy is permissible if the woman was willing and desired it, although the positions might have to change. Only "the menace of miscarriage, and the eminence of birth" would preclude it.[74]

Van de Velde reported that though some of the Oriental encyclopedias of love described a hundred different positions, there was no need to know them all. Still, knowledge of a variety of positions was important and had practical value, because it increased sexual pleasure, sometimes prevented hygienic dangers or injuries, and could promote or prevent conception. He erroneously believed that the most intense orgasm possible occurred when both partners reached climax simultaneously or almost simultaneously. Similarly erroneous was his belief that such an orgasm increased the probability of conception, although it was just such beliefs that allowed him to argue the necessity of experimenting with any positions that might promote simultaneous climaxes.[75] He listed ten different positions, six in which male and female are face to face and four with the male behind the female. He included a summary chart that indicated when such positions should not be used (for example, in pregnancy), the type of stimulation afforded (such as clitoral stimulation), and whether it was more likely or less likely to lead to conception (here he was in error).[76] He also felt that by exploring and engaging in these various coital positions, the desire to achieve variety by a change in coital partners would be curbed and marital fidelity fostered. Van de Velde, in fact, was often in error not only in terms of today's knowledge but in terms of the knowledge of his own time. He taught there was danger in the male thrusting too much, since it might rupture the vagina. He also believed that the human penis could become locked in the human vagina, a somewhat common occurrence for dogs but an extremely rare occurrence for humans. His belief in the importance of simultaneous orgasm was probably the belief that caused the most difficulty to his readers, at least according to modern sex therapists.

EMANUEL HALDEMAN-JULIUS

Collectively, the writers of marriage manuals were changing attitudes toward sex, particularly those held by and about women, and in the process eliminating some of the stigma associated with sex research. The new concepts about sex were not confined to members of the middle class who could afford a hardcover book by Stopes or van De Velde; they also appeared in pamphlets and books aimed at a larger public. Probably the major American disseminator of information aimed at the working classes was Emanuel Haldeman-Julius, a publisher from Girard, Kansas, who issued more than three hundred million copies of volumes in the series Little Blue Books, initially selling at five and ten cents apiece, and later for twenty-five and fifty cents. Sex books furnished a significant portion, perhaps as much as 20 percent of Haldeman-Julius's business. During 1927, for example, some sixty-six thousand copies of Sanger's *What Every Girl Should Know* were sold. Similar titles were aimed at boys, newly married couples, married women, married men, expectant mothers, and so forth. Most of the books for males were written by William J. Fielding, while Sanger wrote many of those for women. Some rather esoteric books by A. Niemoller were also included in the series, including translations of some of the Islamic and Hindu classics. Haldeman-Julius was eclectic in his sex titles and published books about homosexuality, transvestism, incest, rape, and prostitution. There were six titles by Ellis available in 1928 as well as two books on contraception, others on venereal disease, and some on sexual hygiene. In most cases, Haldeman-Julius published the best available information in English on many of these topics, and in his miscellaneous topics book he included discussions of such topics as homosexuality and transvestism, which made his publishing company often the only popular source available.

The Little Blue Books had started out as offshoots to the socialist newspaper *Appeal to Reason* but soon dominated the business. In addition to the sex pamphlets, Haldeman-Julius published works on socialism, free thought, and vast quantities of the literary classics, including all of Shakespeare's plays in individual booklets. His business began to decline in the early 1940s with the advent of large-scale paperback publishing in United States, but during the height of his activities (from the late 1920s through the 1930s), he was perhaps the leading disseminator of sex information in the United States.[77]

CENSORSHIP AND SEX

The battle to bring sex information to the American public had not easily been won. In fact those disseminating sex information were able to do so only with the aid of a series of decisions by the U.S. courts.[78] One of the key cases dealt with whether the public had the right to read about lesbianism. Edouard Bouret's play *The Captive*, which dealt with lesbian love, opened in New York in 1927 and was closed down by city officials as obscene. This action was quickly followed by the passing of a law in the New York State Legislature prohibiting the performance of any drama that dealt with sexual perversion. The immediate question was whether such a law also applied to a novel, particularly to *The Well of Loneliness* by Marguerite Radclyffe Hall. This early classic of lesbian love had been published in England in 1928 where it had been denounced as obscene by a reviewer in the *Sunday Express*, in part for what was regarded as special pleading for lesbians as victims of nature as well as an appeal to God for justice and recognition. The home secretary responded to the review by requesting the publisher to withdraw the book from circulation. An English edition was quickly published in Paris and sent to England, but customs seized a shipment of this printing and, after getting a court order, destroyed it.[79] Fearful that such events would also occur in the United States, the original American publisher decided to drop the book, and its publication was quickly taken over by Donald Friede of Covici. Friede was determined to fight the case through courts if necessary, even if it meant his conviction for publishing pornography. Inevitably, the book was seized, and in the magistrate's court, the judge ruled that the novel "tends to debauch public morals, that its subject-matter is offensive to public decency, and that it is calculated to deprave and corrupt minds open to its immoral influences and who might come in contact with it." On appeal, however, in the appellate court on April 19, 1929, the court ruled the book was not obscene, and it was given American publication.[80] The importance of the decision is that because the book was allowed open circulation, no theme in itself could be declared obscene. Censors had to find other reasons for suppressing the written word rather than the topic with which it dealt.

This decision by no means ended attempts at censorship of sex-oriented books. In 1930, Mary Ware Dennett, an activist in the legal battles for contraception, had been convicted of mailing obscene material, namely a pamphlet she had originally written for her two sons (aged eleven and fourteen) to read; it was titled *Sex Side of Life*. She had allowed some of her friends to read the pamphlet; one of them brought it to the attention of the owner of the *Medical Review of Reviews*, who asked if he could read it. He found the pamphlet informative and proceeded to publish it in his journal without any interference from legal authorities. About a year later, Dennett decided to publish the article in pamphlet form herself and sell it through the mail for twenty-five cents a copy to anyone who wrote and asked for it. It was this for which she was prosecuted.

At the beginning of the pamphlet, Dennett had what she called an introduction to elders (parents and others), in which she says that she had been unable to find a pamphlet on sex matters that she would willingly put into hands of her sons without warning them about the misinformation that was contained in it. She then said her pamphlet contained material "far more specific than most sex information written for young people. I believe we owe it to children to be specific if we talk about it at all." In the pamphlet, Dennett emphasized that sex involved both physiological and emotional responses. She described the sex organs and their operation, the way children are begotten and born, and masturbation. In her mind, the greatest danger of masturbation was the guilt feeling built up about it, since the act itself was not particularly harmful. She also said that venereal disease was curable and noted that the sex impulse was not a base passion but a normal one and that its satisfaction could bring about great and justifiable joy, especially when accompanied by love between two human beings. Dennett warned against perversion, venereal disease, and prostitution and argued for continence and healthy mindedness until marriage and spoke against promiscuous sex relations.[81]

It was such frankness that led to her prosecution. In what turned out to be an extremely important decision, Dennett was acquitted of disseminating obscene material by the U.S. Court of Appeals for the Second Circuit on March 3, 1930. The opinion written by Augustus N. Hand proved to be extremely influential, because it made sex education materials legal, even if those aimed at children dealt with sex in a positive way.

Stopes also added to American case law. She early had been involved in a notorious trial in England over her establishment of a birth control clinic in London in 1921.[82] In this case, she herself had brought suit when an English writer had denounced her as teaching a "harmful method of birth control" and being one of the leaders in a "monstrous campaign" for birth control. Though her attorneys advised her against suing, as nothing came of the review except disparagement of her views, sue she did. She lost, but her loss was more to her pride than to the birth control movement. The American cases came about in 1931 when her book *Married Love* had been seized by customs officials in two different ports of entry, New York City and Philadelphia. In both cases, however, the courts found the book not to be obscene.[83]

It is worthy of note that all of these cases involved women, and two of them were active in the movement for contraception and better sex education. Though these decisions dealt with scholarly or scientific discussions of sex, they set the background for the most important censorship case in American history up to that time, namely that of James Joyce's *Ulysses*, which had been published in English in Paris to avoid British censorship problems. Bennett Cerf, president of Random House, wanted to print the book and devised a stratagem to have the courts rule on the book before he published it. After consultation with his attorney, Morris Ernst, he imported a copy of *Ulysses*, which was then duly seized by the customs inspectors. Cerf brought action in federal court for its release and asked that the action be heard without a jury and before a single judge. The judge turned out to be John Munro Woolsey, the same judge that had been involved in one of the cases involving Stopes's *Married Love*. In December 1933, the judge ruled that the book was not obscene, and Cerf immediately began publication of the book, which appeared early in the next year. One of the key elements in Woolsey's decision was that a book must be judged not by its effect on the abnormal or the young but rather on the average man (or woman). The government appealed the case to the next highest court, the circuit court of appeals, where the decision supported Woolsey two to one. The government decided not to appeal to the Supreme Court. The decision in the circuit court, written by Learned Hand and Augustus Hand, who were brothers, emphasized that a book had to be taken as a whole and could not be banned because it contained some obscene passages. The Hands added that it was settled, as far as the circuit court was concerned, that works of physiology, science, and sex instruction could not be judged obscene, even though some persons argue they promote lustful thoughts.[84]

CONTRACEPTION

Gradually, a growing number of scientists became more willing to examine aspects of human sexuality, instead of just that of animals. Because so much of the popular writing about sex had been aimed at women, and much of it written by women, contraceptive research was one of the beneficiaries. Originally, Sanger fit her patients with cervical caps, simply because she had an available supply in her clinics (started in 1916). Later, she was able to get supplies of the Mensinga diaphragm and quickly turned to them, because they were easier to fit and insert. Still, the contraceptive industry was a clandestine one, with little regulation or quality control, since contraceptives were outside the purview of the various government agencies.

The Birth Control Clinical Research Bureau of the American Birth Control League (after 1942 collectively known as the Planned Parenthood Federation of America) filled in the gap and began to conduct clinical experiments in contraception. They gathered individual case histories of use and attempted to determine the success rate of the users. The first report was published in 1924, but it was essentially a preliminary one. A second report was issued in 1928 based on 1,655 cases; it showed that the use of a modified Mensinga diaphragm—known then as the Ramses pessary (it had a coiled spring)—in combination with a spermicidal jelly was most effective, with a success rate of 96 percent.[85]

Although such information was available to medical specialists and was even available in some libraries, most of the general public remained ignorant. Still, a central goal of the birth control movement had been getting medical acceptance of contraception, and the publication of such data was important. One of the first widely available sources of public information about effectiveness was the 1937 report by Consumers Union, the publisher of *Consumer Reports*. It was a sign of the times, however, that rather than publish the information openly in the magazine, it notified subscribers that a report was available to those who signed an application saying they were married and had been advised by a physician to use contraceptives. Even this method of dissemination ran into difficulty, and in 1941, the Post Office Department barred the report from the mails. Though the action of the post office was upheld by a U.S. district court, the decision was finally reversed by the court of appeals in September 1944.[86] Planned Parenthood officials remained leery of legal action, however. Although an increasing number of physicians were providing contraceptive information, the Planned Parenthood Federation was reluctant to support any widespread public education in contraception outside of the confines of its clinics, because of local and state government action against them.

In 1934, Gamble established a research program under the auspices of the National Committee on Maternal Health that was directed toward the "discovery of better, cheaper and more generally available contraceptives for the underprivileged masses."[87] The committee itself, as mentioned earlier, had been set up Dickinson in 1924 to sponsor medical investigation of contraception, sterility, abortion, and related issues. It had been funded mainly by a group of society women led by Gertrude Minturn Pinchot,[88] although part of its research program also received funding from the Rockefeller-sponsored Bureau of Social Hygiene. Gamble's support appeared at a critical time, because the Bureau of Social Hygiene was being closed down, and its funding was being withdrawn; furthermore, the CRPS was not interested in such research.

Later, another Rockefeller-funded agency also put a strong Rockefeller imprint in this area. This time the money came from the third generation of Rockefellers, when in November 1952, John D. Rockefeller III, following the example of his father in supporting sex research, started the Population Council with his own funds. Through its Demographic and Bio-Medical Division, the council continued the research that had been done on a small scale by the National Committee on Maternal Health and in 1958 took over the committee itself.[89] Included in the council's research program were demographic studies. The early ones showed a decline in the number of children per family through the 1930s among cohorts of women (aged fourteen to forty-nine) grouped into twelve-year intervals from 1897 onward. The 1940s, however, saw an increase in family size and the subsequent rise in birth rate that produced the so-called baby boomers.[90] The effect of such studies was to give accurate indicators of the use of contraceptives, the existence of abortions, and other factors so important in long-range planning for a variety of issues.

SUMMARY

The 1920s and 1930s saw significant breakthroughs in the understanding of the physiology of sex. This understanding was primarily based on animal models and resulted from the efforts of the CRPS funded by the Rockefellers. It also saw a growing public willingness to learn more about sex, particularly by women, many of whom took on the role of sex educator in the family for both their children and their husbands. This growing public receptivity led to more scholarly and scientific research, as academic researchers, mainly in the sciences but also in anthropology, gingerly began to explore human sexuality within their own disciplines. It is this subject that is explored in the next chapter.

6

FROM FREUD TO BIOLOGY TO KINSEY

A major factor in bringing about a change in attitudes toward sexuality in the United States was the growing importance of Freud and Freudian ideas. Though Ellis was widely read, as were the writers of marriage manuals, it was Freudian ideas that captured the interest of the intellectual community. Freud's concept of sexuality was widely adopted by anthropologists, psychologists, literary critics, and others who carried his message to a larger audience. His ideas were also important in treatment. While there were some psychiatrists who did not adopt the Freudian viewpoint and others who belonged to variant schools of psychoanalytic thought, it was Freudian-based concepts that dominated American writing on variant sexuality and exercised great influence over American ideas about sex, in general, and about the diagnosis and treatment of patients' sexual problems, in particular. Psychiatrists saw both private patients and those sent to them by the courts. There was, however, a difference between Freud himself and most of his American followers, because where Freud had emphasized nature and biology as much as nurture, the American psychoanalysts put much more emphasis on nurture.

The historian John Burnham argued that in the 1920s and 1930s

enthusiasts of dynamic psychiatry and psychoanalysis, like the Menninger brothers, were asserting that they and their colleagues could contribute to a better world because they knew the causes of human unhappiness. Later a conservative cult of contentment ... may have been important, but for decades coinciding with the high tide of psychoanalytic influence in America, the analysts, the deviant analysts, and many of their psychiatric and lay followers persisted in describing the benefits of psychoanalytic treatment and applied psychoanalytic psychology in terms of glowing promises.[1]

A major result of this rising influence of Freudian ideas to the development of sex research was to make Americans more conscious and less reticent about sexuality, since simply to read Freud was to become aware of sexuality. For a time, in fact, some American journalists argued that it was impossible to describe Freud's theories in magazines that circulated in the home, because the theories included so much about sex.[2] Gradually, however, discussions of sex, which previously had been limited to the topics of prostitution and divorce in most popular magazines, began to be more open. As early as 1915, Floyd Dell, an author and later publisher, wrote that psychoanalysis gave an air of "ponderous German scientific propriety" to subjects that a few years before would have been deemed "horrific."[3] Psychoanalytic talk, he said, spread from Greenwich Village to the suburbs, although there still remained a tendency to avoid mentioning directly pertinent body parts and circumlocutions abounded. Though articles in the popular media about psychoanalysis remained unfavorable through the 1920s, knowledge of Freudian ideas was increasingly widespread. By World War II, the impact of Freudian ideas was widely accepted in popular culture. His ideas had become fundamental in psychiatry and psychology and had spread over into sociology, anthropology, history, and various forms of literary and other criticism.

FREUD AND SCIENCE

The critics of Freud, and there were many, served only to disseminate further the knowledge of Freudian ideas and concepts. As president of the American Association for the Advancement of Science in 1925, J. McKeen Cattell stated that "psychoanalysis is not so much a question of science as a matter of taste." He went on to add that Freud was an artist who lived in a fairyland of dreams among the ogres of perverted sex.[4] One of the reasons for criticism is that Freudian psychoanalysis became a self-justifying system, not subject to scientific verification by the uninitiated.[5] Freud also either suppressed or ignored information about his patients that did not fit into his theories.[6] Even some of his closest sympathizers, such as Wilhelm Fliess, went so far as to accuse Freud of reading his own thoughts into the minds of his patients.[7] This absence of data, lack of intuitive insight, and even deliberate suppression led Sulloway to question why Freud published even the case histories he did if he did not cure most of his patients or, for that matter, apparently did not believe them generally curable. He argues that Freud's purpose was to establish a way of thinking, that he was involved in what might be called the social construction of psychoanalysis. If this was the case, the American version of Freudianism was a great success.

Freud's influence in the United States went beyond his own ideas and concepts and included the views of those who started as his disciples and then broke with him about various parts of his system. In fact, his very system encouraged the development of rival systems, and in the United States, this aroused even greater interest. Freud's ideas, as indicated earlier, originally had been disseminated through a small coterie of believers. Unable to sell his concepts to the medical establishment in Austria, he created his own education institutes, more or less abandoning the established medical apparatus designed to train neurologists and psychiatrists. Later, he deliberately discouraged his followers from trying to establish medical school ties or to set up training institutes within medical schools. The ultimate result was that Freud, sometime after 1900, either consciously or unconsciously abandoned the methods of science, namely the need to replicate theories and practice not only within one's own immediate social group but outside as well.

Training scientists, however, was not his aim, because Freud wanted practitioners who operated within a relatively fixed system of ideas. As Sulloway put it, "Instead of trusting that his methods would withstand critical scrutiny and flourish independent of opposition, Freud privatized the mechanism of their dissemination and trained a movement of loyal adherents. His most talented followers naturally tended to rebel under this totalitarian regime."[8]

The history of Freudian thought inevitably becomes the histories of heresies and schisms, but the net effect was to disseminate more widely a realization of the importance of sexual issues. Some of those who broke with Freud did so because of disagreement with his sexual theories, and most of them picked up American followers of their own.

ALFRED ADLER

Adler (1870–1937) broke with Freud in 1911 and went on to develop individual psychology. Adler tended to downplay sexuality, using it more symbolically, as he did in his concept of sibling rivalry, than biologically.[9] Closely related to sexuality was his concept of gender theory, by which Adler held that individuals generally compensate for underdeveloped or abnormal organs and occasionally achieve a higher than average achievement by so doing. Though Adler originally did not apply his theory to the sexual organs specifically, others did. They implied that organ inferiority might lead either to neurosis or to perversion, an idea that Adler himself came to accept.[10]

From these seeds, Adler went on to develop the idea of a drive toward masculine protest or will to power. Since society calls women the weaker sex, a masculine protest is a common compensation, and women may strive directly for power or use their sex as a device for gaining power over men. Men with feminine tendencies may become exaggerated he-men or, if they are homosexuals, try to dominate through weakness.[11]

WILHELM STEKEL

In 1911, Stekel (1868–1940) resigned as an officer of the Vienna Psycho-analytic Society and eighteen months later Stekel left the society over an editorial dispute with Freud. He continued to accept many of the concepts of psychoanalysis, but gave them his own interpretation. A gifted writer and an energetic reformer (in 1906, he established a branch of Hirschfelds's Sci-entific-Humanitarian Committee to agitate for the removal of legal obsta-cles to homosexuals), Stekel today is most valuable for his reports on his case studies. These empathetically describe the problems of his patients; in fact, his descriptions were so interesting that he was accused of making them up. His descriptions of fetishism, for example, make one almost feel what it must be like to be a fetishist. He continued to emphasize the importance of sexuality but he did not accept the Freudian interpretation of neuroses. Instead, Stekel claimed that all neuroses and sexual disorders rose from men-tal conflict, not from blocked or redirected instinct and were, therefore, potentially curable. He believed that deviants isolated their problems in their minds and chose a tortuous path through life to avoid challenges that they could not handle. Thus, once the source of the conflict is uncovered, a cure can take place, and in a dig at Freud, he claimed he could cure individu-als that the Freudians found incurable.[12]

CARL JUNG

Carl Gustav Jung (1876–1961) broke with Freud in 1912, in large part because of Freud's emphasis on sexuality. He disagreed with Freud about infantile sexuality, the concept of a latency period, and the emphasis on childhood experiences as the essential causes of neurosis, all of which he criticized on biological grounds. Sex, for Jung, was important only in so far as it was made so through the varieties of individual experiences. Jung agreed with Freud and Adler that every person is essentially bisexual. Jung, however, developed the concept in a different way, one that fits more into the gender theory of today. The male, he wrote, frequently represses the female side of his nature, the *anima*, but it survives in the unconscious, as does the repressed male side of the female, the *animus*. The *anima* in men gives a receptive, nurturing alternative to soften the masculine logical, dominating, and forming features. The *animus* in women imparts logical, dominating, and forming features to the feminine nature. Pathologies develop when these opposites are lost in a shriveled, ineffective, unconscious representative. It is also possible, however, that the wrong nature dominates. In some homosexuals, according to Jung, the female *anima* dominates, as it does in excessively weak or tender men. Similarly, the woman who seeks to dominate men without love may be controlled by her *animus*. The healthy person is one in which the correct nature is present, but there is also a strong dose of the opposite. Jung also claimed that the Western male was afraid of the feminine within himself and the more he tried to avoid confronting this, the greater the toll for avoiding it.[13]

WILHELM REICH

Wilhelm Reich (1897–1957) was much more radical on sexual issues than the others discussed above. He argued that Freud did not carry his theories about sex to their logical conclusion, particularly his megatheory about the need of civilized peoples to redirect their sexual energy.[14] After World War I, Reich set out to incorporate Marxism and Freudian psychoanalysis into a new synthesis based on his sexual concepts. He argued that there was a crucial interdependence of social and sexual liberation and that any political revolution was doomed to failure unless it was accompanied by abolition of repressive morality. It was this failure, he argued, that had undermined the Russian revolution.

Reich also felt that the sexual revolution needed to encompass not only adults but children and adolescents. In his mind, the sexual repression of the adolescent in society led to juvenile delinquency, neuroses, perversion, and of course, political apathy. He devoted numerous pages in his writings to the problem of providing adolescents with the private quarters and contraceptive devices necessary for the fulfillment of their sexual needs. If society followed his prescription for a revolution in the attitude toward sex, homosexuality, he said, would disappear in the wake of the revolution, as would all other forms of sexual perversity.[15]

Ultimately, Reich proved a failure in his endeavors to reconcile Marxism and psychoanalysis and was expelled from both the International Psychoanalytic Association and the Communist Party. In 1936, he was in Norway, where he founded the International Institute for Sex-Economy to study the way the human body used what he called sexual energy. Motivating this search was his attempt to find the basic physical unit of energy to replace Freud's generalized concept of libido.

In place of the libido, he developed a concept of energy, an actual physical component of humans that could be measured and harnessed, to which he spent the rest of his life trying to explain, control, and use. In 1939, Reich went to New York, where he established the Orgone Institute in Forest Hills. There, he attempted to teach others how to use the new kind of energy, which could be tapped by body massage, stored in accumulators, and used to strengthen the body against disease and particularly to increase orgastic potency. The result was the orgone box—a six-sided box made of wood on the outside and metal on the inside, a little larger than a coffin—that supposedly collected the orgone energy, transferring it to the patient sitting inside, who could then direct the energy to the genitalia, thereby restoring sexual potency. If the patient was ill, he or she could be restored to a healthy condition.

Reich held that orgastic potency—the capacity of an individual to achieve orgasm after appropriate sexual stimulation—was the key to psychological health. Orgasm, he claimed, regulates the emotional energy of the body and relieves sexual tensions that otherwise would be transformed into neurosis. His theories led not only to his denunciation by the American Medical Association but to an investigation by the Food and Drug Administration, which enjoined him from distributing orgone accumulators. Reich defied the ban; in 1957, he was sentenced to prison, where he died of heart disease.[16]

HERBERT MARCUSE

Starting out with the same intention of reconciling psychoanalysis and Marxism was Herbert Marcuse (1898–1979). He considered sexual repression one of the most important attributes of the exploitive social order. Under capitalism, Marcuse held that sexual love had been stripped of its playfulness and spontaneity and become a matter of duty and habit, carefully circumscribed by the ideology of monogamic fidelity. In fact, the blunting of sensuality was the inevitable by product of industrial labor and had resulted in the atrophy and coarsening of the body's organs. Sex repression itself was a major tool used to maintain the general order of repression.

Sexual repression also correlated with the performance principle, a key to capitalism, and this had resulted in the desexualization of the pregenital erotogenic zones and reinforced the genitalizations of sexuality. In short, the libido became concentrated in one part of the body, the genitals, thus leaving the rest of the body free for use as an instrument of labor. Resexualization of the body was the goal of human fulfillment.[17]

*   *   *

There were many other individuals who developed theories and schools of
thought that emerged from Freudian ideas. The purpose of examining these
five men is not to explain the development of psychoanalysis (impossible to
do in such a short section) but to emphasize that Americans were becoming
increasingly aware of the importance of sexuality, in large part through the
efforts of Freudian disciples and apostates. Academics in the social sciences
and humanities, especially, were intrigued by Freudian and other psychoanalytic theories and used them heavily in their own writings. Many tried to use
various aspects of the theories to explain society, and from the academic
community, the ideas passed into everyday use. Sex, in effect, could no
longer be ignored. Moreover, the very willingness of psychoanalysts to speculate about sex made them the authority figures for sexual information for
the middle years of the twentieth century, at least in terms of variant sexual
activities. Even when their ideas were challenged—as they were by numerous sex researchers during the 1920s, 1930s, and 1940s—the psychoanalytic
view dominated American thinking on sex until the 1960s, again because the
psychiatrists, particularly the psychoanalysts, claimed a cure rate that no
others could match. They helped maintain a medical dominance over sex
that their research could not support.

Though the psychoanalysts dominated treatment, and case studies of
individual patients or groups of patients became publishing staples, a new
kind of research was emerging on the college campuses. Changing ideas
about sex gradually permeated into the college curriculum, particularly in
courses in child development, psychology, and cultural anthropology. One
result was the growth of what today would be called gender studies, although
the term had not yet been invented. Such courses emphasized the social and
cultural differences between males and females. Many of the women's colleges as well as so-called women's departments, such as home economics,
developed special woman-focused courses so women could better understand themselves. The American studies of sex by Davis, Dickinson, Hamilton, and others had fit into this growing tradition. Such studies continued
during the 1930s, although none of the published studies at that time
matched the earlier ones.

One major benefit of the developing college courses in marriage and family was that they gave the emerging social sciences potential research subjects. Increasingly, studies based on college student responses to questionnaires in psychology, home economics, and sociology classes began to appear.
Some studies sponsored by the Committee for Research in Problems of Sex
(CRPS) went beyond the classroom to study special groups as well. Still, it
was animal studies that dominated American sex research, and in a sense,

FROM FREUD TO BIOLOGY TO KINSEY

academic American sex research, once established in biology departments, gradually moved over into other areas. Initially, at least, animal studies not only were politically safe but, in many cases, had practical application for farmers and others.

ANIMAL STUDIES

T. H. Bissonette was one of the researchers supported by the CRPS; he concentrated on the problem of light in relation to sexual periodicity. Though his studies were mainly done on birds, he also investigated ferrets, rabbits, and raccoons and was able to change the breeding season of these animals by reducing illumination in the summer months and increasing illumination during the winter.[18] Such findings led to increased egg and milk production but were not applied to humans, although it can be statistically demonstrated that seasons do have some influence on human reproduction (and sex life). If there were no seasonal effect on human sexuality, then an equal number of births would, on average, occur on each day of the year. This, however, is not the case, and statistics demonstrate that fewer individuals are born at certain times of the year than at others.

Studies on nutrition and sex in terms of reproduction sponsored by the CRPS demonstrated that in rats a diet of pure fats, carbohydrates, and proteins, supplemented by the then-known vitamin groups (A, B, C, and D), was enough for the rats to live well, but the females could not breed. The addition of wheat germ oil cured this difficulty—a fact that led to the discovery of vitamin E (α-tocopherol).[19] Similarly, nutrition and diet affect fertility ratios in humans, but it was not until later that such studies were made.

Often the results were more immediately important. C. R. Moore found that mammalian testes could not form sperm unless they were subjected to temperatures lower than the interior of the body, hence the need for a scrotal sack. Furthermore, if a human male wanted to get a female pregnant, he should not keep his genitalia too warm by taking too many hot baths or wearing fur-lined jock straps.[20]

It was not only on rats, sheep, and goats that experiments were carried out but also on nonhuman primates. It was in this area that Yerkes, the chair of the CRPS, was most active. Yerkes was a comparative psychologist and devoted himself from 1924 to work with nonhuman primates. By 1929, with aid from the CRPS, he had established an experimental station near Orange Park, Florida, later named the Yerkes Laboratory of Primate Biology. One of the questions Yerkes examined was whether primates had a cyclic variation in sexual response similar to that of the human female. Yerkes and his collaborators agreed there was, but they also found that several nonsexual factors were involved, namely the individual personality traits of the consorts, their past experience, and various social and environmental factors, all of which influenced the basic behavior cycle by overriding hormonally conditioned responses. In the female, preference for one male or another could completely overshadow sexual receptivity and lead to the occurrence of copulation throughout the cycle.

Yerkes found that when a male and female chimpanzee were intimately acquainted and congenial, the male, ordinarily dominant in all their competitive activities, yielded precedence to the female during her estrous period. Most important, he observed that variation in mating behavior among females was very great and that the behavior of a particular female differed in reaction to different males. Yerkes emphasized that touching and manipulation of genitals was a common pattern among nonhuman primates and a wide range of other animals and, contrary to traditional thinking, was not against nature. His study of the great apes, which he published in 1929 with Ada Watterson Yerkes, was the standard work on the biology and psychology of these animals for several decades.[21]

A major result of these and other animal studies was to emphasize just how many of the body systems (muscular-skeletal, circulatory, neurologic, and endocrine) were involved in sexual activity. Though much of the copulatory pattern is mediated by nervous circuits that lay within the spinal cord, these are regulated by the hypothalamus and the cerebral cortex. It was also found that various sensory-motor adjustments were involved in coitus and the stimulation of various cutaneous receptors located in the genitals and elsewhere contributed to the intensification of sexual excitement. In the primates especially, learned patterns of social interaction are also influential in forming relationships. In sum, if it was not known before, the animal studies of the 1920s and 1930s proved that sexual activity involved a whole series of neuropsychological and sociocultural factors.[22]

ANTHROPOLOGICAL STUDIES

Giving emphasis to sociocultural factors hinted at in the animal studies were a growing number of field studies in anthropology, many of them influenced by Freudian concepts. Bronislaw Malinowski (1884–1942), the father of anthropological fieldwork, set the tone for the future study of such factors through his emphasis on the necessity of the meticulous investigation of a culture and its social organizations. He believed that it was important to determine how each specific aspect of cultural behavior contributed to the functioning of the group.

In Malinowski's mind, sex was more than a simple physical connection between bodies. He argued that understanding the sexual practices in their sociological and cultural contexts was essential to understanding any cultural grouping. Malinowski's insistence on studying sex linked the professional anthropologist with the amateurs, such as Richard Burton, of an earlier generation and the new Freudian concepts. Sex behavior was defined broadly by Malinowski, and The Sexual Life of Savages (1929), one of his major studies, dealt more with courtship and marriage and family than with descriptions of sexual behavior.[23] Still, Ellis, who wrote an introduction to the book, emphasized its importance as a comprehensive picture of a society that integrated sexuality within its culture and noted that Malinowski did not stress the aberrations as many earlier studies had done. Ellis pointed out that although the Trobriand Islanders made up only a small community and lived in a confined space it was evident that these comparatively primitive people were very much like their modern European counterparts and had the same vices and virtues. Malinowski's critics, upset at his emphasis on sex as a key to life in primitive societies, accused him of generalizing from too narrow a base of experience as well as failing to catalog and describe his native informants.[24] Though these criticisms are not without merit, his importance lay in institutionalizing sex and gender issues into anthropological field studies.

A number of other anthropologists sought to answer specific questions that had sexual overtones. Margaret Mead, a student of Franz Boas (1858–1942), focused on the nature of adolescence and sexuality in Coming of Age in Samoa.[25] Boas and Mead, at this stage in her life, were both cultural relativists. Human nature consisted entirely of such physical needs as food, water, and sex, and how different cultures coped with these needs was seen as enormously varied. On the basis of data furnished by her informants, Mead concluded that adolescents in Samoa had complete sexual freedom and that it was the ambition of every adolescent girl to "live with as many lovers as possible" before she married. The book was extremely influential in its time and was praised by Bertrand Russell, Ellis, Mencken, and others, although not all observers of Samoa agreed with her. In fact, many anthropologists as well as the Samoans themselves now argue that traditional Samoa was just the opposite of what she said, that it was a culture of strict parental controls and unbending sex taboos, a society in which female virginity was so highly prized that girls were tested for virginity before they were allowed to marry.[26] Mead later went on compare sex roles among three New Guinea groups (Arapesh, Mundugumor, and Tchambuli), emphasizing that sex roles were not inborn but a product of learning, a view that even her critics never challenged.[27]

Whether Mead was misled by her informants, and she probably was, is important historically, but such charges came much later, too late to prevent her work from becoming part of the canon challenging traditional sex and gender roles. She was regarded as giving further proof that ethnography could furnish a catalog of information on human sexuality, illustrating the malleability and variety of human behavior.[28] One source for testing this malleability, which went beyond the individual reporter such as Mead, was a collection of reports that had been gathered together at Yale University during the 1930s on more than 190 cultures; it has continued since then in what was originally known as the Yale Cross-Cultural Survey and now called the Human Relations Area Files, Inc. This compilation of information about different peoples is classified by both subject matter and geographic area and is based on thousands of books, articles, and papers by a variety of reporters and observers ranging from casual travelers to missionaries to professional ethnographers.

One of the early attempts to summarize these data in terms of sexual behavior was by Clellan S. Ford and Frank A. Beach, whose research was sponsored by CRPS. They concluded that there was such a wide variation among peoples and cultures that no one society could be regarded as representative of the human species.[29] Though they found many cross-cultural similarities, they also found a number of differences. Some societies condoned and encouraged the sexual impulses of children, but others prohibited and punished such behavior. Different societies held widely different rules and attitudes about masturbation. Regardless of whether the attitude was approval or condemnation, they noted that at least some adults in all or nearly all societies appear to have masturbated. Though a number of societies reported the existence of bestiality, most such references were found only in folklore and not in everyday life. Homosexual behavior was not found to be predominant among adults in any of the societies, although some form of same-sex activity was either observed or reported in a significant proportion of societies. The data indicated that though certain social factors probably do incline certain individuals toward homosexuality, the phenomenon could not be understood solely in such terms.[30]

Though the summaries of Ford and Beach are accurate summaries of the data then available, it is important to emphasize that the data they used reflected the biases of the compilers of their source material. This means that such data must be used with caution, because so many of the early reports were made by missionaries or other amateurs who lacked the training of the modern ethnographer. Moreover, many of the early observers tended to approach sexual subjects with preconceived moralistic assumptions.

A good illustration appears in the findings concerning homosexuality. Ford and Beach found references to homosexuality in seventy-six societies. In forty-nine (64 percent), homosexuality was considered normal and socially acceptable, at least for certain members of the community. In the other twenty-seven (36 percent), homosexual activity among adults was reported to be totally absent, rare, or carried on only in secrecy.[31] Because no reference to homosexuality, either positive or negative, was reported for the majority of societies, these groups were excluded from the statistics. This leads to the question of whether the absence of either positive or negative statements about homosexuality means it was unknown in those societies or simply that the informants neglected to mention the subject or the informants felt it was something that the Western observer would not understand or sympathize with. The answer would seem to lean more toward the latter explanation, and if this is the case, the accuracy of definitive conclusions drawn from such observations is open to debate. Some of the investigators reporting the existence of homosexuality might have been looking for proof of its existence, while many of those who did not look for it were unaware of its existence.

Another problem with such studies is a definition of what constitutes homosexual activity. Balinese society, for example, was classified by Ford and Beach to be one of the societies in which homosexual activity was rare, absent, or carried on only in secret. Yet the crossing of sex roles is common among the Balinese, since their religion places a high value on the hermaphroditic figure of Syng Hyan Toenggal, also known as the Solitary or Tijinitja. Tijinitja, according to Balinese cosmology, represents the time before the gods, before the separation of male from female. Thus Tijinitja is thought of as both husband and wife.[32] Obviously, the cross-dressing associated with the god represents transvestite conduct, but is it homosexuality? Ford and Beach did not think so, but then the question has to be asked as to why they classified the cross-dressing among the *berdache* as homosexuality since many of the early studies classified a variety of conduct under this label.[33] Can the classification in the one case be any more justified than in the other? Probably the answer depends on when the data they used were reported and by whom, something they did not take into account.

Part of the confusion stems from the fact that as research into sexuality has developed there has been both a greater awareness of the sexual implications of some customs and a greater attempt at precision in definitions. New categories such as transvestism and transsexualism have been established in recent years that were once classified as homosexuality; the use of the new categories challenges older interpretations. Unfortunately, it is impossible to redo the earlier studies because many of the cultures studied have themselves changed. Since most of our past historical and anthropological data are not usually detailed or precise enough to indicate into which category an individual should fall, whether an individual is classified as homosexual in part depends on the time and place when the data were gathered and the sophistication of the observer.

This difficulty in classification is not a unique problem for anthropologists but pervades the study of human sexuality. Modern physicians, for example, who attempt to read historical descriptions of symptoms have difficulty in determining what disease is being described, because physicians in the past looked at different things than do modern ones and had different diagnostic categories. This meant that they lumped together categories we now separate and gave great emphasis to some factors we no longer think important. The most workable and practical solution for the historical diagnostician often is to avoid analyzing an illness outside the culture in which it existed; even when the data are sufficient to permit some preliminary classifying, definite assignment in terms of current understanding can be misleading if not actually erroneous.

The same caution must be applied to labeling same-sex conduct in other societies as homosexual in the modern sense of the term, since it had a different meaning in that society from the one it has in ours. Still, after all of this has been said, the anthropological data that were being gathered in the period between the World Wars (1918–41) emphasized the variety of possibilities existing in human sexual behavior.[34]

HISTORICAL DATA

Similarly, various historical works appeared during this same period, emphasizing that even within the Western tradition, there had been a variety of different sexual activities that had not only been tolerated but institutionalized in societies. Among the Greeks, for example, homosexual contact between adolescents and young adults were considered normative.[35] Similar studies appeared about sexual life in ancient Rome and in ancient India[36] as well as specialized studies dealing with references to masturbation, sexual intercourse, bestiality, and so forth in classical culture.[37] Undercover editions of pornographic classical works were also translated and distributed,[38] and the major value of such works is that they throw light on different attitudes from what the official sources would allow, not only in ancient times but also in modern times.

A good example is the anonymous work *My Secret Life*, a sexual autobiography of a Victorian gentlemen believed to be based on a real life. In such works, we find that there was widespread interaction between prepubescent young girls and older males. The author records some of the difficulties of having sex with prepubescent girls, and we know from other sources that it was not uncommon in the nineteenth century to chloroform the girls during penetration. The author said nothing about this but stated rather as a matter of fact:

> Verily a gentleman had better fuck them for money, than a butcher boy (fuck them) for nothing. It is the fate of such girls to be fucked young, neither laws social or legal can prevent it. Given opportunities—who has them like the children of the poor—and they will copulate. It is the law of nature which nothing can thwart. A man need have no "compunctions of conscience"—as it is termed—about having such girls first, for assuredly he will had done no harm, and has only been an agent in the inevitable. The consequences to the female being the same, whosoever she may first have been fucked by.[39]

He went on to state that the youngest girl he had ever had sex with was ten years old, but he complained that she could not give the pleasure "that fully development women could." Nevertheless, he admitted he had several orgasms with her.[40]

The existence of such conduct emphasizes just how much the silence about sex in the late nineteenth and early twentieth centuries allowed such abuses to exist. It causes one to question seriously official histories and to challenge the pious hypocrisies of the past. But the problem with the specialized sexual histories, as with the pornographic classics, is that many of these works reflect the same kind of problems as did the anthropological literature before Malinowski. The specialized investigators into historical sex failed to integrate the sexual customs and data into the culture of the society they were examining. They tried to look at it in isolation, and the result was often a prurient, even salacious study, which though informative, probably gives a rather distorted picture of the place of sexual activities in a society.

This emphasis on the salacious and prurient was the same problem that Americans in general faced between the two World Wars. American culture was only gradually losing what could only be called a prurient view of sexuality. Americans often officially denied the existence of a wide variety of sexual activities and yet sought out, in pornography or in sex-segregated audiences, information about forbidden activities. This was particularly true of all-male groups, which gathered to watch one-reel stag films or strip tease shows, with the men becoming aroused at the forbidden.

STUDIES ON PREMARITAL SEXUAL ACTIVITIES

Still, serious quantitative studies reflected a growing change not only in attitudes but in behaviors. Among the best of the studies of the 1930s were those by Dorothy Dunbar Bromley and F. H. Britten on male and female college students and that of Lewis M. Terman, Carney Landis, and others on psychological factors in marital happiness.

The Bromley and Britten study consisted of a questionnaire given to 1,364 college students, 43 percent of whom were male and 57 percent female. Data collecting could only be called haphazard. The two authors were journalists who visited fifteen college campuses: five men's colleges (four in the East and one in the South), five women's colleges (all in the East), and five co-educational universities (four state institutions—two in the Midwest, one in the South, and one in the West—and one private university in the Midwest). They simply walked on campus, introduced themselves to students, and then scheduled interviews with them. Authorities at one women's college asked Bromley and Britten to leave, but officials at the other institutions either did not know they were there or tolerated their existence. In total they interviewed 154 women and 122 men, all of whom were undergraduates. Then Bromley and Britten recruited undergraduates to give or send their questionnaires to other students, paying them ten cents for each returned questionnaire. Of the 5,000 questionnaires sent to forty-six additional colleges and universities, 1,088 were returned: 618 from women and 470 from men. The largest number from any one campus was 169 from a midwestern state university, and the next largest was 77 from an eastern women's college.

The statistical manipulation of the data was scant, and there is no breakdown of the sample. From what we know of other available information, responses to the questions about masturbation and homosexuality were underreported by the male respondents, although this was not necessarily the case with females. Among the questions asked was whether homosexuality was prevalent in their college and whether they knew much about it. Students were then asked if they themselves had ever engaged in same-sex practices. Many men reported knowing nothing about homosexuality, and only a few admitted having such experiences. Furthermore, those men that did report same-sex behavior insisted that it was simply a casual affair.

Probably more accurate data were given on students' self-reports of heterosexual activity. Some 25 percent of the women indicated they had engaged in sexual activities with men, but almost all said they had been in love with their partners at the time, and many said they were engaged. On the other hand approximately 50 percent of the males indicated that they had sexual experience with women, and there was less of an attempt to justify their activities with statements of love.

On the basis of their data, Bromley and Britten argued that there was a social revolution in manners and morals taking place in America in terms of premarital sexual relations, although their respondents overwhelmingly reported they still wanted to marry and settle down. Perhaps because of the pressure put on them for sexual intercourse by the men in their lives, the college women for the most part wanted to marry young, much younger than the males did.[41] In fact, the unconscious pressure to engage in premarital sexual activity, which went contrary to officially sanctioned morality, might have been the reason so many women dropped out of college to marry.

Terman and colleagues examined attitudes toward sex and marriage based on 2,484 subjects, made up of 1,250 married couples, mostly living in California and mostly college educated. The mean age of their subjects was thirty-nine, and within the limits of the sample, the study illustrated changes that were taking place in the United States. Terman and co-workers found an increase in premarital sex among college-educated women, with between 35 and 65 percent engaging in premarital sexual activity, depending on their age cohort. The younger women had engaged in premarital sex more than the older ones. The authors predicted that, if the trends portrayed in their study continued, by 1960 no American girl would reach the bridal bed as a virgin. One of the more interesting findings was that 13 percent of the married women experienced multiple orgasm. Subjects, however, were not interviewed, and data were based on a questionnaire.[42]

Landis studied married couples, and his work was supported by the CRPS (as was Terman's). He found that beneath an "emancipated facade," and even behind "emancipated behavior," traditional attitudes still remained strongly ingrained. Many husbands reported deliberately varying sexual techniques and extending intercourse to bring their wives to orgasm, but the majority of wives either remained unaware of these efforts or felt they were unsuccessful, because the women said that their husbands cared only about gratifying themselves. Miscommunication about sexual matters seemed to be a common problem among the couples.[43] A number of other studies on marriage were done, and Terman was often involved in some capacity or another.[44]

Terman and C. C. Miles pioneered in developing a masculinity-femininity scale, although by today's standards the measurements were very crude. Instead of beginning with a theory of masculinity or femininity as psychological concepts, Terman and Miles followed the more simple strategy of finding empirical sex differences between the responses of men and women to questionnaire items about their interests and preferences. At its most mindless, such an operational definition of *masculinity* equated it with men's preference for taking showers, for seeking work in business or as contractors, and for reading magazines like *Popular Mechanics*. Femininity was equated with a preference for taking baths, for seeking work as dress designers, and for reading magazines like *Good Housekeeping*.

Such an approach, Terman and most of his contemporaries seemed to believe, enabled them to discriminate between men and women as well as separating off homosexual men and lesbian women. They assumed that males and females were opposites and homosexuals were the inverse of heterosexuals. They also assumed that healthy men would be masculine and healthy women feminine. Any cross-gender activity or habit aroused suspicion.[45] Many such simplistic assumptions carried over into profiles that the military, during World War II, used to reject or discharge individuals whom they identified as gay.*

Indicative of the growing awareness of sexual issues are the studies of specialized samples. Landis and M. M. Boles, for example, examined the personality and sexuality of physically handicapped women, but the study should be noted more for its intent than for its results.[46] Earlier, F. M. Strakosch had done a study on the sex life of seven hundred psychopathic women in the New York State Psychiatric Institute and Hospital in New York City. The data were compiled by a number of psychiatrists, raising the potential of possible variation in the standard of recording. Complicating the usefulness of the study was the fact that the ages, education levels, and social levels of the respondents fell within such a wide range, especially since Strakosch compared his data with that of Davis, Hamilton, Dickinson, and Beam, whose subjects were drawn more or less exclusively from the upper educational and social levels.[47]

Many of these social science–oriented studies were examined by Alfred Kinsey and colleagues, who concluded that it was difficult to generalize from such studies because of their limited or class-biased samples. Many were heavily based on material from the New York City area, while most of the others came from the eastern United States. All of those studies attempting to go beyond this limited geographical area were questionnaire studies, not interview ones, and none of the samples attempted to reach a wide range of populations or age levels. Their results were further compromised by the ambivalence the authors had in asking questions about sex, and many of the studies used circumlocutions or judgmental terms, particularly the earlier ones.[48]

---

* Even I thought that something must be wrong with me because, during an induction examination, I said I liked baths. This was regarded as a feminine rather than a masculine characteristic and resulted in further questions about my masculinity, even though my preference for baths was because there was no shower in the home in which I was raised.

HOMOSEXUALITY AND LESBIANISM

Increasingly, however, American sex researchers became ever more daring, and one of the most ambitious efforts to study sex dealt with the issue of homosexuality. This study was carried out by a self-selected group of individuals who, in 1935, formed the Committee for the Study of Sex Variants.[49] Most of the research and writing for the ultimate publication were done by psychiatrist George Henry. The subjects for the study were recruited by Jan Gay, a woman who had contacts in the homosexual and lesbian community. Subjects willing to participate underwent a psychiatric interview, using what Henry called a modified free-association method. With an occasional suggestion or question subjects were led to talk freely and spontaneously about themselves and their family and a verbatim shorthand record was made of the answers. The note-taking instead of inhibiting the subject was reported to encourage the respondents that everything they had to say was of value. The notes were taken by the psychiatrist who showed no reaction to what was divulged other than interest.

After a detailed history had thus been obtained, the study, as far as the subject was aware, was at an end, although many volunteered to return if further information were desired. Two years later interviews were resumed but this time the questionnaire method was employed. This procedure was adopted for the purpose of checking on statements previously made and to supplement information already obtained.[50]

How many were interviewed is not clear (somewhat more than two hundred), but eighty subjects were ultimately selected to have their case histories and analyses published; half of them were female. To report his data, Henry separated lesbians from gays, but put each into the same three categories, bisexual cases, homosexual cases, and narcissistic cases, although narcissistic cases were simply another form of homosexuality for him.

Henry presumed there was a physical basis or predisposition for homosexuality, and thus in addition to the interview, there was an extensive physical examination of the subjects made by Joseph C. Roper. Furthermore, a special pelvic examination was given to the women by Mary Moench. Both Roper and Moench were physicians at New York Hospital. About 33 percent of the group were photographed in the nude, and all of them had x-ray examinations of the head with special reference to various physical features that were thought to be indicative of homosexuality or lesbianism. Males were also asked to give sperm samples. As a result of these examinations, Henry and co-workers concluded that the body-carrying angle of homosexuals of both sexes was intermediate between male and female. Lesbians often demonstrated abnormal pelvic formation and immature skeletal development.[51]

Henry also looked for hereditary factors and made up genealogical charts for his subjects, noting the occurrence within each family of factors such as homosexuality, bisexuality, suicide, psychosis, alcoholism, and tuberculosis as well as "artistic inclinations." He found that sex education had been an ignored subject for his subjects when they were growing up, and that many of them had been told by their parents that they had wanted a child of the opposite biological sex. In traditional psychiatric fashion, Henry and colleagues looked at the parents and found that mothers were in many cases unhappy, almost martyrs, who made their children conspirators with them against the man of the house. Some fathers fought back with angry displays of infidelity, making the mother's judgment come true, whereas others withdrew and became ciphers. In other households, fathers flaunted their sexuality and suffering to win attention. Usually only one parent won the child's solicitude. He documented a relationship between homosexuality and aggression-passivity behavior and showed that this behavior was often passed through several generations. He thought there might be a genetic basis for this behavior but that a neurotic environment made it particularly important in personality development. Henry also found that on tests of masculinity/femininity, male and female homosexuals were intermediates,[52] which he emphasized was an indicator of the existence of an intermediate sex.

Henry's conclusions were strongly influenced by the attitudes of his time and his psychiatric assumptions, but what made his study valuable then and now is the long, detailed case histories that sometimes indicate patterns to the current generation of readers that Henry did not see or ignored.[53] Generally, his homosexual men and women were a varied group, some openly homosexual, some secretive; many were successful in their careers and social lives, while a few had a history of depression, career failure, and suicidal urges.

Henry had the traditional medical-psychiatric assumptions and emphasized that the sex variant could be studied best and most efficiently by a psychiatrist who had specialized in sexual pathology.[54] Ultimately, his study turned out not to be the dispassionate study he had set out to do, because he became deeply involved with his clients and occasionally even expressed skepticism about standard psychiatric assumptions. During World War II, Henry was consulted by the military about the problem of homosexuals. He had the courage to tell them that far more homosexuals served with the armed services than were eliminated before or after induction. In fact, the army had in a sense encouraged homosexuality by making men aware of their sexual orientation. As a result, many men had their first overt homosexual experience while in the army.[55]

It was the feeling of compassion showing through his psychiatric assumptions that led a group of Quakers to approach him in 1945 to set up the Quaker Emergency Committee in New York City to deal with the problems of the young people arrested on charges of homosexuality. A network of clergy, physicians, and educators was established, but differences soon developed between Henry and the Quakers. The Quakers withdrew and set up their own group, the Quaker Readjustment Center, a title indicative of the conflict with Henry. The new director of the Quaker group was the psychiatrist Frederic Wertham, who was far more conservative than Henry. Under Wertham's direction, the center dealt mostly with sex offenders. The earlier committee was reorganized as the George W. Henry Foundation and concerned itself with giving aid, advice, and encouragement to youths troubled with problems of homosexuality. Many of the directors of the Henry foundation became active in organized homosexual groups in New York, and following Henry's death in 1964, the foundation was reorganized into a social work agency to deal with problems encountered by homosexuals.

Henry also published a more popular and rather different version of his psychiatric experience with nine thousand clients (eight thousand men and one thousand women) under the title of *All the Sexes*. This was a pioneering effort to emphasize that each individual was an incalculable complex of masculinity and femininity. Henry pointed out that every man possessed feminine attributes and every woman possessed masculine ones; human beings represented an imperceptible gradation between the theoretical masculine and the theoretical feminine. Though his psychiatric explanations of imbalances were not the traditional ones of the time, his message was an important one, and in a sense, the book was an important contribution to what later came to be called gender studies.[56] Henry even began to retreat from his earlier emphasis on physiological factors and realized that body form itself was not a reliable indicator of sexual competence.

Though Henry changed his mind about physical indicators, other researchers did not. Such concepts were pushed to the extreme by W. S. Sheldon through what he called "constitutional psychology." Sheldon took photographs and measurements of thousands of men and women, which he then divided by physiques into three basic categories: endomorph (dominated by stomach and massive digestive viscera), ectomorph (dominated by brain and nerve endings), and mesomorph (dominated by musculature). He then went on to claim that body build and emotional disturbances were related. Sheldon implied that male homosexuals were weak, fragile, defenseless, and poorly endowed and had physical characteristics typical of women. The reverse was true of female homosexuals. He also argued for a correlation between high sexual activity and perversion.[57] In light of such attitudes, Henry's studies appear to be a model of objectivity.

ALFRED KINSEY

In spite of the criticism made of some of the American social science–oriented studies, such studies were increasing in number and, with some exceptions, gradually becoming more sophisticated. Interestingly, they had no counterpart in Europe. In the Soviet Union, to which many of the earlier generation had looked for new breakthroughs in sex, Stalin had repressed any kind of social science sex studies. Early in the 1920s, before Stalin had consolidated his power, the Soviet Union encouraged studies on the sex lives of factory workers and college students. For a brief time, there was even the Scientific Society for Sexology and Forensic Sexological Expertise, which held a conference in Leningrad in 1928, but after this, sexological studies were more or less terminated.[58]

It in this setting of a growing awareness of the importance of sexuality and an ever-increasing volume of studies on human sexuality that Alfred Kinsey began to do his research. Kinsey was born in 1894 in Hoboken, New Jersey; he was at the height of his career in 1938 when he shifted from the study of gall wasps to the study of human sexuality. He probably also was going through what might be called a midlife crisis, hunting for new fields to conquer. In the summer of that year, Indiana University began to teach a course in marriage, one of the many colleges and universities to venture into this new area. Because no professor on the faculty was considered qualified to teach it singlehandedly, teachers (all men) were gathered together from the departments of law, economics, sociology, philosophy, medicine, and biology to do so. Kinsey ended up as coordinator of the course.

To add to his own knowledge, he soon began taking histories of the students, many of whom came to him for counseling. He sought information on age at first premarital intercourse, the frequency of sexual activity, the number of partners, and similar data. Gradually, he amplified his search for information by including questions about prostitutes, the age of the partner with whom the subject had his or her first intercourse, the percentage of partners who were married, and so forth. Kinsey, a compulsive data gatherer, began an extensive reading program into all aspects of sexual behavior. This led him to build up a personal library, since serious studies on sex were difficult to find in most public or university libraries (some thirty years later, I also had to face this problem). To extend his collection of data beyond the classroom, Kinsey took a field trip to Chicago in June 1939 to conduct interviews. About this time, he also began working with inmates at the Indiana State Penal Farm and their families, compiling their sexual histories. All of this he did in consultation with the university officials, who had ruled that the histories were to be kept completely confidential. His students apparently trusted him, and many of them who had taken the class continued to write Kinsey about their sexual problems long after they had graduated.

Kinsey's expanding research into human sexuality was not without controversy, and one of his most persistent critics was Thurman Rice, a bacteriology professor at the university who had written extensively on sex, primarily from the point of view of eugenics. Rice had long given the sex lecture that was part of a required course in hygiene at the university and for which males were separated from females. Rice was typical of an earlier generation of sex experts, in that he considered moral education a part of sex education. He believed masturbation was harmful, condemned premarital intercourse, and was fearful that Kinsey's course on marriage was a perversion of academic standards. He charged Kinsey with, among other things, asking some of the women students about the length of their clitorises and then demanded the names of students in the class so he could verify such classroom voyeurism. Rice totally opposed Kinsey's questioning in general, because he believed that sexual behavior could not be analyzed by scientific methods as it was a moral subject, not a scientific one.

Some parents also objected to the specific sexual data given in the course, and university president Herman Wells, a personal friend of Kinsey, offered him the alternative of either continuing to teach the course or to conduct his sex research.[59] In any case, Kinsey would continue to teach in the biology department. He elected to do the research and dropped his participation in the marriage course.

To get funding for his research, Kinsey applied to the CRPS in 1940. After Yerkes, the chair of the committee, met with Kinsey, the committee upon Yerkes's recommendation voted to give Kinsey an exploratory grant. Kinsey, to them, seemed to be the ideal person to do a project on human sexuality, for which they felt the committee had originally been set up. He was an established scientist, was married, had a family, and had a history of carrying projects through to the finish. This last was important because the committee earlier had given money to Adolf Meyer to examine sexual attitudes of medical students at Johns Hopkins and to make observations on the sex life of selected groups, but though he gathered considerable data he was never able to bring it together.

Kinsey was not only interested but well prepared. As a bench scientist, he felt the researcher had to be directly involved in the project. He was somewhat disdainful of the work of most of his predecessors in sex research. He was appalled at how Freud and the early analysts, still under the influence of Krafft-Ebing, had looked on masturbation as a sickness. He was also concerned that Freud relied on subjective impressions and did not test them. Similarly, he disagreed with Stekel and, ultimately, with the whole psychoanalytic approach. He had no use for Krafft-Ebing's unscientific cataloging of sexual behavior, but he thought highly of Moll. Kinsey believed that American psychologists and American followers of Freud were not objective scientists and were too highly influenced by traditional moral codes. Though he had good words to say about Ellis, his esteem dwindled when he learned that the British researcher was so timid about his work that he could not talk to his subjects face to face and depended entirely on letters written to him. Kinsey was also offended by Hirschfeld's open proclamation of his own homosexuality, which led him to regard Hirschfeld as a special pleader and not an objective scientist. Similarly, he was disdainful of Malinowski, because in his mind Malinowski was not only afraid of sex but had been taken in by the islanders. He and Mead disagreed publicly, because Mead accused him of talking only about sex per se and not about such things as maternal behavior. Kinsey thought they were different things and said he wanted to study sex, not love.

Obviously, Kinsey was a strong-minded individual—some might call him arrogant; he was critical of most of his predecessors, although he was always careful to cite them in his work if they had broken new ground. Moreover, in spite of his criticism, he recognized that some, particularly Freud and Ellis, had made important contributions for their time. They just fell short of what Kinsey felt was necessary, namely the study of human sexual activity in as detached and scientific a way as possible. He had the commitment and the temperament to do so, since he thought he had to be rigorously neutral and nonjudgmental and let his data speak for him.

In sum, Kinsey seemed to appear at a particularly opportune time, when the committee was hunting for someone to fulfill its long-neglected original purpose, and Kinsey seemed to be the only person the committee had so far come in contact with who was able and willing to look at human sexual activity scientifically.[60]

For these reasons, George W. Corner saw him to be a model candidate. Corner, who visited Kinsey for the CRPS to see if the grant should be renewed, wrote:

He was a full professor, married with adolescent children. While carrying on his teaching duties in the zoology department he worked every available hour, day and night, traveling anywhere that people would give him interviews. He was training a couple of young men in his method of interviewing. Dr. Yerkes and I submitted separately to his technique. I was astonished at his skill in eliciting the most intimate details of the subject's sexual history. Introducing his queries gradually, he managed to convey an assurance of complete confidentiality by recording the answers on special sheets printed with a grid on which he set down the information gained, by unintelligible signs, explaining that the code had never been written down and only his two colleagues, Wardell B. Pomeroy and Clyde E. Martin, could read it. His questions included subtle tricks to detect deliberate misinformation.[61]

Corner added that Kinsey was the most intense scientist he had ever met, and that Kinsey could talk about little else than his research. Corner was sympathetic, since Kinsey had "met so much criticism and so much resistance to his research program from prudish and timorous people that he trusted nobody on short acquaintance, even professional scientists like Yerkes and me."[62] Corner believed that Kinsey trusted him but never quite trusted Yerkes, because Kinsey felt Yerkes was sex shy. Corner believed he won Kinsey's approval by casually reporting to him that he had read a recent book about homosexuality, and the two of them, after a knowledgeable discussion on the topic, found themselves in general agreement. From such factors were grants given and new paths broken. Kinsey received a renewal grant, and by 1946–47, he was receiving one-half of the committee's total budget. In short, after years of skirting around the subject of human sexuality, the CRPS had jumped in with full support for Kinsey. The result was a revolution in sex research. Aiding this revolution was what for a time was believed to be the elimination of the threat of venereal diseases, or as the Centers for Disease Control (now called the Centers for Disease Control and Prevention; CDC) began to call them, sexually transmitted diseases.

The first big step in this direction was the discovery of sulfa drugs in 1935, and this was followed by the development of a commercial process for making penicillin during World War II. Sulfa proved effective against gonorrhea, while penicillin was effective against both gonorrhea and syphilis. Other new antibiotics soon appeared in the postwar period, and for a time at least, the fear of sexually transmitted diseases was no longer an issue and, more important, no longer an inhibitor, in sexual relations. In sum, Americans, who had been among the most sexually inhibited, proved to be a receptive audience for the new findings about human sexuality, which are discussed in the next chapter.

7

FROM STATISTICS
TO SEXOLOGY

*T*he two decades following the appearance of the first Kinsey report in 1948 saw a radical change in public attitudes about sexuality spurred both by the development of the oral contraceptive and by new studies in human sexuality, including additional ones by Kinsey and his team and by William Masters and Virginia Johnson. The results of these studies included the establishment of a new discipline, sexology; the emergence of a new helping profession, sex therapist; and a reorientation of the way sex was taught. Individually and collectively, there was also a changing attitude, more positive if you will, toward sexuality.

KINSEY'S RESEARCH

Kinsey is a good marker of these changes because, unlike almost all previous American sex researchers, Kinsey emphasized the sex part of sex research and held that sex was as legitimate a subject to study as any other. He recognized the many facets of sexual behavior from biology to history and gathered together one of the great resource libraries of the world devoted entirely to sex. He openly challenged the traditional medical dominance of sexual topics and, in the process, opened up the field to many other disciplines. Though some of his statistics can be challenged, it was the combination of all his contributions that make him the most influential American sex researcher of the twentieth century.

His two major works, the male study in 1948 and the female study in 1952, serve as effective indicators of the change taking place in American society.[1] Though Kinsey is known for his diligent interviewing and summation of data, his work is most significant because of his attempt to treat the study of sex as a scientific discipline, compiling and examining the data and drawing conclusions from them without moralizing.

THE KINSEY INTERVIEW

The key to Kinsey's studies (as indicated in chapter 6) was the interview, since Kinsey was convinced that it was only through this means that accurate data could be compiled. More than any previous investigator, he was troubled both by the possibilities of deliberate deceit and willful or unconscious exaggeration as well as by the uncertainty of accurate recollection when it came to sex, problems that he felt a mailed survey could never overcome. His interview technique included a number of checks for consistency, and if inconsistencies appeared, either from attempts to deceive or from faulty memory, the interviewer probed deeper until the apparent disagreement could be explained or eliminated. Kinsey strongly believed he could detect fraudulent answers, and certainly his ingenious coding system was designed to detect the most obvious ones.

Exaggeration proved almost impossible in the system, in which questions were asked rapidly and in detail, because few subjects could give consistent answers. Though he recognized that some subjects might not remember accurately, he felt errors resulting from false memories would be offset by errors other subjects made in an opposite direction. A deliberate cover-up was a more serious problem, but he felt his numerous cross-checks made it difficult. If histories were taken of a husband and wife, the two were cross-checked to see how they conformed; some retakes were conducted after a minimum interval of two years and an average interval of four years to see if people would give the same basic answers.

Kinsey was also concerned with potential bias by the interviewer, and he sought to overcome this by limiting the number of interviewers to four: himself, Wardell Pomeroy, Clyde Martin, and eventually Paul Gebhard. These men engaged in discussion sessions after a series of interviews to see if they agreed on the coding of certain kinds of responses. Collaboratively, the four interviewed some eighteen thousand individuals: eight thousand each by Kinsey and Pomeroy and two thousand by Martin and Gebhard.[2] Kinsey actually hoped to get one hundred thousand sexual histories, but his death ended this long-term plan. Pomeroy continued to gather histories from various individuals long afterward, but none of these was published. In fact, most of them were not taken as part of the Kinsey studies but primarily in an effort to teach the interview method to others.

The interview covered a basic minimum of about 350 items, and these items remained almost unchanged throughout all the interviews.[3] A maximum history covered 521 items, and whenever there was any indication of sexual activity beyond what the basic questions covered, the interviewer could go as he thought necessary to get the material. All the questions had been memorized by the interviewers, and there was no referral to any question sheet. Questions were asked directly and without apology, and the interviewer waited for a response from the subject.

Initial questions were simply informational ones about the informant's age, birthplace, educational experience, marital status, and children. These were followed by questions on religion, personal health, hobbies, special interests, and so on. It was not until 20 minutes into the interview that sex questions appeared, and these started with sex education, proceeded to ages when a person first became aware of where babies came from, and then on to menstruation and growth of pubic hair and various anatomical changes. From here, the questions went on to early sex experiences, including age at first masturbation. Techniques of masturbation were investigated for both men and women. There were questions on erotic fantasies during masturbation and about erotic responses, and next was a series of questions about actual sex practices. The answers to the basic 350 questions could be coded on one page; different spots on the paper represented different questions, and a check in one area meant something quite different from a check in another area. Pomeroy estimated that the code sheet provided information equivalent to twenty-five typewritten pages.[4]

Before any specific questions about homosexuality were asked, twelve preliminary inquiries were scattered throughout the early questions, the answers to which would give the interviewer hints about the subject's sexual preference in partners. If the interviewer thought the subject was not being honest, he told the person so and generally refused to finish the interview. In some cases, the interview continued, but at the end the interviewer then told the subject that he wanted to go through some questions again, so that the subject could answer accurately questions that he or she had not been honest about the first time. In general, the interview ran from 1.5 hours to 2 hours. Children were also interviewed, but a different approach was used and at least one parent was always present.

Some individuals were interviewed for much longer periods of time. For example, those individuals who had extensive homosexual experiences were asked more questions than those who did not; subjects who had engaged in prostitution were also asked more questions. The longest interview was of a pedophile. It took some 17 hours and involved both Pomeroy and Kinsey. Kinsey had heard of the man from Dickinson and, unlike most of the subjects, this man was sought out because he was known to have kept accurate written records of his sexual activity, a not uncommon occurrence among pedophiles. The man had sexual relations with six hundred preadolescent males and two hundred preadolescent females, as well as intercourse with countless adults of both sexes and with animals of many species. He had developed elaborate techniques of masturbation and reported that his grandmother had introduced him to heterosexual intercourse and that his first homosexual experience was with his father.

His notes on his sexual relations with preadolescents furnished much of the information on childhood activity that Kinsey reported, since it included the length of time it took the child to be aroused, the child's response, and other such data. Kinsey's use of these data has been much criticized,[5] in part because Kinsey did not report his subject to the authorities. During the interview, the man was boastful about his ability to masturbate to ejaculation in 10 seconds from a flaccid start, and when Kinsey and Pomeroy openly expressed their disbelief at such a statement, the man effectively demonstrated his ability to them then and there. Pomeroy added that this was the only sexual demonstration that took place during the eighteen thousand interview sessions.[6] There were, however, laboratory observations from which data were derived, but these were separate from the interview and did not necessarily include the same individuals.

KINSEY AND STATISTICS

One of the major criticisms of Kinsey was the way in which he drew his sample. Two difficulties were at the heart of the criticism: (1) it was not random, and (2) it depended on volunteers. His critics urged him to undertake at least a small interviewing project on randomly selected individuals to test the validity of his findings,[7] but he refused. His reason for the refusal is that he believed some of those chosen randomly would not consent to answering the questions, and thus he argued it would no longer be a random sample. Though sampling techniques when Kinsey began in the 1930s were not as advanced as they later became, the issue of Kinsey's sampling concerned the Committee for Research in the Problems of Sex very early in their support. They had concluded, however, that the cluster method he advocated was as good as could be expected. After the first Kinsey volume was published, Corner, the chair of the committee, had called a conference in Bloomington, Indiana, with Kinsey and his staff and six statisticians, three of whom were favorable to Kinsey and three of whom were hostile.[8] Neither side changed its mind, but Kinsey took greater care to explain his sampling method, in his second book and also eliminated some of the more controversial data gathered from interviews with prisoners.[9]

Kinsey's sample is clearly overrepresented in some areas; for example, there are too many midwesterners, particularly from Indiana, and in the male study there is a disproportionate number of prison inmates and perhaps also of homosexuals.[10] Critics also charged that those who volunteered for the project were among the less inhibited members of society, and this gave an erroneous picture of the American public. There probably is some truth in this charge, but Kinsey tried to guard against it through what he called 100 percent sampling. When he turned to organized groups to obtain subjects, all members had to agree to be interviewed about their sexual histories, whether the group was a college fraternity, a woman's club, or the residents of a particular building. About a quarter of his sample was picked this way, and since he found few significant differences between the reports of those who belonged to groups and those he contacted in other ways, he felt he was able to establish the representativeness of his sample. Though this was an ingenious resolution to the problem, his sample was, by any definition, not a cross-sample of the total population.[11] Part of the problem was the way the Kinsey studies were used by others, some with special interests.

One of the problems with any statistical summary of sex life is what is reported and how it is reported. Kinsey, for example, put sexual activity on a 7-point continuum that ranged from 0 to 6; exclusively heterosexual behavior was on one end (0) and exclusively homosexual or lesbian behavior was on the other (6). The effect of this was to emphasize the variety of sexual activity and to demonstrate that homosexuality and lesbianism were more or less a natural aspect of human behavior. This was a partial solution to an impossible question: What is homosexuality, or for that matter what is heterosexuality? Kinsey avoided these questions by defining sex in terms of outlet, any activity that resulted in orgasm. This was something that could be measured with his 7-point bipolar scale.

At the time Kinsey began his research, 5-, 6-, and 7-point scales seem to have been the most popular, and he probably adopted such a scale for this reason. He did not attempt any greater refinement, probably believing as most scale constructors of the time did, that further refinement would simply make it more difficult to score. For whatever reason he chose this method of measure, the decision was a stroke of political genius. Note the term *political*. Although the Kinsey scale can be improved on and although it does not measure all the things that many researchers would now want to measure, it did two things of great importance. It offered comfort to both homosexuals and heterosexuals. Kinsey, in effect, demonstrated that homosexual activity was widespread in the American population: 37 percent of his American male sample had at least one homosexual experience to orgasm sometime between adolescence and old age.[12] This statistic gave assurance to many worried heterosexuals who had experimented briefly with same-sex activities that they were not homosexuals and could relax in their normality.

Homosexuals, on the other hand, found that they were more numerous than the general public (and perhaps they themselves) realized and that many heterosexuals had experimented with homosexuality. It also led many writers on homosexuality to claim a higher percentage of homosexuals in society than probably existed. Reports of the proportion of gays in the population ranged from one person in twenty to one person in ten to even higher ratios, depending on which Kinsey statistic was used.[13] However, only 4 percent of Kinsey's subjects could be labeled as exclusively homosexual; this percentage is close to what has been found in more recent studies. Kinsey noted that the proportion of women engaging in same-sex activity was less than half that of the men.

KINSEY'S DEFINITIONS AND HOMOSEXUALITY

Kinsey's insistence on a behavioral definition of homosexuality has led to speculation about his own potential homosexuality,[14] a question that seems to arise about almost every investigator of homosexuality, including me. There is no evidence for this, but Kinsey did not condemn homosexuality, which might have been the basis for the charge. He also rejected the popular stereotype of the homosexual as effeminate, temperamental, and artistic; instead, he held there were wide variations among homosexuals. To gauge this he turned to measuring sexual activity. He did, however, believe that homosexual relations were characterized by promiscuity and instability, a statement somewhat contrary to his own data, as homosexual contacts accounted for only 6 to 7 percent of all male orgasms. He explained this discrepancy by saying that homosexuals were finicky in choosing partners.[15]

Undoubtedly, Kinsey's findings and the publicity about homosexuality was valuable in assuring many a parent and many a client that one experience does not a homosexual make. On the other hand, his conclusion that a significant percentage of his sample was exclusively homosexual or almost exclusively homosexual allowed American society to come to terms with the facts of life and to recognize the widespread existence of this phenomenon. These are extremely important contributions, and the modern gay movement would probably not have come into being without them, at least at the time it did. Kinsey, in effect, accepted the bisexual potential of humans as a reality and this in itself was a major challenge to existing concepts in the psychoanalytic community, which tended to argue that bisexuals were really homosexuals trying to adjust to societal norms.[16] Kinsey's emphasis on outlet and his bipolar scale not only challenged traditional attitudes about sex but undermined them.

OTHER FINDINGS

Kinsey was also important in emphasizing that there are class distinctions in sexual practices, that highly educated individuals have a different history of sexual activity than do the less educated, and the affluent have patterns that are different from the poor. This finding basically challenged the validity of most of the studies that had gone before his, which for the most part were based on college-educated or upper-middle-class samples. He also found that the younger generation in his male study was less likely to visit prostitutes than the older, suggesting not only that there was a generational change, but that age cohorts also must be taken into account. Kinsey was not the first to recognize generational change; it had been much commented on by others, including Terman, even though the phenomenon had not been measured effectively by his predecessors.

Kinsey challenged all sorts of myths about sexuality. One such challenge had to do with female frigidity, or what is now called anorgasmia. A total of 49 percent of the females he studied had experienced orgasm within the first month of marriage, 67 percent by the first six months, and 75 percent by the end of the first year. More remarkable was the fact that nearly 25 percent of the women in the sample recalled experiencing orgasm by age of fifteen, and more than 50 percent by the age of twenty and 64 percent before marriage. The orgasms occurred through masturbation (40 percent), through heterosexual petting without penetration (24 percent), and through premarital coitus (10 percent). For 3 percent it was through a homosexual experience.[17]

Women varied enormously in the frequency of their orgasmic responses, with some reporting only one or two orgasms during their entire lives, while some 40 to 50 percent responded being orgasmic almost every time they had coitus. Still, 10 percent of his sample who had been married at least fifteen years had never had an orgasm. He also reported cases in which women failed to reach orgasm until after twenty years of marital intercourse. He also documented (as had others) the female ability to achieve multiple orgasm. Some 14 percent of the females in his sample responded that they had multiple orgasms. Several managed to have a dozen or more orgasms while their husbands ejaculated only once.[18] He concluded from his data that the human female, like the human male, is an "orgasm experiencing animal."[19]

Sometimes Kinsey seemed deliberately to flaunt the differences between widely held beliefs about traditional conduct and reality. He showed that fewer than half of the orgasms achieved by American males were derived from intercourse with their wives, which meant, he said, that more than half were derived from sources that were "socially disapproved and in large part illegal and punishable under the criminal codes."[20] He seemed to imply that premarital abstinence was unnatural and argued that nearly all cultures other than the those in the Judeo-Christian tradition made allowance for sexual intercourse before marriage.[21] Similarly, he found that nearly 50 percent of the women in his sample had coitus before they were married, although in a "considerable portion" it had been confined to their fiancé and had taken place within one or two years preceding marriage.[22] He also argued that his data did not justify the general opinion then existing that premarital coitus was of necessity "more hurried and consequently less satisfactory than coitus usually is in marriage."[23] Kinsey, in effect, ended up defending premarital intercourse just as he had masturbation and petting, arguing that premarital experience contributed to sexual success in marriage.[24]

The two reports hit different emotional responses in the American public. For the male study it was the incidence of homosexuality that received much of the headlines, while for the female study it was the generalized premarital and even extramarital activity of the women. Some 26 percent of the women had engaged in extramarital coitus,[25] and about 50 percent of the married male population had.[26] Still, it was the case of the women "adulteress" that roused public opinion.

Kinsey reported that 50 percent of the males who remained single until age thirty-five had overt homosexual experiences, and some 13 percent of his sample had more homosexual than heterosexual experiences between the ages of sixteen and fifty-five, and Kinsey noted that between 4 and 5 percent of the male population were exclusively homosexual.[27] This figure corresponds to some of Hirschfeld's figures and, as indicated, tends to be supported by more recent data. Women in his sample reported considerably fewer homosexual contacts than the men. Some 28 percent had reported homosexual arousal by age forty-five, but only 13 percent had actually reached orgasm. Less than 3 percent could be regarded as exclusively homosexual.[28] The homosexual pattern, however, differed between men and women by social class. Among men it was the lower socioeconomic class that had more homosexual experiences, whereas among women, it was the upper-class, better-educated group that had more homosexual activity.[29] He did not really explain this difference, which might well have been due to the ability of the upper-class women to have more choices in their partners and the economic capability to be independent of a man.

Kinsey openly and willingly challenged many basic societal beliefs. Though there is considerable evidence of Kinsey's commitment to marriage, and he demanded that his interviewers be happily married,[30] his data seemed to many to undermine the belief in marriage and traditional family. Kinsey had questioned the assumption that extramarital intercourse always undermined the stability of marriage and held that the full story was more complex than the most highly publicized cases led one to assume. He seemed to feel that the most appropriate extramarital affair, from the standpoint of preserving a marriage, was an alliance in which neither party became overly involved emotionally. He was, however, more cautious in the female book and conceded that extramarital affairs probably contributed to divorces in more ways and to "a greater extent than the subjects themselves realized."[31] Inevitably, his ideas came under attack, because he seemed to be assaulting traditional religious teachings.[32]

Interestingly, Kinsey ignored what might be called sexual adventure, paying almost no attention to swinging, group sex, and alternate lifestyles as well as such phenomena as sadism, masochism, transvestism, voyeurism, and exhibitionism. He justified this neglect by arguing that such practices were statistically insignificant. But the real answer is probably that Kinsey was not interested in them. He was also not particularly interested in pregnancy[33] or sexually transmitted diseases. What he did, however, was to demystify discussions of sex as much as it was possible to do so. Sex, to him, became just another aspect of human behavior, albeit an important part. He made Americans and the world at large aware of just how big a part human sexuality played in the life cycle of the individual and how widespread many kinds of heterosexual and homosexual activity were.

CRITICISM

Though the general public accepted the importance of the study,[34] many people attacked it, including Harold W. Dodds, president of Princeton University, and the Reverend Henry P. Van Dusen, president of Union Theological Seminary as well as a member of the Rockefeller Foundation.[35] While a significant proportion of the more serious criticisms was based on the sampling method and the statistical reliability of the data, the vast majority of criticism was based on what can only be called moralism and prudery. Kinsey was surprised and upset by the criticism, but since he basically had challenged much of psychoanalytic thinking, disagreed with and criticized the findings of many of his predecessors in the social sciences, and stated that much of Western moral teaching ignored reality, it is difficult to understand why he did not expect severe criticism. Moreover, as Lionel Trilling reported, in spite of Kinsey's scientific stance, his book was "full of assumptions and conclusions; it makes very positive statements on highly debatable matters, and it editorializes very freely."[36] This made criticism easier than it might have been if he had not, either consciously or unconsciously, engaged in editorializing.

Kinsey was particularly concerned by the criticism because he was afraid that the CRPS would cut off his funding. Though in terms of serious criticism, he had as many defenders as he did hostile critics, most of his defenders also had some criticism not only of his results but of his plans. One critic, for example, cast doubt on the wisdom of getting one hundred thousand histories, if only because the continued accumulation of data by the same methods would lead to the point of diminishing returns.[37] Despite the criticisms of the first report, the CRPS continued to fund Kinsey.

Because the response to the first volume had made it a best-seller, the press had eagerly anticipated the publication of the second volume on the female. By the time the book was ready to appear, the advance interest was so great that Kinsey and co-workers were literally besieged by the press, which was engaging in what has since come to be called a frenzy. Some 150 magazines and newspapers had asked to see the book before it was released to the public, and out of this group, Kinsey and his team selected 30 to receive advance galleys, providing that the editors did not publish anything before August 20, 1953. This policy antagonized as many as it satisfied, and it almost immediately came under attack. The center of the assault on the female volume was essentially by the moralists, particularly by the clergy, who seemed to feel that Kinsey had undermined the virginal status of American womanhood. Some who had supported the first study, such as Karl Menninger, joined in the denunciation of the second. In part, some of the criticism was a turf war. For example, Menninger said, "Kinsey's compulsion to force human sexual behavior into a zoological frame of reference leads him to repudiate or neglect human psychology, and to see normality as that which is natural in the sense that it is what is practiced by animals."[38]

CONGRESSIONAL INVESTIGATION

One result of the mounting criticism was that Kinsey lost his financial support from the CRPS and the Rockefeller Foundation. Corner requested additional funds for the CRPS in November 1953, after the publication of the female study, but the membership of the CRPS was changing, as was the composition of the Rockefeller Foundation. A. J. Warren was the new director of the medical division (replacing Alan Gregg), and the new president was Dean Rusk, who was later secretary of state under John F. Kennedy. Rusk had to fight something that none of his predecessors had, namely the threat of a congressional investigation into the activities of the foundation itself, and this threat undoubtedly entered into the decision making.

Kinsey had not made matters easier for himself or his cause. He had, for example, insisted on crediting the Rockefeller Foundation as a source of funding, a policy discouraged by the foundation and for that reason had not been done in the earlier Rockefeller-supported studies.[39] This deliberate decision of Kinsey to do so gave his opponents a larger and extremely wealthy target to attack, and attack they did. The congressional investigation had been initiated by Congressman B. Carroll Reece, a highly influential Republican who later became chairman of the Republican National Committee. Reece had been encouraged by a number of individuals, including Harry Emerson Fosdick and Dodds, who were concerned with the moral implications of the Kinsey reports. Rather than concentrate on this question, which clearly had two sides, Reece seized on the criticism of Kinsey to set up the House Committee to Investigate Tax Exempt Foundations, thereby attacking Kinsey's funding rather than his findings. In announcing the committee, Reece plainly stated that Congress had "been asked to investigate the financial backers of the institute that turned out the Kinsey report last August." Corner wrote:

The atmosphere of the time was so full of suspicion and fear that instead of asking me directly to go to New York to talk to him, Rusk sent an emissary who with an air of profound concern divulged to me that Mr. Rusk would appreciate a letter from me asking for an appointment to see him. I obliged. This brought a cautiously worded invitation to the foundation's office for discussion of unspecified matters. At Mr. Rusk's office I found that he had called in one of his chief legal advisers. I was quizzed at length about our committee's business. When my replies had (as I took it) established that I would show reliability and common sense if I were called on to testify in Washington, the lawyer coached me thoroughly as to how I should reply to the congressional interrogation. Mr. Rusk, I felt, was deeply worried. . . . [Later] we were informed that the foundations' grant to the National Research Council would be continued with a provision that our committee would no longer grant funds to Kinsey's group. I was so upset about this that I proposed that we should dissolve our committee, but cooler heads among us pointed out that this would cut off support of the other non-controversial grantees. As for Kinsey, the royalties from his two books were so large that he could get along without our committee's grants.[40]

Reece's committee began its public hearings on May 10, 1954, and twelve witnesses were heard, all handpicked and all supporting Reece's view. Their testimony was directed against foundations, in general, but the Rockefeller Foundation and Kinsey, in particular. Though two members of the committee, Gracie Pfost from Idaho and Representative Wayne L. Hayes of Ohio, issued a minority report charging that the hearings were carefully staged and witnesses were preselected, a condemnatory report was issued that was little more than a recapitulation of the original charges and with no additional evidence. Later, one of the Republican congressman who had signed the majority report stated that he had done so on the assumption that it would include a long list of exceptions and qualifications, which was not included. In effect, the report was not a majority report but represented only the majority summary of a divided committee.

Two things resulted from the report: the removal of Rockefeller support to Kinsey and the decision of the foundation to give one of the largest grants it had ever awarded to Fosdick, a major critic of Kinsey. The award did not go to Fosdick himself but to the Union Theological Seminary, which he headed. The nominal purpose of the grant was to aid the seminary in the development of vital religious leadership. Though no direct link between the grant to Fosdick and the curtailment of criticism was ever hinted at or suggested, Fosdick certainly remained a strong booster of the foundation.

For public relations purposes, it was announced that Kinsey's support was not renewed because he had failed to request support, but there is ample evidence in the Rockefeller archives that he did. Some of the slack in funding for Kinsey was taken up by the University of Indiana through the effort of its president, Wells. The scope of project, however, was severely curtailed. Kinsey continued to try to gain funding from the Rockefeller Foundation, and on Corner's suggestion he decided to do something on abortion because it was thought that such a study might prove more immediately useful to public policy issues. Kinsey, however, believed he was not the person to do such a study, since he did not regard abortion as a traumatic experience for all women, nor did he believe abortion was a result of a disordered family or a poor social or economic situation. He continued to pursue his research and tried desperately to raise more funds, up until his death on August 25, 1956. In spite of the trauma of his last years and the serious and legitimate criticism of his studies, he was probably the major figure in transforming American public attitudes about sex, helping Americans to come to terms with the existence of real sexual behaviors that had been previously ignored.

KINSEY AND CENSORSHIP

Kinsey also broke new legal ground in disseminating information about sex. This was because he was nothing if not thorough, and typical of his research was his attempt to survey exhaustively the literature about human sexuality. This, among other things, involved collecting materials from all over the world. Inevitably, he ran into difficulty with postal and custom officials. Alden H. Baker, collector of customs at Indianapolis, called some of the incoming materials, "Damned dirty stuff," and held in 1950 it was inadmissible. Kinsey believed that the law specifically granted exceptions to scientists and medical individuals in matters dealing with possible obscenity, and he argued, it was under this category that the materials should be admitted. Washington, D.C., customs officials backed up Baker, saying there was nothing in the materials that was of intrinsic value or that made it valuable to scientists. Rather than destroy the material outright, as they held was their right, they agreed to wait for final court adjudication.[41]

The case, U.S. v. 31 Photographs, was finally decided after Kinsey's death in the Federal Court of New York.[42] Judge Edmund L. Palmieri, ruling in Kinsey's favor, stated that there was no warrant for either custom officials or the court to sit in review of the decisions of scholars as to the bypaths of learning on which they would tread. The legal question was narrowly defined, whether among those persons who sought to see the material, there was a reasonable probability that it would appeal to prurient interest.[43] In this case, Palmieri decided it would not. Harriet Pilpel, a prominent American Civil Liberties Union (ACLU) and First Amendment attorney, served as the lawyer in the case. Indiana University had also filed a friend of the court brief.

The important aspect of the case was that the court, in determining community standards for defining whether a material was obscene, recognized those scientists and scholars interested in studying human sexuality as a community when it could be shown that this was the audience for which the material was intended. Customs decided not to appeal the ruling, and this has allowed various institutions and professionals to collect materials essential for sex research.

Kinsey was determined to make the study of sex a science and had projected a number of projects and book-length reports, including the examination of the sex laws and sex offenders, the development of sexual attitudes and overt behavior in children, the institutional sexual adjustment in the armed forces and similar places, the heterosexual-homosexual balance, the sexual factor in marital judgment, the physiological aspect of sexual arousal, the practice of prostitution, and the erotic elements in art. Many of these studies obviously went beyond his own capacities, but he seemed to visualize a world center in Indiana where, under his direction, sexual studies would flourish. Though some of these studies on which data had been collected were brought to fruition by his successors in Indiana and other new projects were initiated, Kinsey's death led to a greater dispersion of sex research across the United States than might have been the case had he lived.

A good example is the study of the biological factors involved in sexual behavior, for which Kinsey had been gathering data. He had collected photographs and had photographs made of some subjects and had gathered records of what others reported about the physiology of sex, but none of this appeared in his writings. At his death, his collection included, among other things, more than four thousand sets of measurements of penises made by subjects who gave their case histories, and another twelve thousand measurements made by a person who turned his records over to Kinsey. In the Kinsey files, the longest authenticated measurement of a penis was 10.5 inches in erection, although there were unofficial reports of longer ones. The average length was nearly 6.5 inches.

Kinsey had also attempted to measure clitorises, but this was more complicated because the amount of fleshy material and the position of the material in the prepuce. Still, clitorises that measured as long as 3 inches were reported (primarily in black women), and Kinsey noted that peep shows had exhibited women with 4-inch clitorises. Kinsey also turned to gynecologists to determine the extent to which women were aware of tactile and heavier stimulation in every part of the genitalia. He thought that clitoral stimulation was the key to female orgasm.[44] Kinsey, ever the entrepreneur, had grand plans to do much more in this area and had requested funds for a physiologist, a neurologist, and a specialist in the sexual behavior of lower animals, but nothing had come of these requests. Instead, William Masters and Virginia Johnson were the pioneers in this area.

CONTRACEPTION

Before turning to Masters and Johnson, it is important to look at what was happening in the field of contraception in the postwar United States. Simply by having greater control over their own bodies through better contraceptives, women looked on their own sexuality with a different mind-set. Several key developments took place in the 1950s and 1960s that added to the influence of Kinsey. New methods such as the intrauterine devices (IUDs) and the birth control pill appeared on the market. A different kind of research from Kinsey's was needed to develop these technologies, and in a sense, it was a more traditional kind of sex research, but equally important. The histories of both need to be recounted here.

IUDS

Like most contraceptives, IUDs have a long history. One of the early recorded uses was Arab camel drivers who put stones in the uteri of camels to prevent contraception while on long trips. Most of the early IUDs, however, were not technically IUDs but were stem pessaries, as most of the devices were placed in the vagina and had an extension that went through the cervical opening into the uterus. While nominally inserted to correct uterine position, they also induced abortions and prevented pregnancies. Drawings of these devices, many of which were patented, indicate that they also made intercourse difficult.

The first IUD to be widely used in the twentieth century was a ring of gut and silver wire developed by the German gynecologist and sex researcher Ernst Gräfenberg; the device became popular in Germany in the late 1920s.[45] In 1934, Tenrei Ota of Japan introduced gold and gold-plated silver intra-uterine rings into the uterus, which he claimed were more effective than Gräfenberg's devices.[46] Though there was initial enthusiasm for these devices, both quickly ran into difficulty because of the dangers of uterine infection. The Japanese government for a brief time prohibited the use of Ota's rings, while Gräfenberg abandoned his because European physicians objected to the potential of infection.

As the new antibiotics developed in the 1930s and 1940s finally overcome fears of infection, several physicians began to experiment with IUDs. In 1958, Lazar C. Margulies, a member of the obstetrics department at Mount Sinai Hospital, approached the head of his department, Alan F. Guttmacher, a member of the medical advisory committee of the Council on Population, about the potential of IUDs. Margulies argued that some of the dangers of the past could be eliminated by making an IUD out of plastic and that infection could be controlled. Guttmacher allowed Margulies to try out the device and found out that they did work.

Other research on IUDs serendipitously appeared at about the same time. Willi Oppenheimer reported on his long experience in Israel with 866 women who used Gräfenberg rings made of silkworm gut. Atsumi Ishihama of Japan described his experience with nineteen thousand women who used the Ota IUD, which had again been legalized in Japan.[47] It was not so much the data that were available but the timing, since the reports on the successful use of IUDs that were presented at the Fifth International Conference on Planned Parenthood, held in Tokyo a few years earlier, had failed to stimulate any follow-through.[48]

Progress was rapid after 1958, however, and several devices were evaluated over the next few years made from polyethylene, stainless steel, nylon, silkworm gut, and other materials. The polyethylene devices proved particularly effective, because they could be threaded into a catheter or cervical cannula to be inserted while the other devices required a slight amount of dilatation. The more effective devices had a marker that extended through the external os giving assurance that the device was in place and making removal easier.

One of the key figures in the development of a safe, functional IUD, and in bringing about a change in attitude regarding its use, was the gynecologist Jack Lippes, from Buffalo, New York. Influenced by reports of the successful use of IUDs in Japan and Israel, Lippes began using the Gräfenberg ring made from silkworm gut on twenty of his patients. Lippes soon ran into problems, primarily when trying to remove the device, as it lacked a tail and required him to use an instrument similar to a crochet hook for retrieval. Lippes found this a difficult procedure and began experimenting with the Ota ring, to which he attached a string that dangled through the cervix. Although this facilitated removal, in about 20 percent of his cases the Ota ring would rotate and wind the string up into the uterine cavity.[49]

As he continued his experiments, Lippes tried making a polyethylene loop with a monofilament thread of the same material hanging from it. Initially, this caused difficulty because the thread was hard to see in the vagina; he dyed the thread blue and found he could see it and thus easily remove the IUD. The existence of the blue thread also allowed women to check that the device was still in place. Though there were other competitors, such as the Margulies Spiral and the Binberg Bows, the Lippes Loop became the best known and most widely used device in developing countries outside of China. It proved particularly effective. In a twenty-month clinical trial carried out in Korea in 1962–63, some 7,364 women were fitted with the loop, and later 244,450 women received it. Between 2.5 and 3.8 percent of the women involuntary expelled the loop; the lower expulsion rate was for the second group of women who benefited from improved insertion techniques. In addition, 10.8 percent of the first sample and 7.9 percent of the second sample had the IUD removed because of pain, bleeding, and infection. None of the others reported any ill effect, and the success rate (nonpregnancy) was more than 99 percent among those who retained the IUD in place.[50] The loops were made in four sizes, with the smallest being A and largest, D. These sizes soon became the standard for evaluating other IUDs.[51]

For much of their history, the precise mechanisms by which IUDs prevented pregnancy remained unknown. This is partly because in animal studies the way IUDs prevented pregnancy varied from species to species. In sheep and chickens, they blocked sperm transport; in the guinea pig, cow, and pig, they inhibited implantation, while in the guinea pig, rabbit, cow, and ewe, they also interfered with the function of the corpus luteum. Obviously, several things were happening at the same time in some of these experiments. Similarly, research on humans has tended to show that IUDs affect ova and sperm in a variety of ways. They stimulate a foreign body inflammatory reaction in the uterus, not unlike the body's reaction to a splinter caught in a finger. The concentration of white blood cells, prostaglandins, and enzymes that collects in response to the foreign body then interferes with the transport of sperm through the uterus and fallopian tubes and damages the sperm and ova; thus fertilization is impossible.[52]

As research continued into the 1970s, a second generation of IUDs developed as the shift was made from the unmedicated Lippes Loop to copper-releasing IUDs, and to a growing extent, to IUDs that release progestins into the uterine cavity. The copper devices had some of the same advantages as the Lippes Loop, since some varieties of the cooper IUDs are less likely to be expelled, produce less menstrual blood loss, are better tolerated by women who have not yet delivered a baby, and are more likely to stay in place after postpartum or postabortion insertion. The second generation of copper devices also seemed to be slightly more effective than the Lippes Loop, though they needed to be replaced more often and cost more as a result.

The new IUDs include those that release a steroid such as progesterone or synthetic hormones called progestins into the uterus. The effective doses of steroid are substantially lower than doses required for oral administration, and the systemic side effects are less frequent. The only hormone-releasing IUD currently marketed is Progestasert, which contains 38 milligrams of progestin released at a rate that calls for its replacement after one year. A long-lasting progestin-releasing IUD is being widely tested but is not yet available for general use.

Testing Failures

One of the problems with IUDs is that testing procedures for some of them were almost nonexistent. This is particularly true of the Dalkon Shield, a poorly designed, relatively untested device that was rushed onto the market by the major pharmaceutical firm the A. H. Robins Co. to capture a share of the growing IUD market. Robins apparently did little testing of the product itself, but allegedly relied on the testing done by Hugh Davis and his business associates, from whom Robins purchased the device in 1970. Apparently, most of the reports of testing were not particularly accurate.

There were also serious ethical issues involved, because Davis was doing both the testing and the marketing. Though questions were raised about the Dalkon Shield almost as soon as it appeared on the market—insertion was exceptionally painful and there was a high rate of infection—the complaints were ignored by Robins.

By 1976, seventeen deaths had been linked to its use, but Robins stalled until 1980, when it finally advised physicians to remove the shield from women who were still wearing it. The company's failure to act resulted in a large number of additional deaths and so many lawsuits that Robins was forced to declare bankruptcy.[53] In the aftermath of the Robins failure, other companies were also sued, and though lawsuits against other IUD manufacturers were not particularly successful, the liability insurance rates had risen so much that all companies in the United States temporarily ceased to distribute the devices. IUDs were, however, available in Canada and in most of the rest of the world and have now returned to the American market.

One of the reasons that something like the Dalkon Shield case could occur was that no federal regulatory agency had any control over them before 1976. Drugs, including the pill, had to be approved by the U.S. Food and Drug Administration (FDA), which after 1938 had also been required to regulate the quality of condoms. Since the IUDs, however, were considered devices rather than drugs, and no congressional legislation had mandated their testing, FDA approval was not required before marketing them, and the agency could step in to ban such devices only if they later proved hazardous to health. As a result, IUDs such as the Dalkon Shield could be promoted without adequate proof of safety and effectiveness; this meant the public depended on the integrity of the company and the developers. One result of the Dalkon Shield affair was the establishment of FDA rules to ensure that this could no longer happen. When the copper TCu-380A (also known as Copper T-380A) IUD went on the market, for example, it had been extensively tested for several years and the data and the device had been examined by the FDA.[54]

HORMONAL CONTRACEPTIVES

Some of the latest IUDs, as noted, use steroids, much as birth control pills, which ultimately represented one of the more important results of endocrinology research (see chapter 5). The first steps toward a hormonal contraceptive dated from 1936, when an experiment on rats demonstrated that daily injections of progesterone inhibited the estrous cycle.[55] This was followed by studies indicating that progesterone injections inhibited ovulation in the rabbit and in the guinea pig.[56] Following up on this was the Columbia University endocrinologist Raphael Kuzrok, who treated his patients for painful menstruation with large doses of estrogen and found that the hormone also inhibited ovulation. He concluded that the "potentialities of hormonal sterilization are tremendous. The problem is important enough to warrant extensive work on the human."[57]

Obviously, the potential was there, but there were difficulties. One difficulty was the high cost of hormones, and a second was that the long-term effects were not well enough understood to give hormones to healthy women to control fertility. Still, as the Harvard University endocrinologist Fuller Albright (whose research was supported by the CRPS) wrote in 1945, "Since preventing ovulation prevents pregnancy, one could employ the same principles in birth control as in preventing dysmenorrhea. Thus, for example, if an individual took 1 mg of diethylstilbestrol by mouth daily from the first day of her period for the next six weeks, she would not ovulate during the interval."[58] He went on to say that a hormone regimen would have to be worked out that allowed periodic menstruation but that he did not see that as a problem.

The major obstacle to implementing research on this idea was the high cost of hormones, a problem that became increasingly severe as the treatment potentialities of steroids for a variety of illness became apparent. Much of the high cost was due to the complex and elaborate methods used to extract the hormones. Butenandt (see chapter 5), for example, had to start with nearly 4,000 gallons of urine to obtain less than 0.01 ounce of pure testosterone. When Ernst Laquer turned to bull's testicles as an alternative, he had to process nearly a ton of bull's testicles to get a slightly larger amount than 0.01 ounce of testosterone. In terms of female hormones, Edward Doisy processed the ovaries of more than eighty thousand sows to get 0.012 gram of estradiol.[59]

Russell Marker

As chemists improved their hormone production techniques, yields were rising and the cost of manufacture began slowly to decline. However, many of the key process patents were owned by drug companies, most of them European, who used their monopoly to maintain high prices. The key change in this method was due to the efforts of Russell Marker, a maverick scientist who, though probably a genius, was uninterested in almost anything but chemistry.

When Marker received his degree in organic chemistry from the University of Maryland in 1923, he also received with it a fellowship for advanced studies, but it was in physical chemistry not organic chemistry, and he wanted to do organic. He quickly advanced to candidacy and completed his dissertation, which was eventually published, but he was never granted the doctorate degree, because he had not completed the requirement in physical chemistry. Although Marker avoided the course requirement in collusion with his adviser, he still did not officially receive the degree.

Shortly after this, Marker married and went to work as an organic chemist for Ethyl Gasoline Corp. Within a short time, his salary had increased by 50 percent, but he soon lost interest in organometallic compounds and left to join the Rockefeller Institute. Over the next few years, he co-authored more than 32 papers with P. A. Levene, the head of the chemistry department; he soon had a whole laboratory assigned to him. In the spring of 1935, Marker wanted to embark on the study of steroids. When he failed to get a transfer to the pharmacology department where such research was being carried out, he resigned and, taking a cut in salary, accepted a fellowship at Pennsylvania State College subsidized by the pharmaceutical company Parke-Davis, which was anxious to expand its position in the growing hormone field. Over the next eight years, Marker published a total of 163 scientific papers and secured over 70 patents, which were assigned to Parke-Davis.

Gradually, Marker became convinced that the easiest and cheapest way to make sex steroids was from plant materials, particularly plants of the lily family and other tuberous species, because their roots contained quantities of sapogenins. The sapogenin compounds contain the basic four rings of carbon atoms typical of all steroids, and all of the sapogenin molecules had a long side chain extending from the seventeenth carbon, which is located on the fourth ring. In this, they had a remarkable resemblance to cholesterol, a monohydric alcohol important in the synthesis of the natural sex hormones. During 1940, Marker developed an efficient five-stage process of degrading the side chain and converting sapogenins into progesterones. Through the development of three more processing stages, he turned progesterone into testosterone.

Marker spent his summer vacation that year and the next collecting roots of scores of sapogenin-secreting plants, a project in which he was aided by both American and Mexican botanists. He collected over 40,000 kilograms (about 100,000 pounds) of plants from Mexico and the southern parts of the United States, comprising more than 400 species. In analyzing the plants he found new sources for almost all the known sapogenins, including diosgenin, the starting material for the synthesis of certain sex hormones. He also found twelve new sapogenins and identified two new steriods that he believed were progenitors of sapogenins.[60] About half of his plants yielded nothing, but the rest varied from a trace to a substantial amount. He decided the richest source derived from the roots of the *Dioscorea* family of plants, particularly a wild yam that grew in mountains of southern Mexico outside of Veracruz. He then tried to get backing to harvest the plants and mass produce hormones. Since southern Mexico at that time was rough country with no roads and no potential labor force and Mexican politics in rural areas were unstable, it was perhaps understandable that he failed to get backing for his project.

Marker took matters into his own hands by breaking his connections with Parke-Davis and resigning from Pennsylvania State College effective December 1, 1943. Before doing so he had spent considerable time in Mexico collecting the roots of the cabeza de negro plant under incredibly primitive conditions. He had set out for Veracruz, where the plant grew wild; and then after purchasing a mule, a spade, a machete, and a couple of dozen burlap coffee bags, he went yam hunting. Somehow he conveyed to the local Indians that he was looking for the *Dioscorea* vine they called cabeza de negro and though they did not know why anyone would want the plant, they helped him fill his bags with roots, and he returned to Mexico City to go to work by himself.

In just two months of activity during the summer of 1943, he had produced nearly 3 kilograms of progesterone, which was then selling at eight dollars a gram. He then met with Laboratorios Hermona, a small drug firm located in Mexico City. The company had been founded by Emeric Somlo, who had emigrated to Mexico in 1928, and Federico A. Lehman, a 1933 Jewish refugee from Nazi Germany. Marker's choice of the firm was pure luck, although it was not the first one he had tried. He found it because the company had listed itself in the telephone book as dealing with hormones. Marker called the firm, asking to talk to someone who spoke English. It turned out that Lehman spoke English, knew of Marker's earlier work, was an endocrinologist himself, and knew how important steroids were. The result was the foundation of a new firm called Syntex Sociedad Anónima, with Somlo, Lehmann, and Marker as partners. Marker held 40 percent of the stock.

One of the first effects of the production of steroids was to lower the price of hormones, until by 1945 the retail price had fallen by almost half, and physicians everywhere were prescribing them more than ever before. By 1946, a dispute arose between Marker and his two partners, and unable to buy them out, he sold them his interest in the firm. Marker formed a new steroid-producing company, Hormosynth, and continued to publish research papers. He left Mexico in 1949, and disappeared from public view. Actually, he had returned to State College, Pennsylvania, near Penn State, where he and his wife retired. Marker wrote in 1969, "Since retiring from the laboratory 20 years ago, I have never returned to chemistry or consulting, and have no shares of stock in any hormone or related companies. My one appearance in public was recently on April 23, 1969 to accept an award by the Mexican Chemical Society showing their appreciation for the work I had accomplished."[61]

For a time, Syntex enjoyed a virtual monopoly on the production of steroids from plant material because the Mexican government embargoed export not only of the root but of the incompletely processed diosgenin. Because, however, several of the key processes in steroid manufacturer were covered by American patents that had once belonged to Schering Corp., a German-owned drug house whose assets were confiscated under the Alien Property Act during World War II, the U.S. government had some control of Syntex. Under the threat of antitrust prosecution, Syntex stopped opposing the granting of Mexican licenses to other companies.[62] Gradually, the cabez de negro plant was replaced by *Dioscorea* (known as barbasco), whose roots had a higher diosgenin content.[63]

The Pill

One of the chemists who joined the Syntex staff after the departure of Marker was Carl Djerassi, who worked on a commercially feasible process for synthesizing orally active analogues of progesterone. In April 1952, he announced at a meeting of the American Chemical Society his success in producing a 19-norprogestin. The drug was so named because it lacked a side chain of one carbon and three hydrogen atoms (called a methyl group) at the nineteenth carbon in the progesterone molecule. Rapid clinical evaluation of 19-norsteroids followed under the auspices of Syntex. The commercial value of Djerassi's discover was apparent to Searles's chemists, quite apart from its contraceptive possibilities, and by November 1955 they had managed to synthesize an analogue of progesterone that differed enough from Djerassi's to be patented.[64]

The impetus for an oral contraceptive based on Djerassi's discovery came from a coffee table meeting of Margaret Sanger and Gregory Pincus in Abraham Stone's New York apartment in 1950. Stone and his wife, Hannah, had founded one of the first fertility clinics in the United States, if not the world, and he had carried on alone since his wife's death. Among his many duties, he served as medical director and vice president of the Planned Parenthood Federation and vice president of the *Journal of Human Fertility*. Pincus, who had outlived his early notoriety to gain worldwide recognition for his hormone research, was then editor of the annual *Progress in Hormone Research* and co-director of the Worcester Foundation for Experimental Biology.

Sanger, a continual crusader for better contraceptives, ended up urging Pincus to investigate the possibilities of steroids for contraceptive use. Sanger, however, did more than urge. She also had a source of money willing to support such research.

The money came from Katharine Dexter McCormick (1875–1967), who controlled much of the McCormick estate. McCormick graduated from MIT in 1904 and soon after married Stanley McCormick, the son of Cyrus, the founder of the International Harvester Co. Two years later, her husband became mentally ill, and he remained a schizophrenic until his death in 1947 at the age of seventy-three. McCormick, who never divorced, won a legal battle with her father-in-law over the control of her husband's estate, and using both her money and energy, had quickly emerged as an important figure in the women's rights movement. She had established the Neuroendocrine Research Foundation in 1927 and, among other projects, supported Pincus and Hoagland's research at the Worcester Foundation. The death of her husband in 1947, however had led to such a heavy tax bill that she had been forced to close the foundation. She still supported the Robert Dickinson Memorial Fund of the Planned Parenthood Federation, but she wanted more action.

In 1952, at Sanger's request, she promised the Worcester Foundation $10,000 a year; this soon was increased to $125,000 and then to $180,000 a year for the rest of her life. At her death, she willed the foundation $1 million. Pincus and co-workers were off and running in their research effort to find an oral contraceptive, testing various compounds that were originally furnished by Searles's steroid chemists.[65] Those compounds used in the clinical studies, however, were furnished by Productos Esteroides founded in 1955 by the former United States manager for Syntex, Irving V. Sollins, a personal friend of Pincus.[66]

Pincus's team included his long-time associate Min Chueh Chang and the gynecologist John Rock. Apparently, Pincus had considered his friends Stone and Guttmacher for the job of working out a contraceptive regime for women, both of whom were leaders in the birth control movement, but he ultimately chose Rock. Whether religion entered into the choice is unclear, since Rock was one of the few Catholics at that time who said that religion had nothing to do with medicine or the practice of it.[67] Pincus's choice, however, did give the pill a strong Catholic supporter.

Rock tested Pincus's progesterone regimen on fifty infertile women who ovulated regularly to see if the dosages would repress ovulation or perhaps even cause them to become pregnant. One of the batches of 19-norprogestin turned out to be contaminated by a tiny amount of estrogen, but when this defect was remedied the researchers found there was a higher incidence of breakthrough bleeding toward the end of the ovulation-inhibiting cycle of medication. The result was a decision to combine progestin and estrogen in the daily dosages. They used Enovid (norethynodrel) for the progestin component and 1.5 percent mestranol for the estrogenic component.[68] Rock had also found that his patients, once they had quit taking the pill, resumed their normal patterns of ovulation.

The next step was to do a mass study, and Pincus selected Puerto Rico, in part because Edris Rice-Wray, medical director of Puerto Rico's Family Planning Association, was willing to undertake the exacting and demanding task of supervising the series of prolonged experiments. The study began in April 1956 in a suburb of San Juan, and by January 1957, when the preliminary trial records were reviewed, 221 women of proven fertility had taken Enovid as directed for periods varying from one to nine months, and not a single one had become pregnant.

Like any experiment on people, the human element was very much present, and although the women were warned that they might experience nausea, vomiting, dizziness, abdominal pain, diarrhea, or other side effects, they volunteered in great numbers. Not all the women had proved able to endure the side effects, which varied from individual to individual, and some twenty-five had quit taking the pill either because they were frightened by the side effects or because their priest or personal physician advised them against it. Others appear to have been confused about what they were supposed to do. One woman took the tablets only when her husband was not traveling. Another, who became pregnant, complained that the pills had not worked at all even though she had made her husband take them every day.[69]

After the first report, Rice-Wray opened up her enrollment books to hundreds of additional women who were anxious to get in on the trials. A second project was opened up in Humacao under the direction of obstetrician-gynecologist Adeline Pendleton Satterthwaite, and a third project was later opened in Haiti. Though some women in the later trials did get pregnant, it was found in all cases that those women had not followed the regimen of taking the pills. For those who followed instructions, there was a 100 percent record of effectiveness. Trials spread to the United States, and in 1960, the FDA allowed Enovid to be put on the market. Since then there have been various modifications of the pill and the development of other means of administering steroids, including through IUDs, Norplant inserts, and injections.[70]

There were many more uses of the steroids besides contraceptives, but this book is not the place to detail them all. What the search for the sex hormones emphasized is just how much an understanding of human sexuality is related to a basic understanding of human physiology. The implications of this were the major contributions of Masters and Johnson.

MASTERS AND JOHNSON

William Masters and Virginia Johnson from the first were much more practice oriented than Kinsey or the endocrinologists. Masters was a physician
who was concerned with helping his patients overcome their problems. Together, Masters and Johnson thought of themselves as therapists, which
meant they accepted the world as they saw it existing and wanted to help
their clients adjust to it. Kinsey, on the other hand, was a scientist, describing the world as it existed but also emphasizing the contradictions between
actuality and accepted standards. Masters and Johnson conducted their research
for a reason that was entirely different from Kinsey's.

> When the laboratory program for the investigation in human sexual
> functioning was designed in 1954, permission to constitute the pro-
> gram within a university setting was granted upon a research premise
> which stated categorically that the greatest handicap to successful
> treatment of sexual inadequacy was a lack of reliable physiological
> information in the area of human sexual response. It was presumed
> that definitive laboratory effort would develop material of clinical con-
> sequence. This material in turn could be used by professionals in the
> field to improve methodology of therapeutic approach to sexual inade-
> quacy.[71]

Just as Kinsey had challenged, sometimes with considerable hostility, the
psychiatric monopoly on sexual treatment and research, Masters and Johnson offered whole new areas for the gynecologist, urologist, and other medical specialists to extend their services. Ultimately, Masters and Johnson also
helped establish a whole new profession, the sex therapist, which was no
longer restricted to the psychiatrist but included nurses, psychologists, social
workers, and counselors. It should be added, however, that the initial promise
of nonevasive therapeutic techniques for problems of sexual inadequacy was
oversold by some therapists and that, as the years passed, the balance
between medical intervention and nonintrusive therapy changed. The basic
teaching techniques pioneered by Masters and Johnson and their contemporaries, however, still remain important.

Masters was a native of Cleveland and was born to a well-to-do family in
1915. He attended Lawrenceville Prep School and went on to Hamilton College, where he received his bachelor's degree in 1938. He then entered the
University of Rochester School of Medicine and Dentistry, where he worked
in Corner's laboratory. Interestingly, Corner had three of the leaders in what
he called "the practical application of scientific thought to problems of
human sex behavior" as students: Guttmacher, who became internationally

prominent in the family planning movement; Mary Steichen Calderone, co-founder of the Sex Information and Educational Council of the United States (SIECUS); and Masters.[72]

While working with Corner, Masters was assigned the problem of trying to determine how the estrous cycle in the female rabbit differed from or resembled that in the human female. Masters, who had always been more interested in medical research than in the practice of medicine, decided he would like to do sex research when he completed his degree, and went to Corner for advice. Corner essentially gave him three general principles to follow in pursuing sex research: (1) he should establish a scientific reputation in some other scientific field first, (2) he should secure the sponsorship of a major medical school or university, and (3) he should be at least forty years of age.[73]

Masters followed the advice almost to the letter, although he did start his research into sexuality at the age of thirty-eight. After graduation from medical school, he accepted a position at Washington University in St. Louis. Willard Allen, another of Corner's students and an active researcher in endocrinology, helped Masters get the appointment as an intern in obstetrics and gynecology. Masters moved up the ladder through resident to assistant professor and associate professor. He married and had two children. Masters also published a number of papers covering a variety of obstetrical and gynecological topics, although the majority dealt with hormone-replacement therapy for aging and aged women, a treatment that he strongly advocated[74] and that is widely used today.

THE BEGINNINGS

Gradually, Masters turned to studying the sexual act itself. As noted earlier, this was something that had been of interest to others. The pioneer in this respect had been the French physician Félix Roubaud, who had published his account of the female response cycle in 1855.[75] Kinsey had called Roubaud's description unsurpassed,[76] even though the Frenchman had been mistaken on two points, namely the claim that there was direct frictional contact between the penis and the clitoris, and that the semen was sucked up through the cervix.

Much of the early reporting on the female response was serendipitous. One example was the case of the Indiana physician Joseph R. Beck, who, in 1872, treated a woman with a collapsed uterus (retroversion) by fitting her with a mechanical pessary to hold the uterus in place. He reexamined her on August 8, 1872, to see that the device was correctly placed. Before the examination, the patient cautioned Beck to be very careful, because she had such a nervous temperament and passionate nature that she might have an orgasm from the pressure of his finger. Since her cervix was directly visible through her labia as a result of her collapsed uterus, Beck saw an opportunity to describe how the cervix reacted during orgasm. He proceeded to ignore her plea. He wrote,

> [Carefully] separating the labia with my left hand, so that the os uteri [vaginal opening of the cervix] was brought clearly into view in the sunlight, I now swept my right forefinger quickly three or four times across the space between the cervix and the pubic arch, when almost immediately the orgasm occurred. . . . Instantly the height of the excitement was at hand, the os opened itself to the extent of fully an inch, as nearly as my eye could judge, made five or six successive gasps, as it were, drawing the external os into the cervix each time powerfully, and, it seemed to me, with a regular rhythmical action, at the same time losing its former density and hardness, and becoming quite soft to the touch. All these phenomena occurred within the space of twelve seconds of time certainly, and in an instant all was as before. At the near approach of the orgastic excitement the os and cervix became intensely congested, assuming almost a livid purple color, but upon the cessation of the action, as related, the os suddenly closed, the cervix again hardened itself, the intense congestion was dissipated, the organs concerned resolved themselves into their normal condition.[77]

Beck seemingly confirmed Roubaud's observations that the cervix sucked up the sperm at orgasm. Similar observations and conclusions were made by B. S. Talmey,[78] but these conclusions were challenged by Dickinson, who said if the sucking motion was the typical orgasm, then orgasms have not occurred more than a half dozen times in many millions of office examinations nor have they been noted in their wives by doctors familiar with the cervix, or else physicians have consistently hidden knowledge of such actions.[79]

Just as Masters was actively beginning to plan his own program, G. Klumbies and H. Kleinsorge, two physicians at the University Clinic in Jena, Germany, reported on a patient who was capable of fantasizing to orgasm, a fact that made it possible to distinguish the direct effects of orgasm from the muscular exertion that ordinarily preceded it or accompanied it. With the aid of an electrocardiograph and a blood pressure recorder, Klumbies and Kleinsorge recorded physiological changes, including pulse rate, systolic and diastolic blood pressure, cardiac volume, rhythm of heart chamber contractions, position of the heart, respiratory volume, and muscle irritability. The woman identified some of her orgasms as more intense than others, and Klumbies and Kleinsorge noted that the intensity of the orgasm as subjectively reported showed a close relation to the acuteness of the blood pressure peak.[80] Another investigator, Abraham Mosovich, recorded electroencephalograms (brain wave patterns) during sexual arousal and orgasm.[81]

The best and most complete observations made before Masters and Johnson's studies were Kinsey's. He reported that he had access to a considerable body "of observed data on the involvement of the entire body in the spasms following orgasm."[82] Actually, most of the observations had been made on volunteers by Kinsey or his staff, independent of the interview portion of his research. Chapter 15, "Physiology of Sexual Response and Orgasm," in *Sexual Behavior in the Human Female* contained the best, most current, and most accurate data compiled at that time.

When Masters began his studies in 1954, he interviewed at length and in depth 118 female and 37 male prostitutes. Of these, 8 of the women and 3 of the men then participated as experimental subjects in a preliminary series of laboratory studies.

> Suggestions of this select group of techniques for support and control of the human male and female in situations of direct sexual response proved invaluable. They described many methods for elevating or controlling sexual tensions and demonstrated innumerable variations in stimulative technique. Ultimately many of these techniques have been found to have direct application in therapy of male and female sexual inadequacy and have been integrated into the clinical programs.[83]

Ultimately, however, the experimental results derived from the prostitute population were not included in the final published results, because Masters and Johnson wanted a baseline of what they regarded as "anatomic normalcy." To get this, they turned to patient populations and volunteers for data. It was during this phase that Virginia Johnson joined Masters's team, because Masters strongly believed that a woman should be involved in his research.

Born Virginia Eshelman in Missouri in 1925, she had studied music at Drury College and later attended the University of Missouri. In 1950, Johnson married and had a son and daughter before separating from her husband in the late 1950s. She was registered for a job at the Washington University Placement Bureau. Masters was seeking a woman to assist in research interviewing and had specified that he wanted a woman who had experience and interest in working with people. The bureau sent Johnson, and she was hired. The two later married, but were divorced in 1992.

Johnson's work was particularly important in the first two books but played a lesser part in later studies. The fact that Masters and Johnson were a male-female team separates them from Kinsey. Though Kinsey had added a woman to his team shortly before he died, he seems to have not felt it necessary to do so before. Although women worked on the project, they were not regarded as co-researchers. In this he probably reflected the traditional male attitudes of his generation. Masters, in general, gave more emphasis to the female than to the male not only in his team but in his studies. In the discussion of physiology, for example, the female is mentioned first, and though there are obvious similarities in the response, by giving the female data first Masters seems to have emphasized that the female is not just an inferior imitation of the male, an attitude widely prevalent even at the time of his research.

SEXUAL RESPONSE CYCLE

Masters and Johnson held that the sexual response cycle involved much more than a penis and vagina, and they sought to measure heart rate, respiratory functions, muscle tension, breast response, and any other physiological measurement they could think of. A key element in the ability of Masters and Johnson to break new ground was technological. Advances in the miniaturization of cameras and electronic devices meant that they could be used inside of a plastic phallus. This allowed Masters and Johnson to record what took place inside the vagina during orgasm, and they could observe the phenomena in some detail.

This new technology permitted them to give definitive answers to some of the questions about which there had been arguments or on which there had only been subjective data. It allowed Masters and Johnson to determine that there was a moistening of the vaginal lining with lubricating fluid within 10 to 30 seconds of the onset of erotic stimulation and to note that this fluid came from the coalescence of a "sweating" of the vagina's walls. They emphasized that neither the Bartholin's glands nor the cervix, previously believed to be the source of the lubrication, contributed to the fluid. Rather the sweating resulted from the increased blood supply and the engorgement of vaginal tissues.[84]

Masters and Johnson also noted a lengthening and distension of the vaginal walls, while the cervix and the uterus are pulled slowly back and up into the false pelvis (the part of the pelvis above the hip joint). The vagina's walls also undergo a distinct coloration change, from purplish red to a darker purple as a result of vasocongestion, and the wrinkled or corrugated aspects of the vaginal wall (technically called the rugal pattern) are flattened. Gradually, the outer third of the vagina becomes grossly distended with venous blood, and the vasocongestion is so marked that the central lumen (interior) of the outer third of the vaginal barrel wall is reduced by a least a third. All this takes place during what Master and Johnson called the plateau phase, or second phase of the sexual response cycle.

This is followed by the orgasmic phase, during which much of physiologic activity is confined to what Masters and Johnson called the orgasm platform in the upper third of the vagina. Here there are strong contractions at 0.8-second intervals, which recur within a normal range of three to five up to a ten to fifteen times per individual orgasm. The uterus elevates and contracts rhythmically with each contraction, beginning at the upper end of the uterus and moving like a wave through the midzone and down to the lower or cervical end.

These uterine contractions had been long associated with the idea that the cervix sucks up sperm. Master and Johnson, however, theorized that contractions in such a direction would, if anything, expel sperm. They then proceeded to demonstrate the uterine contractions could not possibly lead to a sucking up of the sperm into the uterus. They prepared a tight-fitting cervical cup that they filled with a semenlike liquid in a radiopaque base. Masters and Johnson then made radiograms during the orgasmic experience and found no such sucking action.[85]

To describe what took place during intercourse, Masters and Johnson developed a four-phase description: (1) excitement, (2) plateau, (3) orgasm, and (4) resolution. They found that men responded in terms of basic physiological changes along the same lines as women; in both sexes, there was an increase in heart rate, blood pressure, muscle tension, and in the majority of both men and women a "sex flush" (a rosy measlelike rash over the chest, neck, face, shoulders, arms, and thighs) is observable. At orgasm the heart and respiratory rates are at a maximum and the sex flush at its peak, although the male has what are called ejaculatory contractions during the orgasm. The orgasm phase is followed by the resolution phase in which there is a return to conditions as they were before the sexual excitement phase began. Women were found to have a wider variety of orgasmic responses and many could have multiple orgasms.

Masters and Johnson criticized what they called the "phallic fallacy" of comparing the clitoris with the penis. They emphasized that even though the clitoris might be the anatomical analogue of the penis, it reacts to sexual stimulation in a manner quite different from the penis. It does not become erect during arousal but instead withdraws beneath its protective foreskin, and in fact, its length is reduced by at least half as orgasm approaches. When it is retracted, however, it responds to generalized pressure on the labial hood.[86]

Patient concerns were always present in Masters and Johnson's minds. For example, they reported that the average flaccid measurement of a penis was 7.5 centimeters (about 3 inches) and during erection the penis more than doubled in length. However, they recognized that not all men had the same size penis.[87] To allay the qualms of their readers, they emphasized that the vagina was a "potential rather than an actual space," and was "infinitely distensible."[88] Interestingly, however, there is no evidence that they ever asked any of their female subjects whether penis size made a difference, or if they did, the answer was not recorded.

SAMPLE

All told 694 individuals, including 276 married couples, participated in the Masters and Johnson laboratory program. Of these, 142 were unmarried but 44 had been previously married. The men ranged in age from twenty-one to eighty-nine and the women, from eighteen to seventy-eight. Data included anatomic and physiologic response patterns from seven women, ages nineteen to thirty-four, who were born without a vagina. Volunteers for the laboratory research program were involved in masturbation by hand, fingers, or a mechanical vibrator; in sexual intercourse with the woman on her back, with the man on his back; and in artificial coition with a transparent probe. Also studied were the anatomy and physiology of the aging male and aging female, although the data were not as complete as for the younger ages. Masters and Johnson emphasized, however, that if opportunity for regularity of coitus exists, the elderly woman will retain a far higher capacity for sexual performance than her female counterpart who does not have similar coital opportunity. They reported that even though the postmenopausal woman has lost some of her hormone output, the psyche is as important, if not more important, in determining the sex drive.[89] Similarly, while in the aging male the entire ejaculatory process undergoes a reduction in physiological efficiency, the sexual response remains. Masters and Johnson concluded,

> There is every reason to believe that maintained regularity of sexual expression coupled with adequate physical well-being and healthy mental orientation to the aging process will combine to provide a sexually stimulative climate within a marriage. This climate will, in turn, improve sexual tension and provide a capacity for sexual performance that frequently may extend to and beyond the 80-year level.[90]

SEXUAL DYSFUNCTION

A natural follow-up to the physiological studies of the human sexual response was treatment for dysfunctional clients. For this purpose, Masters and Johnson developed a sex therapy team (a woman and a man) and a methodology through which they said they were treating the "marriage," since the basic foundation of their treatment was that both the husband and wife in a sexually dysfunctional marriage be treated.[91] Such a statement emphasizes Masters and Johnson's marital orientation, something that probably contributed to their widespread acceptance on the American scene. Although a significant proportion of their clients were unmarried, most came to therapy accompanied by a partner.

Because Masters and Johnson always emphasized the therapeutic nature of their research, their aim in effect had always been the development of treatment modalities. In their treatment, they concentrated on specific symptoms rather than generalized disorders. In a way, they adopted some of the concepts of the behavioral psychologists who had begun treating sexual problems in the 1950s,[92] but in the process, they popularized sex therapy and systematized it on a physiological base.

SEX THERAPY

One result was the development of a new specialty in the helping professions, that of sex therapist. Before their entrance on the scene, the predominant treatment of sexual dysfunction, at least in the United States, was through psychoanalysis. What Masters and Johnson essentially did was challenge perhaps the final bastion of the control that psychiatry, and particularly psychoanalysis, had over the sex field. Kinsey had basically undermined many of the assumptions that psychiatry had made about sexual behavior and furnished a new kind of database. He had also attacked psychoanalysis for their unscientific assumptions. With Masters and Johnson, even the treatment option, which psychiatry had dominated, was now redirected to other specialists, many of whom were not physicians. The result was to increase the number of individuals who were not only professionally but economically interested in sex. Kinsey, in effect, had reestablished the concept of sexology. Although sexological research was a somewhat limited field, the rise of sex therapy gave sexology enough other professionals to justify separate sexological societies and journals.

Masters and Johnson were also important because they, although in a much gentler form than Kinsey, emphasized the importance of sex education. For example, in their discussion of the anorgasmic female, they stated that women in general were victims of the double standard, because they more than men had been taught to repress their sexual feelings. Masters and Johnson concluded that repression, in the form of historical and psychological experience, was the most important factor in the development of frigidity.[93] Ignorance and superstition about sex were and remain the major problems in an inadequate sexual response, and when the sexual partners manage to have their prejudices, misconceptions, and misunderstandings of natural sexual functioning exposed, then and only then can "a firm basis for mutual security in sexual expression" be established.[94] In short, for marriage to reach its full potential, and Masters and Johnson were always concerned with marriage, knowledge of sex was essential. This message was seized on not only by a new generation of sex educators to bring about reforms in sex education but by the public in general, who seemed to grow ever more interested in how to have a better marriage, which they, as Stopes had thirty years earlier, believed was highly dependent on sexual performance.

The largest component of the expanding group of sex professionals in the 1960s was the sex therapist, the number of which grew rapidly. Masters and Johnson had established a two-week basic program that involved a male and female sex therapist and a client couple; this program served as the initial model. Their success was phenomenal. Masters and Johnson reported that the two-week session eliminated sexual difficulties for 80 percent of their clients. Not content with these immediate results, they followed up these studies five years later and stated that of those they were able to recontact, only 7 percent reported recurrence of the dysfunctions for which they originally had sought treatment.[95] The result of such claims was a demand by the public for help with sexual problems and an awareness by the various kinds of professionals that they could expand their client base if they could gain some expertise in sex.

Many would-be sex therapists went to St. Louis to take special training sessions with Masters and Johnson. Professionals who entered sex therapy from a slightly different background also offered special seminars. On the West Coast, for example, William Hartman and Marilyn Fithian, who had included sexual therapy as part of their marriage and family counseling, had begun to carry out their own set of experiments on the sexual response in their Long Beach, California, center. As the demand for sex therapists grew, Hartman and Fithian conducted training seminars not only in Long Beach but all over the country, introducing would-be professionals to new trends in sex therapy.

Another important early sex therapist was Helen Singer Kaplan, who tried to combine some of the insights and techniques of psychoanalysis with behavioral methods. She questioned Masters and Johnson's use of two therapists and felt that one therapist of either sex would be sufficient,[96] a finding made by others.[97] Kaplan agreed that many sexual difficulties stemmed from superficial causes, but she believed that when unconscious conflict was at the heart of the problem and involved deep-seated emotional problems the therapist should use more analytic approaches. As a result, her approach is designated as psychosexual therapy to distinguish it from sex therapy, and her entry into the field emphasizes how psychoanalysts themselves gradually adjusted to the new sex therapy techniques.

In the afterglow of success, the sex therapy originally presented by Masters and Johnson did not seem to hold true for a growing number of therapists as the field rapidly expanded. This was perhaps because of not only the existence of deep-seated emotional problems in some clients, as Kaplan had pointed out, but the presence of basic physiological problems such as diabetes. The result was an attack on the success claims of Masters and Johnson, as an increasing number of studies reported much higher failure rates.[98]

The difference in success rates, however, is probably the result of both the changing nature of clients and the differing methods of client selection. Many of the original problems presented by the early clients of Masters and Johnson resulted from a lack of knowledge of basic sexual activity, something that was comparatively easy to overcome. The very success of the books by Masters and Johnson made such clients increasingly less likely to seek the help of a therapist, since they could read about the sources of human sexual inadequacy and adjust their own practices. On the other hand, the physical exam required by Masters and Johnson for their patients undoubtedly eliminated many of those with physiological difficulties that other less knowledgeable therapists attempted to treat and failed to help. The major result of the criticism of Masters and Johnson was to emphasize that sex therapy at its best involved a team, not only of therapists but of medical professionals, particularly the urologist and gynecologist.

PROFESSIONAL ORGANIZATIONS

The increase in the number of therapists in the 1960s and 1970s led to the development of professional societies that were devoted to the study of sex, the advancement of sex education, and the regulation of sex therapy. At first such groups attempted to be inclusive and, in fact, had their American beginning in 1939 with the foundation of the National Council of Family Relations (NCFR). The NCFR threw its net widely enough to include scholars, educators, counselors, and therapists within its membership. Pomeroy, a specialist in sexual matters, was also a member. In fact, many of the pioneering sex therapists had come out of marriage and family counseling back-grounds.[99]

One of the first attempts to organize an American society specifically devoted to the study of sex was by Albert Ellis in 1950, but this resulted in failure. The failure might be explained simply by saying that the time was not propitious for the establishment of such a group, but in fact, Ellis might have succeeded if it had not been for Kinsey's opposition. Whether Kinsey was afraid that such an organization might compete for funds with his own research institute, as some have said, or whether having studied the history of the European groups, he was afraid of having sex researchers involved in campaigning for sexual reforms, as others said, is unclear. Kinsey obviously had his own agenda, and an organization devoted to the study of sex was not a high priority with him. Although Kinsey was critical of the sexual hypocrisy in America and never hesitated to speak about it, he was convinced that the only way he and the institute could function was to avoid any sign of lobbying for any particular sexual cause.

Perhaps for this reason the impetus for the organization of sex professionals took place in New York rather than Indiana. In 1957, a New York group centered around Hans Lehfeldt, Henry Guze, Robert Sherwin, Hugo Beigel, and Albert Ellis began meeting together and planning for such a society. The initial group recruited others, and by 1960, forty-seven professionals joined together to form the Society for the Scientific Study of Sex (SSSS). Albert Ellis was elected the first president.

Ellis was known at that time for his books debunking sexual superstition and the folklore of sex.[100] He had also been a major defender of the first Kinsey report and had stated that the very broadness and comprehensiveness of coverage made "all previous sex surveys look feeble by comparison." Albert Ellis wrote that the Kinsey team had taken care to equip themselves with carefully planned and selected research techniques for a truly scientific adventure.[101] In spite of such support, Kinsey had become critical of Ellis because Ellis had also pointed out what he felt were inadequacies in the book.

Kinsey did not take criticism lightly, and though he did not usually publicly attack his critics, he remembered almost every detail of any negative criticism, even if the overall review had been positive. Undoubtedly, this tendency of Kinsey to remember his critics might have been an additional reason why he opposed any organization of sexologists during his lifetime. One of the first things that the new society did was establish the *Journal of Sex Research*, which by its very existence gave an opportunity for serious scholarly and scientific studies to be published on a wider scale than ever before.

The need to disseminate much of the new information about sex to pub-

lic attention and into the school curriculum was the motivation for the organization of SIECUS. The impetus for SIECUS came from a conversation
between Lester Kirkendall, a long-time family life educator, and Calderone,
then medical director of Planned Parenthood, that took place at a meeting of
the SSSS. Calderone had become discouraged in her efforts to get Planned
Parenthood to enter the field of public sex education. Guttmacher had consistently opposed her on the grounds that it would take away from the
already limited resources of Planned Parenthood that were devoted to
research. Guttmacher also strongly opposed public education programs,
because he felt that any stand taken by Planned Parenthood on any sex issue
would unnecessarily weaken it. Contraceptive planning itself, he held,
remained controversial enough without the organization seeking other
causes.

In their discussion, Kirkendall and Calderone laid the groundwork for
what became SIECUS, which was formally organized in 1964. Calderone
soon resigned from her position with Planned Parenthood and became exec-
utive director of the new organization. SIECUS essentially adopted a public
health approach to sex, focusing on awareness and education; it soon assumed
a leading role in introducing sex education programs in the schools. By the
late 1960s, SIECUS had become the target of virulent attacks from sex education opponents such as the John Birch Society and the Christian Crusade,
and among other things, it was accused of being a tool of the Communist
Party.[102]

The American Association of Sex Educators and Counselors (AASEC)
had been founded in 1967 by Pat Schiller to professionalize the teaching of
sex. Schiller was quick to realize that the developing field of sex therapy
lacked organization. She had the name of her organization changed to
the American Association of Sex Educators, Counselors, and Therapists
(AASECT) and set out to have AASECT become the regulatory arm of sexology.

The efforts of AASECT to control sex therapy was strongly assisted by
Masters and Johnson, who felt the whole field of sex therapy was being taken
over by quacks and that less than one in seven hundred of those who claimed
to be sex therapists could be considered legitimate.[103] The attempt at such
organizational control essentially failed, in part because Masters had in mind
the same kind of regulatory mechanisms that existed in medicine. Neglected
in his calculation was the fact that organizational control of medicine was
backed up by state regulatory agencies. Those flocking to become sex therapists, on the other hand, came from a variety of professions, many of them
certified either by their own professional associations or by state regulating
bodies. They saw no need to pay for another credential from AASECT, par-

ticularly when many of the major figures in the field refused to join AASECT just to be certified.

In attempting to dominate the field, AASECT had become both a membership organization and a credentialing body, and sometimes the one function interfered with the other. The credentialing process also meant that AASECT as an organization grew from the top down, with comparatively little participation of its members who generally joined to get their credentials not because they believed in the organization. Though AASECT had a far richer budget for a time than other sex organizations—in large part because of the number of individuals who did seek accreditation and the high fees they had to pay—the money was not used to develop new programs but to run lavish conventions and give large expense accounts to officers and staff. Most of the other sex organizations relied more on volunteers.

AASECT did, however, begin publishing the *Journal of Sex Education and Therapy*, which though not very scholarly at first, has, since 1990, become an important publishing vehicle in the field. AASECT also seemed to have been more riven by internal fights during its early history than the other organizations, although as it has grown into adulthood it has become much more stable.

Predating the appearance of the AASECT journal was another much stronger one, *Archives of Sexual Behavior*, which was edited by Richard Green. It became the official journal of the International Academy of Sex Research. This small group recruits members by invitation only. Other organizations also formed, including the Society for Sex Therapy and Research (SSTAR), American Academy of Sexology, and The Association of Sexologists (TAOS). As of this writing, a reorganized American Academy known as The American Board of Sexology is challenging AASECT in the total number of individuals it is certifying as sex therapists.

There was also a rapid growth of specialty journals devoted to various aspects of sex research, including *Psychology and Human Sexuality* and the *Journal of the History of Human Sexuality*. Some of these concentrate on special areas of sexuality, such as the *Journal of Homosexuality*, which began publishing in the 1970s, or the more controversial *Paidicka*, which examines adult-juvenile sexual interaction.

SUMMARY

Kinsey and Masters and Johnson were major factors in changing American ideas about sex, at least as much as any research findings can change attitudes. Kinsey certainly ran into more public criticism and critical opposition in the field than Masters and Johnson did. In retrospect, some of this might well have been due to Kinsey's way of presenting data and his own tendency not to suffer fools gladly. Masters and Johnson were much more cautious in their public statements and apparently had a policy of avoiding controversy whenever possible. They accepted the world as they found it and tried to prescribe therapeutic measures to help. Regardless of the difference, however, both the Kinsey group and Masters and Johnson played significant roles in the development of sexology in the United States. Because of the growth of the rival organizations and the entrance of a variety of professionals into sex research, however, no one or two individuals have since dominated the field in way Kinsey, Masters, and Johnson did in their most active years or as Ellis, Hirschfeld, and others had earlier.

In sum, Kinsey's studies challenged traditional ideas, and the development of new contraceptives gave an impetus for change. Masters and Johnson's research set the scene for a new generation of sexologists, who included educators, therapists, and researchers. Some of the results of this new interdisciplinary approach to sex are recounted in the next chapter.

8

THE MATTER OF GENDER

MASCULINITY, FEMININITY, AND CROSS-GENDER BEHAVIOR

*T*he interdisciplinary nature of the emerging field of sexology in the years after Kinsey and Masters and Johnson is best exemplified by the research on what might be called gender issues. *Gender* is an old term that has been widely used in linguistic discourse to designate whether nouns are masculine, feminine, or neuter. It was not normally used either in the language of the social sciences or sexology until John Money adopted the term in 1955 to serve as an umbrella concept to distinguish femininity, or womanliness, and masculinity, or manliness, from biological sex (male or female). In a sense, by using a new term to describe a variety of phenomena, Money opened up a whole new field of research. It was, however, a field ripe for exploration, since it appealed to the increasingly powerful feminist movement, which was concerned with overcoming the biology-is-destiny arguments that had been so long used to keep women in a subordinate status.

A history of any topic poses tremendous challenges both to the reader and to the author as it approaches the contemporary scene, and the history of sex research poses special difficulties. This is because since the 1960s there has been almost a geometric expansion of research into sex (and gender), with the number of articles and books almost doubling every decade. At the same time, I have been deeply involved in some of this research, and I know personally many of the individuals who appear in these last chapters. Inevitably, some individuals who have contributed to the field are not mentioned and not all kinds of research have received equal attention. Such qualifying statements are necessary in any discussion of gender research, which has attracted the attention of numerous individuals interested in bringing about change and many who hope to find scientific justifications for changes that have already occurred. Other researchers are striving to preserve the status quo. To include all the modern issues would necessitate a book in itself. The reader who wants to know more about the development of a particular avenue of research should delve into the endnote references, as what is presented here is an overview of this rapidly expanding discipline.

JOHN MONEY

Money, a psychologist, has been one of the key figures in sex research since the 1960s, and his term *gender* soon took on totally different meanings than he had originally visualized. Much of what appears in this chapter has been influenced by Money's concepts, but many other individuals from a variety of disciplines have made significant contributions.

A native of New Zealand, Money emigrated to United States in 1947 to pursue his graduate studies, eventually ending up at Harvard, where he received his Ph.D. in 1952, with a doctoral dissertation on hermaphrodites.[1] During his study of hermaphrodites, he became intrigued as to how an individual could grow up to live as a woman without the female sex organs or a man without the male sex organs. Recognizing that the existing terminology of sexual discourse was not sufficient to describe what he was observing, he took over the term *gender* from linguistics. By the use of the word *gender*, he believed he could avoid being bogged down by the necessity of having continually to make qualifying statements such as "John has a male sex role except that his sex organs are not male and his genetic sex is female."[2]

Money subsequently differentiated gender identity and gender role, although these constituted an inseparable unity in his mind. The term *sex*, he felt, should be used with a qualifying criterion, as in genetic sex, hormonal sex, or external genitalia sex; whereas the word *gender* was more inclusive since it involved somatic and behavioral criteria, for example, whether one is masculine or feminine and how one conducts oneself personally and socially, and how one is regarded legally. Sex belongs more to reproductive biology than to social science, romance, and nurture, whereas gender belongs to both.[3] Thus he extended the definition of sex research into new areas and gave further authentication to the role of the social and behavior scientist.

As Money realized the value of his original insight, he continued to expand on it, developing such terms as *gender identity*, by which he meant the total perception of the individual about his or her own gender, including a basic personal identity as a man or woman, boy or girl. He also held that the term could be used to make personal judgments about the individual's level of conformity to the social norms of masculinity and femininity. Gender as it is perceived by others, he called *gender role*. The two concepts are tied together, because most people show their perceptions of themselves in their dress, manners, and activities. Clothing and body decoration are the major symbols of gender and allow people immediately to identify the gender role of others. There are, however, other symbols, including mannerisms, gait, occupational choice, and sexual orientation.

Most people are gender congruent, that is, their gender identity, gender role, and all of the symbolic manifestations of gender are harmonious and they will not have a cross-gender sexual orientation. There is, however, a minority who fail to conform in some way, perhaps somewhere between 10 and 15 percent of the population, depending on how one defines nonconformity. Because the definitions vary so much, the diagnostic categories long favored by the medical community simply could no longer be used. Because of the complexity and the newness of the problems of gender, gender-related issues dominated sex research during the 1970s and 1980s.

The label *nonconformist* for those people who do not fit neatly into the gender boxes of male and female implies that they deliberately violated the norms of society. Because this may not be the case, most sexual scientists use the term *cross-gender* to avoid this judgment. Cross-gender means the individuals feel that they do not fit neatly into either the male or female category or their behavior is not totally congruent with the rules and expectations for their sex in the society in which they live. To paraphrase Donald Mosher, another major figure in the exploration of gender: Society has tended to believe in a dimorphic sexual essentialism, that is males and females must and should display congruent erotic sex and congruent gender characteristics of their sex because their biological or God-given nature made it so.[4]

The major area in which a significant number of people depart from societal expectations is in sexual orientation, defined as the personal view of the sexual attractiveness of other persons, including the most basic question as to whether one is sexually attracted to persons of the opposite or same sex (or both) as well as the details of the individual's sexual turn on. The second most common cross-gender behavior is in the area of symbolic expression clothing (including jewelry, tattoos, and other adornments). A smaller group of cross-gendered people seek total identity as a member of the opposite sex.

Though research into homosexuality began in the nineteenth century, the study of the other variants of cross-gender behavior were pioneered by Hirschfeld and Havelock Ellis in the twentieth century. It was not until the 1950s and 1960s, however, that the complexity of the issue became apparent.

Several strains of research seem to have been important. One was the genetic or biological research itself, a second was the study of hermaphrodites, the third was the mass publicity given to the sex change surgery of Christine Jorgensen, and a fourth was the growth of research into cross-dressing. Overshadowing all of these, however, is the change in the conceptualization of gender issues, the change in consciousness as a result of the rise of a new wave of feminist research, and the challenges to traditional attitudes stemming from the research into gay and lesbian issues. Moreover, both women and homosexuals have well-organized constituencies; thus research into gender has also had increasingly political implications.

BIOLOGY

CHROMOSOMES

The development of hormone studies and a better understanding of human genetics were major factors in launching gender studies. Each new discovery made the issue of gender behavior appear more and more complex. When chromosomes were discovered and the science of genetics was founded, scientists assumed that sex was established by the sex-determining chromosomes called X and Y. Two X chromosomes (designated as XX) in the fertilized egg leads to the birth of a female and an X chromosome plus a Y chromosome (XY) leads to the birth of a male. In a sense this is true, but there is far more to sex differentiation than that. Further discoveries were handicapped by the widespread belief that humans had forty-eight chromosomes, when actually there are normally forty-six, something not known until 1956.[5]

Even before the number of chromosomes was confirmed to be forty-six, other combinations of X and Y chromosomes were being discovered and the number of known variations increased after 1956. One result was the realization that there were several viable genetic possibilities, not only XX and XY but also X, XXX, XXY, and XYY. There is also a condition known as mosaicism in which some cells in a given individual have either an extra or a missing chromosome. In terms of the sex chromosomes, embryos with only a single Y and no X chromosome do not live. Embryos with a single Y chromosome and at least one X will become males, unless other biochemical factors intervene to inhibit masculinity.

Individuals with a single X chromosome and no Y have a condition known as Turner's syndrome (designated as 45,X). Such a person is characterized by a female body type, but the ovaries are either nonfunctional or have degenerated entirely. This ovarian deficiency prevents the child from developing normally at puberty and also has a growth-hindering effect—these girls seldom reach a height of five feet. Many people who have Turner's syndrome also have other congenital organ defects, coupled with a strong likelihood of intellectual disabilities. Fortunately, if the condition is diagnosed early enough, the individual can be helped by the administration of estrogen, the female sex hormone, although even after treatment, she remains somewhat shorter than average and is sterile.[6] The embryo with three X chromosomes (XXX) develops into a normal female body type, although her fertility may be diminished, and there is a possibility of mental retardation.

Embryos with a chromosome pattern of XXY have what is known as Klinefelter's syndrome. These individuals have penises that are usually small, and adults have shrunken testes. The output of testicular androgen is so low that female breast formation occurs, although there are individual variations.[7] The individual with one X chromosome and two Y chormosomes is male, is abnormally tall, is usually sterile, and often has some genital anomalies. In the 1970s, it was believed that certain behavior disorders might be more apt to occur with this genetic makeup, but the evidence for this remains controversial.[8]

As of this writing it seems clear that only part of the Y chromosome carries the sex differentiating factors, although the full implications of this are not yet clear. It does mean, however, that the existence of even a partial Y chromosome, if it is the crucial part, leads to the development of a male. In 1993, it was also found that the X chromosome of 64 percent of homosexuals in the study had a distinctive marker (which could be seen).[9] Whether this indicates that there is a genetic component to homosexuality remains unclear. What is still needed is much more research to determine the genetic influence on gender roles.

HORMONES

This lack of definite clarity is in part because even when an embryo has a normal chromosomal pattern, other things can occur that influence sex development in utero. If the child is to be a male, the testes begin to develop in about the sixth week; if the child is to be a female, the ovaries do not begin to develop until about the twelfth week. Experiments on animals have shown that if the embryonic gonads are removed before the sexual anatomy is formed the embryo differentiates as a female, regardless of its chromosomal makeup. This seems to imply that to be masculine it is necessary to add something, or to put it another way, the normal development of the human fetus is toward the female.

Once the testes begin to develop, they secrete a substance (called the müllerian-inhibiting substance) that suppresses all further development of the embryonic müllerian ducts, which is the part of the embryo that becomes the uterus, fallopian tubes, and upper segment of the vagina. If for some reason this substance fails to be secreted in a genetically male embryo, the boy is born with a uterus and fallopian tubes in addition to normal internal and external male organs. The male organs are normal except for cryptorchidism (undescended testicles).[10]

In addition to secreting the müllerian-inhibiting substance, the testes release testosterone, the male sex hormone, which promotes the proliferation of the wolffian ducts, which become the internal male reproductive structures. Via the bloodstream, the testosterone reaches the embryonic sexual organs and affects their development. The genital tubercle becomes a penis instead of a clitoris, and the skin folds wrap around the penis on both sides of the genital slit and fuse in the midline, forming the urethral tube and foreskin, instead of the bilateral labia minora and the clitoral hood of the female. The outer swelling on either side of the genital slit fuse in the midline to form the scrotum, which receives the testes, instead of remaining in place to become the bilateral labia majora of the female.

If testosterone is added to the bloodstream of a genetically female fetus during a critical period in development, a girl will be born with either a grossly enlarged clitoris or, in rare instances, a normal-looking penis with an empty scrotum. In human beings, such masculinization occurs in the fetus through an abnormal function of the adrenal cortex. It may also occur if the pregnant mother has a tumor or other condition that causes her to produce excessive male hormones. This condition is generally rare, but relatively recently the number of cases increased because the synthetic pregnancy-saving progestinic and other hormones, which were taken by pregnant women, led to the masculinization of female fetuses.

Embryologically, the external organs are the last stage of sexual development to be completed. This means it is not uncommon for the external genitalia to be left unfinished, neither fully masculine nor feminine. Because the undeveloped genitalia of the newborn are difficult to differentiate,[11] historically, this condition has often caused difficulties with infant sexual identification. In countries and cultures that put excessively high value on having a male child, doubtful cases tended to be assigned to the male category, and this assignment caused considerable trauma in many individuals when they reached puberty. Even today, in countries such as Egypt, where a male child is highly prized, a number of "boys" have experienced a great shock when they begin to menstruate.[12]

HERMAPHRODITISM

Errors in development of the fetus create only the most obvious problems in gender identity, leaving many questions about the variations in gender and sexual orientation unanswered. As indicated above, it was Money who was a leader in exploring many of these aspects of human nature. After receiving his Ph.D. from Harvard, Money went to Johns Hopkins University Hospital to join Lawson Wilkins, the founder of pediatric endocrinology. There he teamed up with other researchers to study children (and later adults) who presented with hermaphroditism or development anomalies. They used what has been called "experiments of nature" to gain information regarding a wide range of developmental problems.

Money soon was able to found a small research clinic unit, that was supported by grants from the National Institutes of Health (NIH) and private foundations for some thirty-five years. As his reputation grew, patients were attracted to the clinic from all over the United States. Thus he was able to acquire a rich clinical experience and was able to study in depth groups of individuals with rare conditions. Before his clinic, such conditions would have been reported only as isolated cases, if they had been reported at all. In the process, Money and others were able to explain many of the factors involved in hermaphroditism.

One of the conditions he studied in detail was adrenogenital syndrome, which produces precocious puberty in boys and varying degrees of pseudo-hermaphroditism and virilization in girls. When Money began his work, many of his patients were women who had grown up exposed to high levels of androgen throughout childhood. This allowed him to examine the influence of such hormones on development, something that is less likely to occur today in the developed world, in part because of his work, which allows for medical intervention at younger ages when diagnosis is now possible.[13]

Also important were his studies of androgen insensitivity syndrome, or testicular feminization.[14] Androgen insensitivity syndrome is the result of a sex-linked deficiency of androgen receptors, a condition that makes the subject insensitive to androgens in the genital area. It only occurs in genetic males and results in sexual differentiation along female lines, except for the crucial effects of the müllerian-inhibiting factor. The end result is apparently normal female external genitalia, but internally there are intraabdominal testes, a lack of a uterus, and a short and blind vagina. The child usually is reared as a girl and develops breasts, but she never begins to menstruate, which usually leads her or her parents to seek medical advice. This, in turn—at least in recent years—leads to the discovery of her abnormal condition. In spite of their lack of ovaries and uteri, these XY women develop as normal females in their capacity for sexual response and sexual desire.[15]

One of Money's early collaborative studies was with J. G. Hampson and J. L. Hampson, a husband and wife team at Johns Hopkins. They examined seventy-six hermaphrodites, nineteen of whom had been assigned a sex that was not congruent with their chromosomal sex. In spite of this error, all of them had developed a gender role and sexual orientation that was consistent with their sex of rearing. An additional twenty persons had exhibited some sort of contradiction between their gonadal sex and their assigned sex, yet all of these people also conformed to their sex of rearing.

Less conforming to their sex of rearing were the twenty-seven persons who had hormonal stimulation that was not synchronized with their sex assignment. Four of these people (three women and one man) changed their sex to conform to their hormonal patterns. Twenty-three members of the group developed external genitalia that were at odds with their sex assignment and were grateful to have plastic surgery to make their genital sex look more plausible, but they did not change their sex from their sex of rearing.

Because only four of the seventy-six patients (as some individuals fit into two categories, the numbers do not add up) changed their sex from that assigned at birth, the authors concluded that the assigned sex and the sex of rearing were more accurate prognosticators of gender role than any biological factor. They argued that gender imprinting starts early in infancy and reaches a critical period by about eighteen months; by the age of thirty months, gender role is established and change after that time is difficult for the individual.[16] In simple terms, if a child is early acclimated to one sex for the first two to three years of its life, it will identify with that sex, or gender, in later life.

TRANSSEXUALISM

Matters are never so simple as they seem, and they became more complicated with the case of Christine Jorgensen, who though biologically a male and raised as a boy decided that he wanted to be a female. Though both transvestism and transsexualism had been noted and studied by Hirschfeld and Ellis, this kind of cross-gender behavior had then been more or less ignored by those studying sexual behavior until the Jorgensen case again brought it to public and scholarly attention.

In 1950, a young man complaining of severe depression and asserting a conviction that he could not continue life as a man consulted with Copenhagen surgeon Christian Hamburger.[17] George Jorgensen told Hamburger that he had acquired a set of female clothes, secretly put them on, and shaved his pubic hair to shape it like female's. His work as a laboratory technician in New York City had given him access to estrogen, which he had for a time administered to himself. Hamburger and associates, after conducting a thorough physical and psychiatric examination, decided to treat Jorgensen with additional female hormones (parenteral estrogen) in doses much larger than he had given himself.

The large dosages changed the shape of Jorgensen's body to a more feminine contour and his behavior, gait, and appearance also became more feminine. As his beard became sparser, electrolysis was used to remove the remaining hair. Hamburger and colleagues, using some of Hirschfeld's concepts, diagnosed the case as one of genuine transvestism and differentiated it from two other types of transvestism: the fetishist who, as a consequence of a neurotic obsession, concentrates on one or more articles of dress and thus develops an interest in cross-dressing and the homosexual man of the passive type who desires to dress in women's clothing.

After hormone therapy, Jorgensen, who later took on the name of Christine (in honor of Hamburger), was then castrated under provisions of the Danish Sterilization and Castration Act of 1924, which had permitted castration when the patient's sexuality made him likely to commit crimes or when it involved mental disturbances to a considerable degree.[18] In 1952, when Jorgensen continued to express an ardent wish to have the last visible remains of his detested male sex organs removed, his penis was amputated, just one year after he had undergone orchiectomy. No attempt was made to construct a vagina or make his urethra opening appear more female, although Jorgensen underwent such surgery later. It was in this later surgery that a vagina was constructed out of intestinal tissue for Jorgensen, but this operation was done in the United States.

In December 1952, news of the sex change reached the media, and in a few short weeks, Christine Jorgensen became world famous. Seizing the opportunity, Jorgensen sold her story to journalists from the Hearst newspapers. And on her return to United States, she went on the stage. Much later, she also wrote her autobiography.[19] Though Jorgensen was not the first person to attempt a sex change,[20] the media publicity resulted in a flood of letters to Hamburger and associates from people requesting a sex change.[21] The surgical change also created a further turf battle in the field of sex research, and much of the medical controversy over Jorgensen was set off by this conflict.

As indicated earlier, psychiatry and psychiatrists had claimed much of sex research as their own specialty, especially in the United States, although gynecologists and obstetricians were recognized as also having an interest because of the contraceptive issue. Psychologists, sociologists, anthropologists, and historians during the 1930s and 1940s had begun to conduct investigations into areas of human sexuality from their own disciplinary interest, and Kinsey, whatever else he had done, had mounted a full-scale attack on the psychiatric domination. Transsexualism now brought surgeons and endocrinologists into an area that psychiatrists had long claimed as their own, and to complicate matters, these professionals were making decisions about sex change without even consulting the psychiatrists. The psychiatric reaction at such an incursion was almost immediate, some of it perhaps justifiable, and some of it an effort to preserve turf.

The New York psychiatrist G. H. Wiedeman was one of the first in the field. He criticized Hamburger for not doing a more thorough psychological workup and for not treating the patient with psychotherapy instead of surgery. He concluded that the "difficulty of getting the patient into psychiatric treatment should not lead us to compliance with the patient's demands, which are based on his sexual perversion."[22] Other psychiatrists felt, as did Mortimer Ostow, that the real treatment for the Jorgensen case should have been "intensive, prolonged, classic psychotherapy."[23] Gradually, a kind of truce emerged, at least in the more formalized sex and gender clinics whereby a psychiatric examination was usually required before surgical intervention. Though not all surgeons, particularly those outside of the formalized clinics, followed this procedure, some sort of consultation was usually solicited to cut down on the possibility of a dissatisfied patient bringing a lawsuit.

To categorize such individuals, the label *transsexualism* was soon applied, and the operation came to be called sex reassignment surgery (SRS). The term *transsexualism* had been coined by Hirschfeld, who described one of his patients as a psychic transsexual, but the term was not used again until 1949 when David O. Cauldwell described a case of "psychopathia transexualism" in which a girl wanted to be a boy.[24] Others had picked up the term,[25] but it entered the scholarly and scientific literature through the work of Harry Benjamin, who in 1963 reported that he had been using the term for about ten years to describe the phenomenon in some of his patients.[26]

In explaining the term *transsexualism*, Benjamin created three categories of cross-gender types. The first group included transvestites who led reasonably normal lives. Most of them were heterosexual men who could appease their feelings of gender role disharmony by cross-dressing. They derived erotic satisfaction from cross-dressing, but this might decrease over time. Though these patients sometimes displayed neurotic symptoms, they were seldom seen by physicians. Their clash was with society and the law. Transvestites in the second group were more emotionally disturbed and required psychological guidance and endocrine therapy. The third type of transvestism was essentially transsexualism. This type, he said, represented a disturbance of the normal sex and gender role orientation. The individual wanted to be a full-fledged woman and have a male sex partner, or vice versa. The condition could be present as fully developed transsexualism, or it could gradually appear after short or long periods of cross-dressing. However, the male transsexual was much less interested in the symbol of female attire. He wanted to be a woman and function as a woman. Transsexuals, Benjamin noted, were often very unhappy people.[27]

By the time he collected his material into a book, Benjamin had seen 200 cross-dressers, 125 of whom (108 men and 17 women) he diagnosed as being transsexuals. Most patients described the onset of their condition as being "as early as I can remember." Benjamin remained puzzled by the etiology. A moderate sexual underdevelopment was present in 30 percent of the patients (for example, a small penis in men or lack of breasts in women), as had been the case with Jorgensen, but the chromatin tests all conformed to the anatomic sex. Some observers of transsexualism had speculated that the phenomenon was due to unfavorable conditioning, and though Benjamin found possible evidence for such an explanation in 21 percent of his cases, he could identify no causal sequence for at least 50 percent of the cases. This led him to speculate about such factors as imprinting, which is common among animals. Benjamin wondered if imprinting somehow took place among humans and that transsexual patients had been imprinted with the wrong gender. Finally, he speculated about prenatal exposure to hormones. What led to this hypothesis was the fact that 28 of the first 91 patients he had examined exhibited hypogonadism, even though he could detect no current hormonal abnormality.[28]

Benjamin, who was not a psychiatrist, held that the administration of hormones was an important part of the management of transsexuals. He prescribed estrogen for male-to-female (m/f) cases and androgen for female-to-male (f/m) cases before surgical intervention could be recommended. He concluded that since the mind cannot be adjusted to the body the adaptation of the body to the mind seemed to be not only indicated but recommended.

To further study the issue, Money and others joined together to establish the Johns Hopkins Gender Identity Clinic in 1965, and the first sex reassignment surgery was performed there in 1966. Soon afterward, gender identity programs were established at the University of Minnesota, Stanford University, the University of Oregon, and Case Western Reserve University, all of which included teams of psychiatrists, psychologists, and surgeons to diagnose and treat transsexuals.[29]

Not all would-be transsexuals went to such centers, however, and many freestanding surgeons moved into the field. For a time the most skillful surgeon was George Burou of Casablanca in Morocco. He is credited with the development of the surgical technique for SRS on m/f clients that is now almost universally used. This technique involves using penile tissue to create a functional vaginal wall. Before his innovation, attempts to create a vagina using intestines or other tissue (as in Jorgensen's surgery) had not been particularly successful, and many of the early m/f subjects went without this aspect of the surgery.[30] Until his retirement in 1992, the major specialist in m/f surgery was Stanley Biber, a Colorado surgeon who had done SRS for more than two thousand clients. He did a much less elaborate workup than the gender identity centers and struggled to keep his fees reasonable.[31]

Surgery for f/m cases is not as well developed as that for m/f cases, and it is far more complicated because it involves removal of breasts, uterus, and ovaries. The scrotum is constructed from the labia while the penis is constructed from various kinds of skin grafts. Fully functioning surgically constructed penises that can be used for both urination and sexual intercourse have not yet proved to be a possibility. Often prostheses are used that fit into a vagina. In most female-to-male surgery both the vagina and the clitoris are left intact.

Transsexualism, perhaps because its solution remains primarily in medical hands, has created a growing medical literature. The phenomenon became a recognized diagnostic category in the 1980 edition of the American Psychiatric Association's *Diagnostic and Statistical Manual*.[32] Such recognition in this case was important and necessary, because it could then be officially treated with surgical intervention. Transsexuals, however, represent a comparatively small minority of those individuals who are involved in cross-gender behavior. By far the most numerous are the transvestites.

TRANSVESTITES

Transvestites pose a different set of issues, which sexologists are just beginning to face, because many members of this cross-gendered group live traditional lives as men but also spend part of their time living in the gender role of women. Interestingly, they are almost all men, and in terms of partner preference, most of those in the organized groups are heterosexual. What makes the issue more complicated is that transvestism seems for the most part to be a fairly recent phenomenon.

Bonnie Bullough and I have used a combination of disciplines, including history and sociology, to demonstrate that though cross-dressing has been widespread in the past, most of those who did it for any length of time were women trying to escape from the role limitations put on them. This was because the masculine role gave not only higher status but greater freedom. Though there were occasional instances of males cross-dressing to serve as prostitutes and of homosexuals playing at being feminine, the heterosexual male cross-dresser seems to be mostly a twentieth-century phenomenon, and one that has escalated rapidly since 1960. We speculate that this growth is the result of the changing nature and greater attractiveness of the feminine role in society. Furthermore, we hypothesize that the male role itself is being visualized by many men as more confining than it was in the past.[33]

Before 1960, individuals who cross-dressed existed more or less in isolation. A few sought psychiatric help, by which they were usually lumped with homosexuals.[34] Changing the pattern of cross-dressing was a man who adopted the name of Virginia Prince and who in 1960 founded a magazine and, soon after, an organization dedicated to heterosexual cross-dressing. By 1970, she had decided to live full-time as a woman and the pronoun used to identify Prince is a feminine one, since this is how "she" lives. Prince took as her mission the bringing together of previously isolated heterosexual cross-dressers. She was also interested in educating the professional sexological community about the phenomenon. One result was the appearance and growth of organized transvestism, which allowed scholars to investigate this gender behavior on a scale never before possible.[35]

In the process, she further emphasized the complexity of gender manifestation by emphasizing how a socialization process, which she introduced into organized transvestism, can both shape and reinforce the decision of individuals to depart from their assigned gender roles. One of the major functions of an organization for stigmatized individuals such as transvestites is to furnish a subculture for people with common problems to meet and share experiences.[36] Inevitably, such an organization of cross-dressers not only allowed but defined a kind of group behavior. Such groups not only help give solace and end the isolation faced by these individuals but, in the process, set new definitions of appropriate behavior. A social constructionist would describe this as "the process by which people come to describe, explain, or otherwise account for the world (including themselves) in which they live." In short, such a view attempts to articulate common forms of understanding.[37]

Other groups of cross-dressers groups soon sprang up, challenging the kind of characteristics that Prince herself had defined as necessary to being a transvestite. It seems clear that the organizations that Prince, her followers, and her rivals established furnished transvestites with a sexual and *gender script*, a term coined by John H. Gagnon and William Simon, two sociologically oriented sex researchers. They emphasized that sexual activities were social activities guided by what they called scripts.[38] They emphasized that though there was a public discourse about sex, not all individuals opted for the same sexual categories, since there were a variety of alternatives. Using this term, we can see that what Prince's transvestite script did was give cross-dressers a group identity that shaped their own self-concepts. As new groups came into being, they sometimes challenged the Prince interpretation, and what had been seen as a fairly unitary kind of behavior has become ever more complicated, with a variety of scripts.

Male cross-dressers especially needed the support of a group, because by dressing and acting as women they lost status, at least in the eyes of their male colleagues. Despite the growing attractiveness of the feminine role in society, males and masculine values remain dominant. It was the perceived desirability of the feminine role, coupled with the eroticism of doing the forbidden (which gives the cross-dresser a high) and the accessibility of women's clothes (which could be purchased through the mail) that tended to encourage experiments in cross-dressing.

This development would seem to emphasize a social constructionist perspective about human sexuality over the biological one, an issue that finds sexologists on both sides of as well as straddling the fence. Sex for the social constructionist is an ever-changing psychosocial construct. The roots of such a perception about sexual discourse are derived from symbolic interactionism, a theory that describes the development of symbolic social worlds that allow people to interact with each other.[39] In examining any aspect of human sexual behavior, interactionists see their primary task as describing the process by which sexual meanings are constructed.[40] In terms of cross-dressing, the act itself is not nearly as significant as the social meanings the participants attribute to the act and the resulting interactions. This significance is created through public discourse, which not only imparts meaning but establishes categories that individuals use to organize and classify their world and themselves.[41]

One of the cornerstones of cognitive psychology, which also has been useful for some sexologists, is the belief that each individual will have a different set of experiences and cognitive structures that will determine how any new information, such as that about human sexuality, is internally processed and ultimately used. If this is the case then the mind of each individual gives meaning to sexual and self-concepts through the complex interaction of external discourse and social relations with the existing power structure.[42]

Self-concept involves both a personal identity acquisition and a social one. Both terms need definition. A social identity in this case is defined as the individual's knowledge that he or she belongs to certain social groups and membership in such groups has emotional and value significance for him or her. Social groups can be based on sex, nationality, religion, or any number of other categories. On the other hand, a personal identity refers to specific attributes of the individual such as feelings of competence, bodily attributes, ways of relating to others, psychological characteristics, personal tastes, and so on. Although personal and social identity usually function side by side as the self-concept, it is also possible that the social identity can on occasion function nearly to the exclusion of personal identity, particularly among groups who experience discrimination, such as cross-dressers do.[43]

GENDER IDENTITY

While transvestites and transsexuals pose different problems to the researcher than do hermaphrodites, or for that matter homosexuals, each seems to be closely associated with challenges to traditional notions of gender roles. Another way of getting at the problem is to examine cases of babies assigned to one sex at birth and subsequently reassigned to the other sex later on. Money developed the concept of *critical periods*, periods at various stages in development from the embryo through early childhood during which crucial gender decisions could be influenced. He eventually came to hold that in terms of gender self-identification the critical period in the human child begins when it reaches the age of eighteen months, if not sooner, after which any change or reannouncement of sex would prove difficult.

"By the ages of twelve to eighteen months, if not earlier, parents have so completely habituated themselves to having a daughter or a son, both in their relationship to the baby and to their relatives and friends, that they need an expert's counseling and guidance if a sex reassignment should prove desirable or obligatory."[44] This statement does not mean that Money failed to recognize biological factors, but that he felt both biological and social psychological factors were involved in gender behavior. He called this an interactionist perspective, something that most sex researchers now subscribe to, but over which there is considerable disagreement as to which factors are most influential. The developmental critical periods for the fetus are less controversial than the critical periods in the life of the infant, and even those researchers who accept the existence of such a concept are not certain when it occurs and what its limits are.

For this reason, one of Money's most famous early reports became quite controversial. This was the case history of what happened when two identical twin boys were circumcised at the age of seven months. The physician used an electric cautery gun of the wrong type to do the circumcision and during the brief procedure of circumcising one twin an accident occurred, burning the penis so badly that it soon necrosed and was sloughed off, leaving nothing but a urethral opening and the scrotum. The parents sued for damages, of course, but the real issue for them was what they should do to help their child grow up as normally as possible. It was deemed impossible to construct any kind of satisfactory penis then or after puberty, which would be fully functional. One of the surgeons consulted by the family recommended that the child be reared as a girl and to start this process the parents agreed to have the scrotum and testes removed.

The child by that time was seventeen months old. Its name was changed to that of a girl and the child was dressed in girl's clothing and given a girl's hairstyle; the process of being culturally feminized was begun. Construction of an artificial vagina was to be delayed until postpuberty, at which time female hormones would also be given. It was after all of this had happened that the child was brought in contact with Money, who supported the parents in socializing the child into the feminine role, the ultimate outcome to be judged on the criterion of romantic and erotic orientation in teenage and adulthood. There were early reports from Money that socialization as a girl was progressing well, but soon the case disappeared from the literature.[45]

NATURE VERSUS NURTURE

Even as Money was putting forth his ideas about the influence of sociopsychological factors—nurture if you will—during the critical periods, he was strongly criticized by Milton Diamond, another psychologist active in sex research. Diamond indicated that the gender decisions in the cases of hermaphrodites, from whom Money had originally drawn his data, were perhaps not as clear-cut as Money implied. Diamond hypothesized that an individual hermaphrodite might be receiving mixed biological signals, which allowed him or her to conform to the assigned gender rather than change it. Many such individuals, in fact, wanted surgery to make them better conform to their assigned gender. Diamond later hypothesized that it would be difficult to predict the behavior of the twin who was assigned to the female role, because unlike the hermaphrodite whose body was giving mixed signals, this child had only one such signal until its sex was changed at seventeen months. He charged that Money was deemphasizing biology, or nature, and overemphasizing nurture.[46] Later, rumors and popular reports came out about the child's failure to develop into the feminine person (including a special presentation on Canadian television), but now, more than twenty-five years after the case was first reported, no additional information is available.

The argument over nature versus nurture still continues, although both sides recognize the influence of both factors, and it remains an argument over degree. Both Money and Diamond were and are interactionists, that is the two of them emphasize the importance of both biology and social learning. What seems evident, and which is of both theoretical and practical importance, is the plasticity of gender identity and the number of variables at work.

At their scientific best, probably most biologists and social or behavioral scientists agree that the coding of gender is multivariate, sequential, and developmental, reflecting a complex interaction across the boundaries of disciplines and across so-called biological and social variables. Still, each discipline is tempted to bolster its status by claiming to reduce human complexity to its specialties. Sometimes it seems that the struggles over the medical model that existed at the turn of the twentieth century have simply reasserted themselves in the nature/nurture controversy.

Historians, such as myself, have found that every society produces cross-gendered individuals. There are a variety of roles and functions such people carry out in their particular society, and their stigma or place of honor also varies.[47] I have argued that the historical data furnish a powerful argument in support of biological influences on cross-gender identities. Frederick L. Whitam has arrived at a similar conclusion using cross-cultural data. After investigating homosexuality and cross-gender behavior in many different contemporary cultures, he noted, "Although all people in all societies with rare exceptions are socialized to be heterosexual, the predictable, universal appearance of homosexual persons, despite socialization into heterosexual patterns of behavior suggests not only that homosexual orientation is biologically based but that sexual orientation itself is also biologically derived."[48]

Richard C. Pillard, Jeannette Pomadere, and Ruth A. Carretta, in a comprehensive review of family studies, found substantial evidence that sexual orientation was familial.[49] Among the studies they examined was that of Henry (see chapter 6), who drew up elaborate family trees for his subjects, although no attempt had been made to check the subjects' claims about their homosexual or bisexual relatives.[50] Similar subjective data results were reported by Margolese and Janiger, who asked about homosexuality in the families of two groups of men who had been recruited for an endocrine study. Among the heterosexual men, two out of twenty-four reported a homosexual relative; among the homosexual men, seventeen out of twenty-eight reported homosexual relatives.[51] This might simply mean that homosexuals were more likely to recognize homosexual relatives than heterosexuals were or that the homosexuals were more likely to identify doubtful or unknown cases as homosexual, because again there was no evidence presented that other efforts were made to determine the sexual orientation of the relatives. Pillard and co-workers interviewed thirty-six homosexual men (the subjects were rated as either a 5 or 6 on Kinsey's 7-point scale) concerning the sexual orientation of their relatives. Among the thirty-five sisters of the men only one was homosexual, but among the forty-five brothers, most were homosexual.[52] Again, however, there was no attempt to interview the relatives.

In a study of twins, Franz J. Kallman found that in forty sets of monozygotic (identical) twins all were concordant for sexual orientation, while most of the dizygotic (fraternal) twins were discordant.[53] An ongoing twin study by Diamond, Whitam, and J. E. Dannemiller is examining thirty pairs of twins—eighteen monozygotic and twelve dizygotic. They have found there is a 72 percent concordance for homosexuality among the identical twins and a 33 to 50 percent concordance among the fraternal twins.[54] None of these studies is definitive, because the work is usually based on small samples and twins are usually raised in the same environment. Still, such studies are indicative of possible biological factors.

ENDOCRINOLOGY

But what biological factors are involved? Currently there is a vast and growing literature documenting a hormonal influence on gender behavior and the impact of prenatal hormones on the brain. The sex hormones are the messengers that cause reactions in the body that are interpreted as masculine and feminine. Their influence is not only important in children and adults but may be equally or even more important in the prenatal period.[55] As indicated above, without hormones from the testes, the embryo remains female.[56] As the roles of additional hormones have been identified, the matter becomes ever more complex.

Progesterone, for example, has been shown to counteract the effects of androgens and thereby protect the brain of female fetus from their masculinizing influence. It has also been demonstrated that androgens are converted to an estrogen compound to exert their decisive masculinizing effects on the target cells. The gonadal and adrenal hormonal responses are controlled by the pituitary gland, which is in turn controlled by the hypothalamus, which is part of the brain. The brain can itself be altered by the hormonal influences, most particularly during the early months of gestation when the neural tissues are developing.

The research that is particularly relevant here are Money's pioneering studies of children with congenital adrenal hyperplasia. Current practice is to treat females born with this syndrome with cortisone as soon as possible after the child is born. This has allowed researchers to examine the effect of excess androgen on the female fetus in utero and later treat the infant. Anke A. Ehrhardt and Heino Meyer-Bahlburg report that these girls differ significantly from siblings and other controls in that they enjoy rough-and-tumble play, associate with male peers, and are identified by self and others as tomboys. They show low interest in role rehearsal for the wife and mother roles. The boys who had extra male hormones before birth exhibited higher energy expenditure in sports and were somewhat more likely to initiate fights than their peers. The fact that the masculinizing influence remained after the hormonal stimulation ceased suggests that the neural pathways controlling masculinity and femininity have been affected.[57] Thus we not only have recognized the importance of hormones in gender behavior but now know that neural pathways are factors and that these too can be influenced by hormones.[58]

Gunter Dorner and colleagues in Germany studied rats and hamsters and identified two different mating centers in the hypothalamus: one for male behavior and one for female behavior. Sex hormones injected into the appropriate areas stimulates sexual behavior and destructive lesions inhibit such behavior. Genetic males that experienced a temporary androgen deficiency during the hypothalamic developmental period but who had normal androgen levels in adulthood were sexually aroused by same-sex animals. The higher the androgen level during the hypothalamic differentiation period, the stronger the male behavior and the weaker the female behavior, irrespective of genetic sex. For humans, Dorner believed that the critical hypothalamic differentiation phase may occur between the fourth and seven months of gestation. Dorner's experiments on humans have demonstrated different responses to injections of Premarin (estrogen) in homosexual and heterosexual men,[59] but replicating his results has been difficult. Still, the hypothalamus seems to be an important factor in sex behavior and Simon LeVay has tentatively identified differences between homosexual and heterosexual men in a portion of the anterior hypothalamus, although what the differences mean is unclear and whether they are widespread beyond his small sample is not known.[60]

Diamond holds that the growing physiological data are impressive. He has argued that prenatal hormones exert influence on neural pathways and the neural endocrine axis (the link between the hypothalamus, the pituitary gland, and the other endocrine glands). These neural pathways control future hormonal production and consequently influence sexual behavior. He has argued that there are separate neural pathways for sexual identity as a male or female, sexual object choice, sexual pattern for maleness or femaleness, and the sexual response pattern.[61] Certainly, some behavior patterns are related to genetic differences in males and females, and one of them is aggression. Janet Hyde conducted a metaanalysis of 143 studies of the genetic differences in males and females, which she summarized: "Out of the massive research literature on psychological gender differences, a few behaviors have emerged as showing reliable gender differences. One of those is aggression. The greater aggressiveness of males compared with females is generally regarded as a consistent and significant phenomenon.[62]

BEHAVIOR

The reason aggressiveness is so important in the understanding of gender is because this has given power to men, and particularly to the more aggressive men. It has established gender characteristics in women, including the relative servitude that has marked their history as the subordinated sex.[63] It is this physiological component of the male makeup that has led many feminists to argue, erroneously I think, that the whole concept of biological influences is a social construction, because so many norms are seen to be self-serving for men. Certainly, some aspects of hypermasculinity are socially scripted, as Mosher and others have shown,[64] but the behavior is based on what is observed and what we know, and testosterone is a factor.

If aggressiveness does turn out to be hormonally factored, and it seems that it is, it would be the male aggressiveness that has allowed males to call the shots. Aggression provides power, and the importance of this power to men helps to explain the threat some men feel when they see cross-gendered males who are gentle and nurturing, as such men are seen as a threat to continued male dominance. We also know from other data that male individuals kept in a subordinated position, as for example in slavery, certainly adopt many of the patterns of what has been regarded as feminine behavior.

Pillard and James Weinrich suggest that there should be further differentiation of the neural influences on gender behavior, because there are separate components of the nervous system that control masculinization and defeminization. They believe, following Dorner, that masculinizing and defeminizing of the hypothalamus probably occur at four to five months of gestation.[65] As of this writing, the masculinizing hormones are known but the defeminizing agent(s) are not yet identified, although it may be the müllerian-inhibiting substance. Some investigators have argued that the masculinizing and defeminizing agents are the same but that they operate at different critical periods. By differentiating between these two processes, Pillard and Weinrich argue that they can place all of the gender transpositions on a periodic table. Male-to-female transsexuals and certain passive lesbians are both defeminized and most masculinized. Most homosexual men are unmasculinized and defeminized. Transvestites are masculinized and not defeminized.[66]

This hypothesis pushes the concept that homosexuality, lesbianism, transsexualism, and transvestism are all influenced by biological factors. There is other scattered but accumulating evidence of physiological factors involved in cross-gender identities. Muriel Wilson Perkins studied 241 lesbian women and found that the members of her sample had narrower hips, increased arm and leg girths, less subcutaneous fat, and more muscle than a sample of 1,260 adult women measured between 1960 and 1968. She divided her sample into groups by the women's dominance in the sex act and noted that the dominant group was significantly taller than other lesbian women.[67]

A group of researchers have studied the metabolites of testosterone to see if the sexual orientation of the individual was related to the levels of these hormone derivatives. The liver reduces testosterone to androsterone and etiocholanolone, which are then secreted into the urine. These two substances are measured when urine is tested for 17-ketosteroid (most of the other compounds in this particular urine test come from the adrenal cortex). Margolese and Janiger found that the ratios of androsterone and etiocholanolone were different in male homosexuals than in heterosexuals, with a relatively low androsterone level being associated with a same-sex preference.[68] The problem with this study is that these same patterns also occur with stress and certain illnesses.

In 1983 Brian Gladue, Richard Green, and Ronald Hellman found an estrogen feedback mechanism related to sexual orientation. Men and women with lifelong heterosexual orientations and men with lifelong homosexual orientations were administered an estrogen preparation that is known to enhance the luteinizing hormone levels in women but not in men. The pattern shown by homosexual men was intermediate to that shown by the heterosexual women and men. Furthermore, testosterone was depressed for a significantly longer period in the homosexuals than in the heterosexual men.[69]

What such data seem to show is that there is strong evidence accumulating for the influence of biology on homosexuality and other cross-gender behavior, although the evidence does not eliminate social psychological, social, or cultural factors. Still, there has been no shortage of critics who have questioned various aspects of the work cited here. In 1980, Garfield Tourney reviewed sixteen studies linking hormones and homosexuality and suggested more carefully controlled studies were needed before definitive conclusions could be drawn.[70] Other critics have argued that research on biological factors overemphasizes the role of biology in predicting complex behavior and that it supports a medical or illness model of sexual variation.[71]

A comprehensive review of all of the literature suggesting biological determinants of sexual orientation was made by Louis Gooren, Eric Fliers, and Keith Courtney in 1990. Though they personally were skeptical of most of the biological data, they were unable to discount arguments that certain prenatal biological factors could facilitate a homosexual orientation later in life. They emphasize, however, that irrefutable evidence is presently lacking.[72] Readers should keep this caveat in mind.

SOCIOBIOLOGY

Obviously no gene or genes have been isolated that cause men and women to be cross-gendered. Moreover, if such a gene existed, biologists would have to ask what survival value it had for the human species that would have kept it in the gene pool throughout the evolution of human beings. Edward O. Wilson hypothesized a possible genetic predisposition for homosexuality (and cross-gender behavior) in certain humans by using a theory that he called *inclusive fitness*, defined as the sum of the individual's own reproductive success plus the reproductive success of others who carry that person's genes. He theorized that there are homosexual genes that exist not only in the individual who is homosexual but in his or her relatives. Homosexuals contributed to the survival of the family, because by not having children they were available to support and help the family altruistically and to serve in roles such as aunt, uncle, shaman, or medicine man. Thus genes for a homosexual orientation increased in frequency, not because they aided the homosexual person in his or her own survival but because they aided his or her relatives who shared the gene pool. Wilson called this gene the gene of altruism, because it enabled the species to better survive.[73]

Wilson's approach is called sociobiology, and when his first book on the topic was published in 1975, it was widely criticized by people at opposite ends of the continuum. People who were hostile to gays felt that the theory took away the blame that homosexuals somehow deserved, while friends of gays and gays themselves argued that it took away their freedom to be what they wanted to be.[74] There was also a real fear among many observers that if such a gene could be discovered, there would be a massive attempt to eliminate it through gender genetic engineering.

Researchers who have studied the behavior of nonhuman primates have found that homosexual behavior often takes place. In some monkey species, the social organization consists of dominate males who each exercise control over several females and a group of so-called bachelor males who are excluded from the females. These bachelors often live and forage with each other, engaging in homoerotic play, until one of them is able successfully to challenge a dominant male. In captive colonies of chimpanzees, lesbian behavior has also been noticed. Franz de Waal, who studied chimpanzees in the Netherlands, reported on a female who not only occasionally mounted other females when they were in estrus, thrusting as a male does, but also served as a kind of umpire when a number of males gathered around a sexually receptive female. This chimpanzee supported those females who refused the advances of a particular male.[75] Homosexual behavior has long been known to exist among monkeys as well,[76] and the existence of such behavior gives support to a sociobiological factor in cross-gender behavior. This, however, might not be as strong a biological argument as it appears, since those emphasizing the social factors in sexual orientation argue that in such primates as the chimpanzee and rhesus monkey, experience and social group life have substantial influences on the display of sex behavior and sexual activity.[77] Such studies, however, are often criticized because they rely on animals in captivity and not in the wild.

In terms of homosexual behavior, Frank Beach early pointed out that the term *homosexual* has been used in animal research in two different contexts: (1) as a description of individuals that exhibit a coital pattern typical of the opposite sex and (2) as a description of individuals that exhibit coital responses typical of their genetic sex but do so in response to a same-sex partner. He cautioned, however, that similarities in the appearance of behavior in different species are not necessarily manifestations of the same phenomena, and traditionally it is assumed that there is a mutually exclusive and inflexible separation of the copulatory motor patterns of males and females.[78] Again, many gays object to animal comparisons, because they feel homosexual behavior is more than mounting and insertion or intromission and the use of animal models to emphasize the inherent biological base of homosexuality denies components of sexuality such as eroticism, imagery, sexual fantasies, and pair-bonding.[79]

INTERACTION

Current research emphasizes that even though there are strong biological factors, gender identity is apparently the product of a complex interaction among three factors: (1) genetic predisposition; (2) physiology; and (3) the socialization process, which involves a complex of psychological, sociological, and cultural factors.[80] There are critical periods during which sociopsychological factors have greater influence than the others, but it is not always clear what agents are at work.

One of the more controversial reports was the discovery of a group of thirty-three individuals in the Dominican Republic who had been born with an extremely rare condition: 5α-reductase deficiency. This is an inborn inability to convert 5α-dihydrotestosterone, which results in an impaired masculinization of the external genitalia, so that such individuals are usually labeled female. At puberty, however, when testosterone kicks in from the testicles, such individuals experience an increase of muscle mass, the growth of the phallus and scrotum, and a deepening of the voice. Nineteen of the males had been raised as females up to the age of puberty, and of these information could be obtained on eighteen. Sixteen of the eighteen gradually adopted a masculine gender identity and erotic interest in women. One of the remaining two changed his gender identity but continued to dress and live as a woman, and the other married at sixteen and retained a feminine gender identity and role. This suggests that even though the individuals were raised as girls, the biological factors were ultimately dominant in the majority of cases, at least in the Dominican Republic.[81]

This last is an important caveat, because the role and status of women might well have been a determining factor in the decision making, and a different choice might have been made in a society in which the status of women was higher. Further investigation emphasized that the children were not reared as girls but as a locally known type of pseudohermaphrodite (*guevedoche*, or "penis at twelve"), because the phenomenon was not uncommon in the villages. Here, the social psychological attitudes of the villagers helped prepare the individuals for a possible change.[82] This need for support seems to have been supported by Gilbert Herdt and J. Davidson, who studied a comparable sample in New Guinea.[83]

One of the problems with such studies, however, is the potential bias of the observer. Gender differences, as emphasized, seem to occur early, and in the past, those who studied groups of children in elementary schools noted that girls tend to prefer to be with girls and boys with boys. In a 1993 study, however, Barrie Thorne reported that gender barriers are not what they seem even during this childhood period. She emphasized that most studies have looked on children as the next generation's adults and have relied on statistical norms, rather than looking at individual exceptions. Moreover, schools, the environment in which most children have been studied, encourage more gender segregation than other settings, such as neighborhoods or families. She found much more gender mixing and individual variation than previous studies have acknowledged.[84]

Such studies only emphasize the growing importance of interdisciplinary research into sex and gender issues. It is important to go beyond the traditional biological view, even beyond the growing social and behavioral sciences, and to include under the rubric of sex research those from the humanities and arts. Since the 1970s, scholars in the humanities have expressed a growing interest in ambivalence and ambiguity, influenced strongly by the writings of the French psychoanalyst Jacques Lacan.[85]

Marjorie Garber's work is perhaps the best example of this. She attempted to escape the bipolar notions of male and female and advocated a third category, not a third sex as it was conceptualized by Ulrichs, but rather a mode of articulation, a way of describing a space of possibilities. She regarded transsexualism as a distinctly twentieth-century manifestation of cross-dressing and the bipolar thinking of which cross-dressing itself is a tradition. Bipolar approaches create what she calls a "category crisis," a failure of definitional distinction, resulting in a border that becomes permeable and permits border crossing. Category crises threaten the established class, race, and gender norms. This means that cross-dressing is a disruptive element undermining not just a category crisis of male and female, but the crisis of category itself. In this sense, cross-dressing is a commentary on the failure of our own stereotypes.[86]

Somewhat different arguments have been made by Bonnie Bullough and me when we held that the gender boxes of male (masculine) and female (feminine) were much too rigid to deal with the reality of gender behavior, which we believe is best described as two overlapping bell-shaped curves. There is an extreme at one end of each scale that can be labeled "masculine" and "feminine" and most males fall somewhere along the male curve, while most females hold similar positions on the female curve. A number of individuals, however, fall where the two curves overlap, perhaps as many as 20 percent of the population. This means that some females are much more masculine than many males and vice versa.[87] This also fits in with what we know about the human potentiality of bisexuality, something emphasized by the Kinsey data and by other researchers.[88]

MASCULINITY AND FEMININITY AND GENDER IDENTITY FORMATION

In 1973, Ann Constantinople questioned the assumption that masculinity was the opposite of femininity and suggested that the identification of masculine traits might be independent from rather than opposite of the identification of feminine traits.[89] The "both/and" concept of psychological identification quickly replaced the "either/or" notion that had been dominant since Terman. Sandra Bem developed a gender identity measure, the Bem Sex Role Inventory, that treated identification with masculine traits independently of identification with feminine traits.[90] Using published scales as well as their own, J. T. Spence and R. L. Helmreich found wide variations in gender traits, although they also found that stereotypically masculine personality traits in males are correlated with self-esteem,[91] which serves to emphasize just how much influence society and culture have on self-esteem. Still, the difficulty is that the scales are based on observable patterns without any real attempt to think through whether there are behaviors that must be distinctly limited to females or to females, a point that Mosher has emphasized.[92] Obviously, there is a tremendous variation in gender orientation, but there is also compulsive pressure to conform. This is why organized groups have assumed such importance.

A current theory for the formation of gender identities and sexual preferences was developed by Bonnie Bullough, who holds that it includes three steps:

1. There is a genetic predisposition for a cross-gender identity, including high or low levels of activity and aggression.
2. Prenatal hormonal stimulation supports that genetic predisposition and perhaps indelibly marks the neural pathways so the pattern that produced the cross-gender identity is continued after birth.
3. The socialization pattern shapes the specific manifestation of the predisposition.[93]

This theory posits the certain children are born with a gender identity that leans to the other side of the gender continuum to varying degrees. These children are not born with a specific identity as a homosexual, cross-dresser, or transsexual, but these patterns are shaped by the socialization process. The socialization process, however, has a different impact on children who have a cross-gender tendency than it does on children whose tendencies are clustered around the male or female pole. This explanation perhaps puts more emphasis on biological factors than some, but it also recognizes the importance of psychological, sociological, and cultural variables.

Research related to the pattern of feminine boys helps show how the socialization process specifies the paths that can be taken by children who are born with the tendency to develop a cross-gender identity. Although folk wisdom has long linked feminine behavior among boys with later homosexuality, Irving Bieber and colleagues were the first researchers to emphasize the point with hard data. Bieber studied 106 male homosexual patients who were being treated by him or other psychiatrists and found that cross-gender behavior as children was the most common element in their backgrounds. He thus realized that the phenomenon that became homosexuality among adults actually started very early, long before the hormonal surge at puberty focused the attention of the young man on sex. Bieber then went on to explain this in psychoanalytic terms by emphasizing a strong binding relationship with a mother and a weak or absent father.[94]

A. P. Bell, M. S. Weinberg, and S. K. Hammersmith gave a different interpretation to this existence of childhood feminine behavior. They conducted a detailed path analysis of the lives of a sizable sample of homosexual and heterosexual men and women from the San Francisco Bay Area. They found that homosexual men and women were more likely to report poor relationships with their fathers than heterosexual members of the study group, but it is not clear whether the rejection was because many of them were gender nonconforming or whether this was a part of the causal sequence. The most common element in the childhoods of both lesbians and homosexual men was gender nonconformity. Many of the boys had developed a homosexual pattern in their teenage years, although there was no evidence that this was due to a lack of opportunity for heterosexual interaction.[95] Thus Bell and co-workers emphasized the early gender nonconformity found by Bieber, but they were hesitant to ascribe it to a single causal factor.

Other studies also interpret the family dynamics differently from the psychoanalytic school, perhaps because more is now known about the socialization process. Frederick Whitam and Michael Zent, for example, found that in countries where certain forms of homosexuality are less stigmatized, such as in the Philippines and the Latin America countries where the penetrater is not regarded as a homosexual in popular belief, the relationships between homosexual children and their fathers are good. In the United States, where the cultural patterns attach more stigma to homosexuality, parents tend to be more disturbed when they note that their boy is feminine or find out he is homosexual, and a rift develops between fathers and their homosexual sons. Whitam and Zent perceive the hostility as a parental reaction and a withdrawal rather than the cause of homosexuality.[96]

Using a structured questionnaire, Michael Newcomb asked male and female homosexuals, male and female heterosexuals, and male transvestites to describe and evaluate their parents. There were no differences in the male homosexual and heterosexual samples, but there were difference among the transvestite and lesbian samples and the others. These two groups perceived their parents as less sex typed; their mothers were more dominant and independent than their fathers.[97]

Additional data from longitudinal studies of children who are severely cross-gendered from early childhood are also important in evaluating the social and psychological influences on their later gender identity and sexual orientation. The study by Green in which fifty feminine boys were studied over a fifteen-year time span is the most comprehensive. The boys were decidedly feminine, so much so that their parents sought professional help at a University of California at Los Angeles center that was interested in gender studies. The children consistently cross-dressed very early (94 percent by age six), played with dolls, preferred girl playmates, and indicated they wished they had been born girls.[98] Approximately 75 percent of the feminine boys became homosexual compared with only one homosexual man in the fifty-member control group. One member of the feminine group was considering sex-change surgery at the time of the last visit, but there were no transvestites in either group.[99]

In addition to the Green study there have been several other smaller-scale longitudinal studies of feminine boys. The psychiatrist Bernard Zuger followed sixteen boys who were referred to him for their effeminate behavior, which included cross-dressing, wearing lipstick, the use of feminine gestures, the desire to be a girl or a woman, and a lack of interest in or dislike of sports and boys' games. In all cases, these behaviors appeared before age six. When Zuger first reported on the group in 1966, he had followed them for ten years, but only half of them were old enough to have settled on a sexual orientation. He indicated that four were confirmed homosexuals, three were probably homosexual, two were heterosexual, and one was left unclassified.[100] Ten years later, Zuger reported that ten of the sixteen were homosexual, one was a transvestite, one was transsexual, two were heterosexually oriented, and no decision could be made for the remaining two. Zuger indicated that he could not have predicted the differential paths from the early behaviors. The child who became the transsexual had cross-dressed as early as one to two years of age, but the same symptoms appeared in some of the other cases with different outcomes.[101]

At the University of Minnesota Hospital, Phil S. Lebovitz studied sixteen young men who had been seen as children for effeminate behavior. At the time of the follow-up study subjects were between sixteen and twenty-seven years old. The group included three transsexuals, one married transvestite, two homosexuals, three other married men, and seven unmarried men. Lebovitz argued that there were different patterns observed in the early and late cross-dressers. Members of the group that started cross-dressing before age six were much more likely to have a "deviant" sexual identity as an adult.[102]

Money and A. J. Russo reinterviewed a group of feminine boys that Green had originally worked with and reported that all five were homosexual.[103] Charles Davenport reported an eight- to ten-year follow-up study of ten subjects who were referred to him for feminine behavior, including cross-dressing; four were heterosexual, two were homosexual, and one was transsexual. The outcome in three cases was uncertain.[104]

Homosexuality is quite clearly the most common outcome of cross-dressing and other effeminate behavior during childhood, but some boys also ended up as transsexuals and transvestites, indicating perhaps that these phenomena all represent different forms of gender nonconformity. Still other such boys turn out as heterosexuals. There have been studies of transvestites that have suggested that a strong mother may be a part of the picture in some cases;[105] other studies of transvestites have emphasized that they are less cross-gendered as children than transsexuals or homosexuals and more interested in sports and other masculine activities. In fact Bullough, Bullough, and Smith reported that transvestites as children conformed to the image of the typical boy, except for the secret cross-dressing in which many engaged in.[106] Because transvestites tend to come from higher socioeconomic groups than transsexuals and do not cross-dress as early as transsexuals, they perhaps are better able to keep their behavior a secret or perhaps it is less of an issue in the family. As adults, they score high on femininity scales, but they are not necessarily low in masculinity, except when they are cross-dressed. The key element in their socialization pattern is that they somehow learn to develop a double persona that allows them to conform outwardly but not necessary inwardly.

Probably most cross-dressers are also cross-gendered in other regards, in their sexual orientation, in their selection of occupation, and so on. However, a special group of heterosexual cross-dressers has emerged, primarily due to the club movement that followed out of Prince's effort. The social influence forming the behavior patterns of this group of respectable middle-class men is very clear, since the heterosexual identity is highly valued by the dominant members of the group.[107] This high value for heterosexual identity, to the point of homophobia for some, may be another key factor in the socialization process that helps some young boys who feel cross-gendered to choose to cross-dress but to avoid homosexuality.

Transvestites seem to be underrepresented as a whole in the studies of feminine boys, but cross-dressing at a level that is not labeled transvestism might be there and not reported. It might also be, as some studies have found, that the people who later become heterosexual transvestites are not as severely cross-gendered in childhood as those more likely to become homosexuals or transsexuals; as a result, they are not brought in for treatment as children. Moreover, the feminine boy thesis is only a partial explanation for adult gender nonconformity, since as many as one-third to one-half of the boys who later became homosexuals were never regarded as feminine boys.

One of the more interesting facets of the studies of feminine boys is the fact that most of them are not overtly feminine as adults. They apparently go through a defeminization process during adolescence, and this process has been particularly well described among homosexual men. By the time they are adults, their outward gender behavior is conventional, with only a minority of the male gay community remaining overtly feminine.[108] Here it seems that social class might be a major factor in determining which of the feminine boys go through this transition. Joseph Harry divided his sample of 686 homosexuals into groups based on social class; he found that the blue-collar respondents were more likely to remain feminine than the men from a higher socioeconomic level. The blue-collar respondents also generally had same-sex sexual experiences at a younger age than the higher-class men. Harry interpreted his data as showing that the early sexual activity, before any defeminization process could take place, was reinforcing to the young blue-collar boys, thus they were more likely to remain effeminate as adults.[109]

It might well be that by becoming and remaining feminine they were tolerated as sissies by the boys in their group. Middle- and upper-class boys had greater freedom as well as greater privacy, and because many of the things that their parents did would be judged by the lower-classes as effeminate, they could better survive the growing-up process and escape the label of effeminacy.

Certainly there is a sexual script that boys have to follow, although the script varies according to social class. In the United States, probably the most extreme is among the lower socioeconomic males for whom feminine behavior among certain boys is tolerated, because they give service to the machismo leaders. Exceptions can be carved out for the "brain" or other needed members, but in the gang culture, males either have to conform or find a specific role in order to belong. Being a sissy is a low-status but accepted role.

To date there have been almost no study of girls who are tomboys, probably because cross-gender behavior in girls is considered less problematic. It is also because, as will be discussed later, male problems in general have been of more interest to sexologists than female ones. Whitam and Robert Mathy, however, have done retrospective studies of homosexual and heterosexual women in four societies: Brazil, Peru, the Philippines, and the United States. All of the subjects were volunteers who were located through clubs, bars, professional groups, student groups, and even friendship groups. Childhood behavioral patterns of the lesbian women were significantly different than those of the heterosexual women. The homosexual women were much more likely to have been called tomboys, played with boy's toys, and played dress up in men's clothing. The heterosexual women were more likely to have paid attention to women's fashion and played dress up in women's clothing. The authors consider these cross-gender childhood behaviors to be precursor patterns of a later-developed lesbian lifestyle.[110]

SUMMARY

When all these findings are summarized, it seems evident that not all cross-gendered children become homosexual, transsexual, or transvestite. About all one can conclude is that the data suggest cross-gender identity behavior might be a precursor to the three types of adult patterns. The strength of the urge for the cross-gender behavior and the social learning that takes place help to determine the pattern of adult behavior. The biological factors seem to be strongest in the transsexual children. In a society such as ours that highly values macho men and feminine women and punishes gentle men and aggressive women, children who do not conform outwardly are likely to be either unhappy or protected by others, often their mothers. In the comparative study by Bullough and co-workers, only 16 percent of the transsexual sample reported happy childhoods compared with 38 percent of the transvestite sample, 64 percent of the homosexual sample, and 60 percent of the undifferentiated control group.[111] The transsexual child knows he is deviating from the norms of society; he hears the admonitions of his parents and the taunts of his peers, but he is unable to cope in a better way. He does not feel that he belongs to his assigned gender group. Psychotherapy may help him somewhat, but it is often unsuccessful, because the therapist tries to eradicate the cross-gender behavior instead of channeling it or trying to build up his self-esteem.[112]

It might well be that in societies that have adopted strict dimorphic gender patterns, change is more traumatic than those that allow somewhat more ambiguity. Generally, the United States and Europe have adopted strict dimorphic gender patterns; thus cross-gender behaviors have been stigmatized and punished. Since the nineteenth century when the medical model came into prominence as a means of diagnosing nonconforming sex and gender behavior, significant departures from a dimorphic model of masculinity and femininity have been labeled as illness. Similarly, sexual orientations other than exclusive heterosexuality were considered as illness until a 1973 vote of the American Psychiatric Association dropped ordinary homosexuality from its *Diagnostic and Statistical Manual*.[113]

If current studies are on the right track, it shows an etiology of development for transsexualism and transvestism somewhat similar to homosexuality. All three are gender nonconforming behaviors, and although many variables are probably involved, one of the major ones seems to be biological proclivity with genetic, hormonal, and neural components.

9

OTHER VOICES, OTHER VIEWS

*D*iscourse about sexuality does not take place in a vacuum but is a reflection of the concerns of the time and society in which it exists. Research itself has tended to reflect this, with priorities on what is important depending on what issues seem of most concern at any one time to significant segments of the population. Sex research since the 1960s has been greatly influenced by the growth of what might be called the second wave of feminism, which arose in the 1960s, and by the appearance of AIDS in the 1980s. The erroneous assumption that AIDS was primarily a disease resulting from homosexual exchange of body fluids also gave further impetus for research into homosexuality, research that also benefited from the growing strength of organized homosexuals and lesbians. The same period saw the growth of a mass consumer market eager for information about sexual pleasure. The 1960s and 1970s also saw considerable public experimentation with alternate lifestyles, which was manifested by group sex, the growth of transvestite clubs, the recognition of the widespread existence of sadomasochism, and an awareness of and interest in the varieties of human sexuality. That time period also saw demands by previously ignored groups, such as the physically disabled, for the right to seek sexual pleasure. Many of these forces were at cross-purposes with each other, and the result of ongoing research has not always been to clarify issues but sometimes to make them more complicated, perhaps because there were no simple answers in the first place.

FEMINISM

Probably the most influential force at work in directing sex research since the 1970s has been that of feminism. Feminist issues were not new to sex research, and in fact much of what went by the term *sexual reform* in the international congresses held in the 1920s became issues for feminists in the 1970s. Still, the growth of an organized feminist movement gave added strength to this aspect of sex research. It also opened up new topics for investigation that had not appeared to be issues in an earlier generation.

Sounding the clarion call for a new wave of feminism was Betty Friedan, who in *The Feminine Mystique* (1963) reported deep reservoirs of discontent among American housewives.[1] At the time of her writing, most American women had accepted the notion proclaimed by the sex radicals of the 1920s, such as Marie Stopes, that the happiness of the marriage was proportionate to the sexual magic generated between husband and wife. Unfortunately, many women had found that husbands and wives held widely divergent views as to the meaning of sex. As many popular women's magazines of the time stated, women hoped for love and affection, whereas their male partners simply sought orgasmic relief. Moreover, the companionate ideal so popular in the 1920s had not led to equality between spouses, because wives remained economically dependent, conscious that a failure in marriage might spell financial disaster. Although the development of the pill and the dissemination of effective contraceptives had encouraged a new ethic of "permissiveness with affection," the rules had changed, which made it much more difficult for women, who traditionally had used their virginity to guide a male into marriage, to just say no to premarital intercourse. Adding to the discontent were the increasing numbers of women in the work force who found that despite their qualifications they often played second fiddle to men.

More formally, the second wave of feminism began with the foundation of the National Organization for Women (NOW) in 1966. Much of the energy and drive for the movement came from women moving out of what is called the New Left, which in turn had its start in the civil rights movement of the 1960s. The struggle to remove racial barriers in society had engaged the energies of both women and men since the end of World War II, although women often found themselves pushed into subordinate roles in the movement, a role that the more radical did not accept. Inevitably the Civil Rights Act of 1964 did not end the struggle for civil rights but rather extended it to include other groups that felt themselves disenfranchised, including women, gays, and the physically disabled.

Just as the first wave of the women's rights movement had grown out of the movement for the abolition of slavery and followed on its heels, the second also had deep roots in the growing consciousness that even in the civil rights field women were regarded as inferior by the male establishment. The so-called New Left movement, the announced goal of which was human liberation, often treated women, in spite of its rhetoric, as second-class citizens, prized more for their sexual availability than their brains. Many of the more radical woman began abandoning gender-mixed organizations, devoting themselves to a rather loosely structured autonomous women's liberation movement. One aspect of the new wave of feminism was consciousness raising, or making the public aware that women were important and had a role to play in society at large. It also gave rise to feminist theory, which emphasized that traditional institutions such as marriage, family, and motherhood needed redefining to stop them from being used as institutions designed to maintain the oppression of women.

The feminist response was to politicize sexuality, that is, to demand legislation for corrective action, a response that gays, lesbians, and the handicapped also soon adopted. While sexual issues had always been political, the essential difference was that organized groups, many of them from previously stigmatized groups, now turned to politics to demand change. The necessary first step in any demand for political change was seen to be the need to publicize the problems, to educate the disenfranchised group or groups themselves about the existence of basic inequalities. In terms of sexuality, one of the important texts was the Boston Women's Health Book Collective's Our Bodies, Ourselves, which was designed to better acquaint women with themselves and to emphasize the need for change. It was a mixture of cutting-edge research and political rhetoric, giving special attention to the ambivalence women had in recognizing themselves as sexual beings.

We are simultaneously bombarded with two conflicting messages: one from our parents, churches and schools—that sex is dirty and therefore we must keep ourselves pure for the one love of our lives; and the other from Playboy, Newsweek, etc., almost all women's magazines, and especially television commercials—that we should be free, groovy chicks.

We're learning to resist this double message and realize that neither set of images fits us. What really has to be confronted is the deep, persistent assumptions of sexual inequality between men and women in our society. "Frigidity" in bed is not divorced from the social realities we experience all the time. When we feel powerless and inferior in a relationship, it is not surprising that we feel humiliated and unsatisfied in bed. Similarly, a man must feel some contempt for a woman he believes not to be his equal. This male-dominated culture imbues us with a sense of second-best status, and there is no reason to expect this sense of inferiority and inadequacy to go away between the sheets.[2]

Manifestos such as this made certain that the women's liberation movement of the 1970s was not antisexual, as much of the earlier feminist movement of the late nineteenth century had been. Instead, the new feminists concentrated on attacking the sexual objectification of women and the traditional belief that biology inevitably was always destiny. The more radical members of the movement felt that only when women achieved total autonomy could sexual freedom have any meaning. Inevitably, such views led to a challenge to traditional research into sexuality and a demand for reorientation, and the result was a new kind of sex research.

MICHEL FOUCAULT

Important in this redirecting research was the use of history, and in a sense, historical study became a weapon in the hands of many of those urging change. Historians have always recognized that each generation writes its own history; this is because each generation looks at history from a different perspective and wants different kinds of answers. Some of the feminist critics of what they called the male-centered view of sexuality, however, demanded more than a change in viewpoint and turned for it to deconstructionism and the philosopher Michel Foucault, who applied his methods to the study of sexuality.

Foucault had originally planned a six-volume study of sexuality in civilization but had only published three volumes before he died of AIDS in 1984. Two of these volumes dealt with Greco-Roman civilization and have been generally ignored, but the first volume, which was more general, proved particularly attractive to those seeking to challenge established assumptions about sex.[3] Volume 1 presented little research about sexuality and no conclusive evidence. Foucault assumed that from the beginning of the seventeenth century the history of sexuality had been marked by a continuous increase in the "mechanisms" and "technologies" of power. During the course of this history, the locus of power shifted from the confessional to the research laboratories and clinics, where sexuality had become the subject of scientific investigation. He emphasized four mechanisms of "knowledge and power" of sex, (1) "the hysterization of women's bodies," (2) "the pedagogization of children's sex," (3) the socialization of procreative behavior, and (4) "the psychiatrization of perverse pleasure." These mechanisms were directed at four figures: hysterical women, masturbating children, Malthusian couples (using contraceptives or curtailing births), and perverse adults.

Much of the recent emphasis on sexuality has focused on these issues, although not always in a direction that Foucault would have wanted. Foucault's summation was easy to grasp, but it was also simplistic history. Foucault, however, was simply not interested in history, in explaining how such assumptions developed, or in the reasons for them, but rather he was engaged in the process of constructing his own myths of the past, which might or might not be true and which, in my opinion, would help him to come to terms with his own homosexuality and to justify homosexuality itself. To do so, he ignored what had gone before and asserted his own belief about what had happened without explaining why. He tried to experience through his own life, particularly through his own search for sexual pleasures, the kind of salvation that a traditional Christian might well have sought by immersing himself or herself in religious devotion.[4]

Though Foucault's history might be erroneous, one of his positive contributions was to emphasize to historians that their historical researches were really a centerpiece to the understanding of the place of sex in the contemporary world.[5] This call to historical research came at the same time that there was a rapid escalation of studies into the history of sex, and history in fact became the cutting edge of forcing the kind of reassessments that both the feminists and the gays demanded.[6] There was also a more negative consequence, namely the attempt to create new mythologies by the more radically discontented with the status quo. This, for a time, seemed easy to do, since traditional historians had written little about homosexuality or feminism or about sex in general, and nonhistory specialists often felt free to speculate about what might have happened. Historians, for their part, were not so much interested in rhetoric as exploring new areas of history. The results were contradictory. On the one hand, there was an increasing politicalization of sex research based not so much on hard data but on what the more vocal spokesperson of the group said had happened. On the other, there were a growing number of serious and scholarly studies. Occasionally, however, a diligent researcher could find relevant historical data to back up a call for a new interpretation and to demand action.

RAPE

A good example is the case of rape, or what is now increasing called sexual assault, which emphasizes that it is an assault in which sex is involved. Rape had received comparatively little study by sexologists or, for that matter, by historians. There were legal discussions of it and there had been attempts to separate the crime of assault from the sex act, but almost no one had paid attention to the person who was raped.[7] Clearly, any feminist examining rape laws could argue that they seemed to be written by men for the protection of men. Laws in most states demanded corroborating evidence for rape, permitted questions about the victim's sexual history, and assumed that many, if not most, women invited it. In North Carolina only a virgin could claim rape.

Susan Brownmiller concluded in her historical, literary, sociological, and political study of rape that the phenomenon was a natural consequence of the theory that male domination over women was a natural right, a belief deeply embedded in both the cultural value system and the law. She urged that police forces, courts, and other such agencies be fully gender integrated and noted that a massive effort was needed to change attitudes.[8] Historians who then directed their attention to the subject with this new perspective reported that rape was not the isolated criminal act so widely believed but rather was rampant and widespread. Some historians went so far as to argue that on the American frontier, most sexual contact between white males and Native American and Mexican women took the form of rape. A good example was the case of Lieutenant General Custer who, after winning a battle in 1869, allegedly invited his officers to "avail themselves of the services of a captured squaw" while he indulged himself with a Cheyenne woman.[9] Victorious armies generally took it as their right to rape the women, and there was a widespread assumption among many men that women in fact wanted and invited rape.

Obviously, one of the first goals of the feminists and their male allies in the United States was to change the rape laws and most states in the 1970s rewrote their rape laws along somewhat more feminist lines but also incorporated the findings of those investigating the topic.[10] Changing the law, however, was just one step; general public education about sexual assault was a necessary second step, and the education of law enforcement officials was particularly stressed. Rape awareness training became important; and significantly, a number of researchers turned to analysis of the psychological effects of sexual assault on the victim, including males raped by other males or, more rarely, by females.

The effects of rape include strong emotions and feelings of shame, humiliation, rage, personal violation, fear, confusion, and a sense of having lost power or control over one's life. Often there is a compulsion to wash the body, to wash away the degradation and shame. Rape victims often go through a period of being afraid to be alone, of having nightmares, of disrupted sleeps patterns, and of an unwillingness to discuss the attack with others; in some cases, victims deny that anything even happened. One of the more common and difficult aspects of the rape trauma syndrome is the feeling of the victim that she (or sometimes he) somehow caused or unknowingly invited the attack.[11]

Based on these studies, as well as a growing political consciousness, the meaning of the term *rape* was gradually redefined, until it became synonymous with an extreme form of sexual coercion, a much more generalized term than even sexual assault. Though much of this redefinition reflected an empowerment of women, sex researchers—mostly women—joined in furnishing a database, emphasizing that rape did not just involve forced sex with a stranger but could be perpetuated by relatives, friends, and even husbands. One result of such research was the appearance of new concepts and terms such as *date rape* and *marital rape*. These terms did not represent new forms of conduct, but rather the application of new terms to old patterns of behavior. In 1969, E. J. Kanin, for example, found that 25 percent of the male university students he interviewed admitted that they had tried to force intercourse on a date.[12] In an update of his study, he found that though the males involved did not use weapons or their fists, they engaged in what he described as a "no-holds barred contest."[13] Another study found that 43 percent of 201 men questioned said they used a "sexually coercive method" at least once or twice, and 15 percent said they had physically forced intercourse.[14] In general, researchers reported that it was a widespread male belief that no "good girl" would ever say yes to engaging in sex until she was engaged.

What the feminists emphasized was that when a woman said no, she meant no. This fit in not only with feminine empowerment but with an increasing recognition by women themselves that they could say yes if they wanted to have sex. Once this happened, involuntary sexual intercourse was found not to be confined to women on dates but to occur to men as well. Research by C. L. Muehlenhard and his co-workers found that college men who became involved in intense petting and kissing on dates often felt the need to go further than they themselves wanted to because of the culturally based perceptions that their activities were supposed to culminate in intercourse. Some 93.5 percent of the 507 men involved in Muehlenhard's study reported they had experienced unwanted sexual activity, including unwanted kissing and petting, and 66 percent of them had experienced unwanted sexual intercourse. On the other hand, only 46 percent of the women studied had done so.[15] What the date rape studies have emphasized is the need for more effective communication between both partners on the date,[16] and *communication* has become the buzz word among an increasing number of sex researchers.

Carrying the redefinition of rape even further was the concept of marital rape, which entailed a change of laws about marriage. The first state to eliminate the use of marriage or cohabitation as a defense against the charge of rape was Oregon in the late 1970s. This was quickly followed by other states as rape statutes were modified to include sexual intercourse with one's spouse if accomplished under force, violence, or threat of great and immediate bodily harm. During the debate on just such a bill in the California legis-lature, one California senator is reported to have said, "If you can't rape your wife, who can you rape?"[17] As of this writing, a majority of the states have passed laws making it a felony to force a spouse to have sex against her or his will, although usually the statues have the wife in mind.[18] Again the issue becomes one of communication, a problem that sex researchers believe can only be solved by large-scale educational efforts. The evolution of the new concept of sexual coercion emphasizes just how influential social and cul-tural traditions have been in forming sexual attitudes and how much such attitudes have been inculcated into our laws, in short, just how many of our societal attitudes have been socially constructed. Inevitably, as the feminists forced rethinking, they demanded changes in the law, forcing changes in conduct that to many males seemed normative parts of being male. Femi-nism, in effect, was challenging male bastions of power, and feminists among sex researchers forced a rethinking of traditional concepts.

SEXUAL HARASSMENT

This challenge to male power was somewhat disguised in the case of rape but was much more obvious in the case of sexual harassment, which first entered into public consciousness in 1975 at a speakout held on the Cornell University campus in Ithaca, New York. The term was defined at that time by the Working Women's Institute as "any sexual attention on the job which makes a woman uncomfortable, affects her ability to do her work, or inter-feres with her employment opportunities. It includes degrading attitudes, looks, touches, jokes, innuendoes, gestures, and director propositions. It can come from supervisors, co-workers, clients and customers."[19] This original definition is similar to the one later adopted by the Equal Employment Opportunity Commission, which had been authorized under the Civil Rights Act of 1964 to eliminate gender discrimination. At the end of the 1970s, harassment was defined by the commission as one aspect of gender discrimination. It was thus moved from a subject of discussion and the focus of a modicum of research (much of it historical) to a matter of law. Sexual harassment was defined as "unwelcome sexual advances, requests for sexual favors, and other verbal or physical conduct of a sexual nature."[20] The Equal Employment Opportunity Commission was the same agency that Clarence Thomas later headed, and it was he who was charged with harassment by Anita Hill when he was nominated for the U.S. Supreme Court in 1992.[21]

Inevitably, these redefinitions lead to examination of the socialization of males and females. One of the pioneering studies was by Mosher, who looked at the use of sexual coercion as an instrument of power. A total of 36 percent of his male respondents agreed with the statement "You have to fuck some women before they know who is boss."[22] He defined a cult of macho men, afflicted by hypermasculinity. Though there were variants of the script with several distinct prototypes (macho, king, magician, lover, and so on), the macho man used his aggression, courage, callous sexuality, and cool self-control to demonstrate that he was a real man. This entailed a belief that whenever anyone encroached on his rights or insulted him, that person must be stopped. Even sexual aggression could be ideologically justified as men's right and women's place. In this script of masculinity, prestige accrued to those few men who rose to the top, and if such a success was denied to the macho man, preserving his manhood still remained all-important. He could do this by being tough, destructive, and dominant in his own family or home. His ideal remained the hypermasculine heroes in sports or the Rambo character in the movies.[23] Traditional male socialization also has included the belief that women are not particularly interested in sex, but with enough persuasion and seductive power, they can be "awakened" sexually.

This idea led some feminists to argue that gender role rigidity for males and sexual assault or violence were correlated, thus they sought sources for such gender rigidity. Some of the more radical feminists argued that the most influential source in inculcating male stereotypes of women was the kind of reading material boys and men chose, such as *Playboy* and other more sexually explicit material, all of which some militant feminists labeled as degrading to women. As a result, they urged that such material should be labeled as obscene and banned.

PORNOGRAPHY

One result of this radical feminist critique was to cast the ongoing debate over pornography in new terms and bring about an alliance between the radical feminists and the traditional forces of moral and sexual censorship. The new perspective was advanced by radical feminists who formed the organization Women Against Pornography in New York City in the 1970s. This group subscribed to the dictum that "pornography was the theory; rape is the practice." Andrea Dworkin, a leading theorist of the cause, argued that there had been an explosion of pornography as a reaction to fears of feminist power. She argued that "a new campaign of terrorism and vilification" was being waged against women, and that pornography was the propaganda of sexual "terrorism."[24] By the 1980s, Dworkin had joined with the feminist lawyer Catharine MacKinnon to draft model obscenity statues, and Dworkin and her militant feminist supporters allied themselves with forces in the traditional antipornography groups to introduce such legislation in Minneapolis and Indianapolis. Pornography, they argued, using the language of the civil rights movement, was sex discrimination, and where it existed, it posed a substantial threat to the health, safety, peace, welfare, and equality of citizens in the community.

Pornography, it was claimed, represents the systematic practice of exploitation and subordination based on sex that differentially harms women. The harm of pornography, they claimed, includes dehumanization, sexual exploitation, forced sex, forced prostitution, physical injury, social and sexual terrorism, and inferiority presented as entertainment.[25]

Little of the claims summarized above were based on any research on the topic. It represented the rhetoric of feminism at an extreme, and though they were unsuccessful in Minneapolis, initially they were successful in Indianapolis. Ultimately, the courts ruled the Indianapolis statute unconstitutional. Dworkin herself represents an extreme feminist position, and her stand on pornography is only the tip of the iceberg of her basic gut hostility to males. She has argued elsewhere that "intercourse is the pure, sterile formal expression of men's contempt for women," and that men used intercourse to "occupy," "violate," "invade," and "colonize" women's bodies.[26]

Though sexologists had tended to support some of the rhetoric of the feminist movement, on this issue they argued that the research did not support the militants. Ira Reiss, a prominent sexologist, found, for example, that individuals who attended X-rated movies between 1973 and 1983 were more, not less, egalitarian in their gender role attitudes than those who did not.[27] The same seemed to be true of patrons of adult movie theaters.[28] Others found that while there were short-term changes in the viewpoints of students who viewed pornography, these were not deeply ingrained.[29]

There had been long been an unresearched assumption that continued or repeated exposure to erotic stimuli resulted in satiation or a marked decrease in sexual arousal and interest in such material, but this conclusion was challenged by the work of D. Zillman and J. Bryant in the 1980s.[30] The problem essentially was how to carry out serious research on pornography, because, though most existing research was supportive of benefits of pornography, none of it was as yet conclusive. Serious research cost money, and no government agency was willing to be labeled as supporting pornography—for that matter, neither were most private foundations.

Dworkin played on the lack of definitive research by turning to writers such as Foucault, and assumed that her claims were valid because, in effect, she believed they were. In essence, polemics had replaced research, and an extreme male personality, the machismo man, was claimed to represent all males. Inevitably, there was a strong feminist opposition to the extreme feminist position represented by Women Against Pornography. Some of these women organized into the Feminist Anti-Censorship Taskforce, joining with the ACLU and other groups to fight such censorship efforts under traditional free speech guarantees of the Constitution.[31]

Even before Women Against Pornography had begun its campaign, the issue of the dangers of pornography had reached national consciousness. Giving impetus to this was the growing challenge to traditional attitudes emphasized by the removal of bans against sexually explicit material in such countries as Denmark and the Netherlands. To examine the issue, three national commissions have been appointed, one by the British government and two by different sections of the U.S. government. They represented different approaches—political and scientific—to the problem.

In 1968, President Lyndon B. Johnson appointed what came to be known as the President's Commission on Obscenity and Pornography; in July 1977, the British home secretary appointed the Committee on Obscenity and Film Censorship, usually referred to as the Williams Committee after its chairman; and in May 1985, Attorney General Edwin Meese announced the formation of the Commission on Pornography. Though each of these groups asked whether pornography was harmful, they came up with different answers. The President's Commission said no, the Meese Commission said yes, and the Williams Committee was ambivalent. The only commission that relied on any research was the President's Commission, and it concluded that empirical research as yet had not found any evidence to support the claim that erotic materials caused sex delinquency or crimes.[32] The Williams Committee, holding that harm to others was enough to initiate legislative action, said it was impossible to reach a conclusion about pornography and sexual crime. The Meese Commission totally ignored any research and, in keeping with its political aim, relied on carefully selected witnesses to furnish anecdotal data. It concluded that the available data supported the belief that pornography led to antisocial violence. Two women members of the committee, however, refused to sign the final report and issued their own dissenting statement, charging that what data the committee did have did not support the conclusions. Mosher was allowed to testify before the Meece Commission on behalf of the Society for the Scientific Study of Sex, almost the only research expert permitted to appear, and he emphasized the lack of conclusive evidence. He also laid out a possible research program for the committee to follow, if it was interested in determining the harms that might come from pornography. The committee, however, indicated it was not interested in any such research,[33] since it already knew pornography was harmful.

One reason research data on the subject are confusing, as Mosher pointed out, is because the conclusions depend on which research model is being used. Mosher labeled the model most often used in the past by sex researchers the catharsis model. It assumes that pornography and other sexually explicit material provide a safety valve in a sexually repressive society. Such a model recognizes that sexually explicit materials are not always good literature or art; in fact, sometimes they are disgusting. However, they are still useful in diverting tensions that otherwise might trigger aggressive antisocial behavior. Much of the studies printed in the supplementary volume to the President's Commission report went on this assumption.

A second model adopts a different hypothesis and holds that sexually explicit books, pictures, and movies provide powerful role models that can, by conditioning and scripting of the readers or observers, promote antisocial, sexually aggressive behavior.[34] Evidence for this is inconclusive, although it is strongest for sexual violence. Even here it seems that violence is the major factor, not the sexual activities themselves. It was this view that was adopted by Women Against Pornography and by Attorney General Meese, who ignored any evidence not supporting their case.

Another model addresses the personal and societal uses of sexually explicit materials in different cultures. Using such a model, for example, it allows the researcher to look on pornography as a product designed to serve as an alternative source of sexual arousal and as a way of enhancing masturbation.[35] This view is as yet little explored, although there is much anecdotal information.

There are also models of pornography based on communication, Marxism, psychoanalysis, feminism, and religion.[36] This last would simply condemn all sexually explicit materials. Mosher emphasized the need for researchers to distinguish between violent and nonviolent pornography, or erotica, as many sex researchers are now calling it to avoid the traditional question of what is or is not pornographic or obscene. So far, however, no one has carried through on any of Mosher's suggestions for research.

The state of the research on the topic was summarized by Richard Allgeier and Elizabeth Allgeier:

Clearly, violent erotica degrades both women and men. Further, such aggressive depictions are arousing to assaultive men—and to about a third of "normal" males who report that they might rape if they were sure they wouldn't get caught. . . . Many men and women report sexual fantasies that involve force, and obviously most do not carry out their fantasies. Further we have no definitive evidence indicating that exposure to aggressive erotica *causes* men to victimize women. Before censorship could be instituted, someone would have to determine what kinds of violent material should be prohibited. Who should make that determination? Another problem is that the effect of censorship in the past has been to increase the profits of producers of banned material. If violent erotica is found to pose a danger to women, we believe that withdrawal of consumer support would be more effective in stopping production of aggressive erotica than would attempts at censorship.[37]

The issue of pornography, however, has never been limited to writings and productions aimed at males as Women Against Pornography tried to emphasize. In fact, they tended to exclude from the discussion of pornography that especially designed for women and lesbians. This exclusion only emphasizes the political nature of their charges, since one result of the growth of feminist consciousness has not only been a conscious awareness of female arousal patterns but an increase in erotica written by women for women. Though erotica aimed at women has a long history, in the past much of it has been a soft-core reflection of women's fantasies rather than hard core, that is, the depiction of actual sexual intercourse.

Erotic romances, often known as gothics, are even written to a formula in which a young woman is aroused by the attentions of an older man who continues his persistent attention until marriage results. In the struggle between the two for dominance, the man's eventual triumph and his ultimate domestication are consistent plot elements. In one study, readers of these novels reported having sexual relations twice as often as nonreaders, emphasizing their potential for erotic arousal.[38] Recently, feminine erotica has become much more sexually explicit, for both heterosexual and homosexual women, and mail-order markets have developed to deal with the increasing interest.

Joani Blank, a sexologist concerned about the antipornography movement, joined with a group of women in San Francisco to start the Sexuality Library to give women who liked to read erotica an opportunity to purchase it without having to enter the so-called adult bookstores, generally an all-male preserve. As her customer list grew, she extended her sexological business by establishing Good Vibrations, set up to feature sex toys, vibrators, oils, dildos, and so forth. Among the books sold are the *Cunt Coloring Book*, *Sapphistry*, *Masturbation*, and a series of erotic fiction that includes depictions of intercourse. Erotic magazines edited for women (and for men) such as *Libido*, *On Our Backs*, and *Yellow Silk*, and a variety of X-rated movies are also part of the catalog.[39] There are now even sadomasochistic materials aimed at a female market, and women are involved in the writing, production, and distribution of these materials. In short, there are feminists against pornography, feminists against censorship, and feminists involved in pornography, with the result that on this issue women present the same kind of division as men do. What research there is would support the right of the public to read pornography.

One reason that pornography has existed is that males knew little about female sexuality. Most of the writing about sex in the past had been done by men, and often what they wrote was as much fiction as fact. This was a problem that women researchers such as Katharine Bement Davis set out to correct, and later generations of women researchers have continued to do so. Though the Freudian myth of the vaginal orgasm had finally been laid to rest by Masters and Johnson, some researchers wondered whether the vagina was as devoid of nerve endings as the research of the 1960s implied. Again, in turning to historical literature, some researchers found mention of sensitive areas in the vagina, a phenomenon that some lesbians reported as well. Beverly Whipple and John Perry hypothesized that there actually were sensitive areas, which they named the G spot after Ernst Gräfenberg (see chapter 7). William Hartman and Marilyn Fithian reported that they had observed sensitive spots in the vagina at the ten, two, and four o'clock positions, but other researchers insisted there was no evidence of a discrete anatomical structure that could be called the G spot, although there might be a diffuse area of vaginal sensitivity.

Whipple and Perry also reported at the same time that some women ejaculated, and though they had originally connected this "ejaculation" with the stimulation of the G spot, the two concepts became separated. Female ejaculation had also been reported earlier by other researchers, but the Whipple and Perry claim focused new attention on the phenomenon. Though almost all investigators agree that many women do have fluid expulsions during their orgasms, the question remains as to what it is and what its source is. Many women do have tissue analogous to the prostate surrounding their urethra close to the vagina and this might well secrete fluid into the vagina.[40] Whipple and others are still investigating both phenomena, and as of this writing both factors seem to have some empirical basis; some women apparently do have sensitive areas in their vaginas and some women do ejaculate fluid that is chemically different than urine.[41] Beyond that, however, there is no current agreement among researchers.

PROSTITUTION

Similarly, feminist-oriented studies have revived and reinvigorated the study of prostitution simply because different questions have been asked. Traditionally, studies of prostitution were dominated by questions of why women went into prostitution and how it should be regulated and controlled. Such questions result in a different sort of studies than if the questions centered

around why men went to prostitutes and what alternative economic opportunities were available to women.[42]

Many of the challenges to traditional attitudes about prostitution came from
historical examination, including those by Bonnie Bullough and me.[43] Some of
the more interesting studies were done by the historian J. R. Walkowitz, who
found that though women in the nineteenth century supported the right of
working-class women to adopt prostitution as a career, they were ambivalent
about sexuality. This ultimately led them to engage in a campaign against
white slavery and to support a single standard of chastity for both males and
females. Walkowitz concluded that there was a tendency for feminist campaigns against commercial sex to be transformed into repressive state policy,[44] a lesson that an increasing number of the current generation of feminists are just beginning to learn.

Martha L. Stein did a quite different kind of study, namely of the clients who
visited call girls. She observed encounters of 1,230 men with call girls. Using
these data, she categorized nine types of clients: opportunists, fraternizers, pro-
moters, adventurers, lovers, friends, slaves, guardians, and juveniles. Regardless
of type, however, she found that all wanted their sex needs met conveniently,
professionally, and without obligations other than the monetary one. Some,
however, enjoyed the temporary illusion of love or friendly involvement.[45]

As in many other areas of sexuality, for a time psychoanalytic studies dominated the studies of prostitutes, the best of which was by Harold Greenwald, who found that the primary predisposing factor to prostitution was a
history of severe maternal deprivation.[46] The difficulty with many of the
studies was that there was an assumption that once a woman entered prostitution, she left the world of respectable women. Obviously, prostitution is an
occupation, it fills a social need, and it offers financial reward to its partici-
pant even though it is a deviant or stigmatized one. In the past, it has been
highly dependent on the existence of a double standard, and in this sense, it
can be looked on as having the same basic objective as marriage: The woman
fills the sexual and social needs of the man. The occupational explanation,
however, is not inconsistent with modern sociological theory that accepts the
fact that both deviant and accepted behaviors and conditions can be built
on the same social structures and values.[47]

Most contemporary recruits probably drift into the occupation, at least in
countries like the United States. Usually, they experiment hesitantly with
accepting payment and then ultimately decide to become professional. Though
they go through a process of working into their identities, the crucial factors
in their coming to think of themselves as prostitutes are the reactions of
society toward them and the labels that others attach to their work. The
prostitute identity is not sought and can be painful at first, but eventually it

can also be supportive as the subculture provides an explicatory worldview to the prostitute that defines her work as significant and allows her to develop friends in the life. Many prostitutes feel that their occupation serves an important public service, more useful than many other occupational groups. Since, however, it is an occupation in which mobility tends to be downward as the woman ages, most women tend to seek other roles. If the legal and other barriers are not too rigid, they ultimately leave the life for marriage or other work. Many enter and leave prostitution several times, depending on the economic situation in which they find themselves. The stigma under which they suffer is one that society imposes on them. Interestingly, their customers suffer from no such stigma.

Increasingly, prostitution in United States has been decriminalized, the double standard has been weakening, and the nature of the customer has changed. Many men go to prostitutes to get sexual services that they feel they cannot receive from their wives or lovers, or because they simply do not want to become involved. Many of the recent generation of feminists who have been concerned with prostitution have concentrated on eliminating the stigma associated with prostitution and have argued that it represents a female response to the larger social forces that have fostered and maintained the practice of sexual inequality and oppression of women.[48]

There has also been a growing number of studies about male prostitutes, most of whom serve homosexual men. Male prostitution seems to be less hierarchized than female, and there seem to be basically three subcategories: the professional, the amateur, and the runaway. The professional is typically in his late teens or older and has a good deal of experience with commercial sex and is able to make a steady living or to supplement his earnings from other sources. The amateur performs only sporadically when he needs the money or the thrill or adventure involved in the activity. The runaway may be quite young, may have been disowned by his family, and finds himself struggling to survive on his own by selling his body. Generally, the career of the male prostitute is brief, much briefer than the female because the emphasis is even more on youth.[49] Interestingly, most of the male prostitutes do not regard themselves as homosexuals and in fact often regard their customers with hostility.[50]

ABORTION

Perhaps the most political of the changes wrought by the growing feminist movement was changes in abortion, a cause that had been advocated at some of the world congresses sponsored by Hirschfeld and his allies back in the 1920s and 1930s, as well as by the founders of the SSSS in the late 1950s.

Kinsey had included questions on abortion in his sample and had begun extracting the data for publication, largely at the urging of Calderone, then the medical director of the Planned Parenthood Federation. The Kinsey data were published after his death in what Gebhard labeled the third report of the Institute for Sex Research.[51]

The study indicated that of the pregnancies that ended before marriage, 6 percent resulted in live births, 5 percent ended in spontaneous abortions, and 89 percent ended in induced abortions.[52] Of the married white women in the sample, between 20 and 25 percent had induced abortions. In the total sample of 7,074 women (single, divorced, black, white, and in prison), the researchers concluded that the majority of all induced abortions came from pregnancies in marriage, although induced abortion of a premarital pregnancy was a fairly common event.[53] Included in the study were data on the costs, methods, and consequences of illegal abortions. They also looked at studies of abortion in other countries, some of which had legal abortion and some of which did not.

These Kinsey data proved to be the opening wedge in what became a struggle for abortion rights. Even before the feminists became involved, changes had been taking place. The American Law Institute in its 1962 suggestions for a model penal code had advocated making the termination of pregnancy legal for cases of conception from rape, incest, or other felonious intercourse; when the child might be born with a grave physical or mental defect; and for any girl who was under the age of sixteen.[54] Like other demands for changes, this one was echoed by a historical study of abortion. Lawrence Ladder concluded that medically induced abortions were a privilege of money; that for the less affluent there was an underground network of abortionists, most of them unskilled; and that safe and simple procedures were known and available.[55]

Many in the feminist movement joined to help form the National Association for the Repeal of Abortion Laws in 1969, and in 1973, the U.S. Supreme Court ruled that abortion was a constitutional right.[56] Almost all sex researchers support the right of abortion, and there is little research that demonstrates any negative effects. But, as is the case in many of the other sexually related issues, neither research, a court decision, nor a change in law ended the controversy. Abortion, and even contraception, remained politicized,[57] a struggle between conflicting assumptions rather than conscience.

LIMERENCE

Much less controversial has been the study of limerence, a term coined by Dorothy Tennov in 1979 to describe the intense emotional state of falling in love and being love smitten.[58] To describe something as having limerence means that it is beyond conscious control. An individual is so preoccupied with the loved one that he or she is oblivious and blind to reality, at least temporarily.[59]

There is, however, a distinction between romantic limerence and the lust of physical passion. Weinrich has suggested that limerent sexual attraction eroticizes the personality and physical traits of the particular person with whom we fall in love. A lusty sexual attraction, on the other hand, produces erotic arousal when a new object of lust appears. It is likely that both men and women experience both kinds of love, but in our society, most women experience limerence as a general desire and physical passion, while lust arises mostly as a reaction to a particular person. Most men in our culture experience a general lusty desire toward most women but limit their limerence to a particular person. The gender of the person with whom we fall into limerent love may not be important, but lusty attraction is.[60]

Traditionally, men have believed that they, and not women, are the sexual initiators. Research data tend to suggest that women are not the reluctant, hesitant, or coy besieged person that myth holds but are probably the initiator in communicating sexual interest by either touching or holding eye contact.[61] H. T. Remof has gone so far as to argue that *female choice* is an evolutionary mandate given to women so that they can select the best mate, thus ensuring the survival of the fittests. According to him, women are the natural initiators of the majority of encounters that lead to sexual contact.[62] A number of researchers have tested the idea, among them Timothy Perper and Monica Moore.

Moore observed some two hundred randomly selected women to gain an insight into the wide range of nonverbal solicitation signals women use to get the attention of a man they find interesting. These signals include smiling, glancing, laughing, hair flipping, head tossing, whispering to the man, licking lips, primping, holding hands, leaning toward a man, and soliciting his aid.[63] To Perper, eye contact seemed to be important in female choice, although other nonverbal means were also used. He found that in almost every stage in the early contact phase, women can either encourage or discourage by nonverbal signals and remain in control of the situation.[64]

CHILDREN

Just as the first wave of feminism at the end of the nineteenth century led to a greater legislative concern for children, so has the second wave of feminism led to renewed interest in children, although much of this has also been politicized, particularly in terms of child sex abuse. One aspect of childhood sexuality, namely the developmental stages, was a major focus of the psycho-analytical school of researchers who followed in the steps of Freud. Of these, the most influential was Erik Erikson, whose developmental consciousness phase is still widely used and has not been supplanted.[65] Studies like Erik-son's were done mainly by observation and did not involve any intrusive questioning of children. These nonintrusive kinds of studies are still con-ducted and have looked at sex differences in developmental phases.

Perhaps the most comprehensive is that of June Reinisch and collabora-tors. Until 1993, Reinish was the director of the Kinsey Institute for Research in Sex, Gender, and Reproduction at Indiana University. She suc-ceeded Gebhard and the institute founded by Kinsey was renamed and somewhat refocused under her direction. Reinisch did not collect the data herself for her study but turned to the Copenhagen Consecutive Perinatal Cohort study, which examined 9,181 children who were born between Sep-tember 1959 and December 1961 in Copenhagen, Denmark, from 9,006 pregnancies.[66] Each mother of these children was instructed to record the day of the first occurrence of each of ten developmental milestones in a diary, which were collected when the child was twelve to eighteen months old. Reinisch's data include information on 4,653 children, all of whom were born within the normal gestation range of thirty-eight to forty-one weeks.

Reinisch and collaborators found sequences of behaviors that, although identical for males and females, took place at slightly different times. They found that agenetic/instrumental behaviors—those behaviors that relate to the advancement of the individual, or self-preservation—were more charac-teristic of males, whereas communal/expressive behaviors were more charac-teristic of females. Reinisch argued that at least some of the identified early sex differences reflect the divergent prenatal hormonal milieu of males and females, although she also recognized that some differences were responses to the variation in attitudes of the caretakers to males and females.[67]

To go beyond the kind of developmental studies that Erikson and Reinisch have done is difficult in today's world, since the topic is surrounded by taboos. Permission is needed from parents to ask questions about sex when a child is involved, and obtaining enough such permissions to form a large sample is difficult, unless the children are undergoing treatment. More-over, getting such a project through a human use committee on almost any university campus would be impossible.

Many researchers in the past who even dared to write on the topic came under attack. Calderone, one of the founders of SIECUS and its long-time director, for example, was accused by *Time* magazine of believing anything was possible in childhood sexuality, and the magazine implied she condoned sexual interaction between children and adults.[68] What Calderone had done to bring on this accusation was to teach that it was important for parents to socialize their childhood's sexuality by providing them with adequate and appropriate information. She was highly upset that anyone would imply that she condoned pedophilia, and in fact, she had a long record of stating that sex between adults and children was inappropriate and indefensible.[69] This did not stop the smear campaign,

Inevitably, this has meant that much of what we know about childhood sexuality is acquired from retrospective research, of individuals looking back on their childhood. Havelock Ellis had included recollections of childhood sexuality by many of his subjects,[70] and of course, psychoanalysis is in part based on such recollections. Kinsey had in part relied on such retrospective recollections and, as noted earlier, had come under attack.

One of the few Americans to study childhood sexuality in detail was Floyd M. Martinson, who combined retrospective studies of his college students with interviews from young mothers along with case material on children in six different communities. He divided his data into three groups: infants from birth to three years of age, children three to seven, and preadolescents from eight to twelve. He found that most children have the capacity for self-stimulation and orgasm by age five. He also found that often the older children taught the younger, and for the vast majority, initial sexual encounters were with a family member, a relative, a neighbor, or a babysitter.[71] In spite of his caution both in gathering data and in interpreting it, Martinson had difficulty in finding a publisher and eventually published the work himself. Publishers seemed to be leery of studies of children's sexuality, fearing that they might be sued or that the author who investigates such subjects might be charged with pedophilia.

Ernest Borneman, an Austrian sex researcher, has done the most significant studies on the topic, although his work has not yet been translated into English. Though Austria furnished a much more hospitable climate for such studies than the United States, Borneman was accused by some of being a pedophile, because he asked children under eight years old questions about their sexual development.[72]

Inevitably, retrospective studies seem much safer to do, although these also present special problems. In David Finkelhor's study of New England college students, 13 percent reported having sexual experiences with their siblings. A total of 40 percent of the students were under the age of eight at the time of the experience, and 50 percent were between eight and twelve. One problem with Finkelhor's study is how sexuality is defined and another is how much information is actually remembered and how much is really fantasy. Finkelhor defined sex activities to include fondling and touching the of genitals of a sibling, mostly of the opposite sex, which may have even been encouraged by a parent. Force of some type was only used in 25 percent of the experiences, but even these did not necessarily involve penetration.[73]

Asking different kinds of questions of college students, James Elias and Gebhard reported that more than 50 percent the boys and 33 percent of the girls in their sample reported having engaged in homoerotic activities between ages four to fourteen.[74] Generally, such childhood sexual activities involved mutual masturbation, fondling or touching of the genitals, and exhibitionism, although occasionally oral and anal contact occurred.

Much current discussion, however, is not focused on sex between or among children but on adult-child sex. It has in large part dealt with the adult perpetrators. Contrary to popular opinion, however, it has been estimated that as many as 95 percent of the cases of sexual abuse of girls and 80 percent of the cases of sexual abuse of boys involved heterosexual men, not homosexual men.[75] Kinsey had pioneered in gathering data on the adult sexual abuser as part of his ongoing studies, and after his death, these data were added to and published as part of a study on sex offenders by Gebhard and others.[76]

Much of the Kinsey team's study focused on sexual activities between adults and minors. The researchers reported that sexual abusers are usually not violent, and physical damage to the child occurs in only about 2 percent of the cases, although it should be kept in mind that such statistics are based on the statements of the sexual abusers themselves.[77] Convicted pedophiles tended to be older than other convicted sex offenders, with an average age at conviction of thirty-five. About 25 percent were older than forty-five. Many turned out to have low intelligence, with about 20 percent rated as mentally retarded and another 5 percent as senile. No clear-cut psychiatric disturbance was undebatably present in most cases, although many showed emotional immaturity. Interestingly and paradoxically, convicted child molesters tended to be conservative, moralistic, and frequently quite religious. They hold very strict attitudes about female sexuality, and those who were married had demanded premarital chastity in their wives. Women tended to be classed as either good or bad by these men. Most of the convicted pedophiles were not primarily interested in children; their contacts with children occurred during periods of stress, frustration, lack of other sexual outlets, or unusual opportunity. Most convicted sex offenders had traumatizing developmental experiences, including having been sexually abused as children.[78]

The Gebhard and other studies of sex offenders have been challenged as not giving a true picture of these people. It is argued that the convicted offenders are perhaps not the best group from which to generalize, because they are more compulsive, repetitive, blatant, and extreme than the undetected abuser, and they suffer from more conspicuous psychological abnormalities than nonconvicted child abusers.[79] More recent studies of adults who engaged in sex with children have shown that, again contrary to popular opinion, the sexual abuser is likely to be acquainted with the child; he or she is either a relative, sibling, family friend, or neighbor and not a passing stranger. Most adult-child sex takes place in the child's home or in the home of the perpetrator rather than in an alley or the woods.[80]

Much of the data about the victims of child abuse are in dispute, including whether specific incidences really did occur. There are a large number of studies, in part because government money is available for such studies, but there is as yet no agreement on who is defined as an adult or child and even what is defined as abuse. There is also considerable evidence that, at least in retrospective analysis, suggestions of abuse as a child can be implanted so strongly by professionals that the individual believes it even though such abuse never happened.

To emphasize the danger of such implanting, a group of concerned professionals, psychiatrists, psychologists, and sex researchers formed the False Memory Syndrome Foundation in Philadelphia. Not so much false memories as implanted ones are known to be a possibility among young children as well. Unfortunately, this phenomenon is something that many of those involved in prosecuting alleged child abusers tend to ignore. There are also other variables at work. Is a person five years older than the victim an adult if the victim is eight and the abuser is thirteen? How much age difference should there be between an adolescent and his or her abuser to be called an adult-child sexual interaction? In part, the definition of adult-child sexual interactions is both cultural and historical. In many countries, twelve is the age of consent, and in many parts of the United States thirteen was, until actions during the last fifteen to twenty years have moved to raise the age of consent to seventeen or eighteen.[81]

Even long-term effects seem to be an area of disagreement. The Dutch sexologist Theodor Sandfort published a report of his research on twenty-five boys, ages ten to sixteen, who had been involved or were involved in pedophiliac relationships with adult males. The boys were located through their adult partners and interviewed. Most of the boys described their relationship as predominantly positive, and they did not perceive their sexual contact with the adult males as representing abuse by adults of their authority.[82]

Systematic surveys of normal populations suggest that child-adult sexual contacts do not inevitably lead to long-term problems of adult functioning.[83] This was true of men as well as women, although women who had sexual relationships with a parent or relative who used pressure, force, or guilt to obtain sexual contact did show somewhat more impairment.[84] The more invasive the act, the more troubled the person was likely to be, although there are exceptions. Paul Abramson encountered two sisters, one of whom he named Sarah, who had both been sexually and physically abused by their father, stepfather, and stepbrothers during childhood. Although Sarah had engaged in a considerable amount of delinquent and self-destructive behavior during adolescence, by her twenties, she had become a healthy, self-directed women and had formed a satisfying relationship that culminated in marriage. Her sister, on the other hand, had exhibited considerable more psychopathology, and her chances of developing into a fully functioning, health adult appeared remote.[85]

Finkelhor found that sexual activity per se, whether it involved exhibitionism, fondling of the genitals, or sexual intercourse, had almost no relationship to the degree of trauma experienced by the child. The two factors that contributed the most to trauma or negative reaction was the use of force and a large difference in age.[86] Also important was the reaction of the adults to the incident; if they overreacted, children felt they were guilty of some unspeakable act and blamed themselves for what occurred.

Like so many other issues in sexuality, the issue of child abuse is mainly a political issue. It is also a media issue, since few issues incite greater public concern than innocent children being sexually abused. Though sexual abuse of children has existed, for much of American history it was ignored by both the social work and legal establishment, because the emphasis was on holding the family together rather than trying to intervene in family disputes. Specific attention to the issue picked up as a result of a 1962 article by C. Henry Kempe and colleagues on the battered child syndrome.[87] Attempts to deal with the battered child syndrome proved too difficult legally, and the result was a gradual change in state laws.

This change culminated with the passage of the Child Abuse Prevention Act of 1973, of which Senator Walter Mondale of Minnesota was the key mover. The emphasis for political reasons was on gross physical abuse rather than neglect, and it brought the power of the federal government directly into the affairs of the family, more so than any earlier legislative action had. This only emphasized the ambiguities of what constituted gross physical abuse, since it was essentially unclear. One of the results of the act was to provide $85 million over the following four years to establish the National Center for Child Abuse and Neglect.[88] Thus there was a bureaucracy to deal with child abuse, and like any other bureaucracy, there was the possibility of expanding its power and influence far beyond what the original sponsors intended.

Each revision of the legislation added more power to the bureaucracy, particularly in 1984 with the passage of the Federal Child Abuse Act.[89] Little discretion was left to the professional in terms of reporting abuse, and when it was reported, the bureaucracy went to work relentlessly. While there was real child abuse, little research had been done on abused children or, for that matter, on children as expert witnesses.

The focus on child abuse was coupled with a changing role for women, by which many were leaving the home to work, something they could do only by leaving their children in child care centers, which left many women feeling guilty. The change was strongly opposed by many conservatives in the country, who felt women's place was in the home. Inevitably, child care centers came to be a center of focus, and those opposed to the changing role of mothers could play on these women's anxieties and could hit at their weakest point.

All of this was compounded by the fact that there was basic lack of research on childhood sexuality or even on childhood talk about sexuality, and the result was a kind of mass hysteria. Neutrality in such investigations was more or less eliminated, and simple charges were often enough to have children removed from their families, if the suspicion centered on the family, or to have professionals charged, if the abuse occurred in a school or similar institution. Special law enforcement task forces involving social workers and police, among others, were set up to investigate allegations. The result was an escalation in charges and in anxiety among parents.

The most notorious of the child care cases was the McMartin case in Manhattan Beach, California, although the Jordan, Minnesota, case received almost as much publicity. In most such incidents, those accused of child abuse were not found guilty or, if found guilty, were later found to have been convicted on false evidence. Still, many are serving long terms behind bars on the basis of suspect testimony. The fear of possible child abuse in child care centers only added to the guilt of working mothers and raised their anxiety levels to new highs. This led them to suspect even the most harmless incident as a sign of possible abuse. Some of those who have most emphasized the fear of child abuse have even revived the age-old myth of satanism and witches to explain what they call the epidemic of child sexual abuse.[90]

In a brief period, the United States went from a country that had swept child sexual abuse under the rug, to one in which the slightest hint of child awareness of sexuality might be enough to charge sexual abuse. The term was ambiguous enough also to be used as a basis to attack television and movies for their increasing emphasis on sexuality and as a need to return to the moral standards of an earlier generation. The fear of child abuse proved to be the weak point of the sexual revolution, namely the American insistence on the innocence of the child. It was hoped that the clock could be turned back to the Victorian past, when sexuality was not mentioned. Interestingly, it was the very ability of the media to discuss what had been previously forbidden that allowed child sexual abuse to become such an issue.

In this sense, however, the campaign against child abuse represents a breakthrough in that the previously forbidden could be talked about. Juvenile court records in the 1920s, 1930s, and 1940s, for example, demonstrate that social workers usually discounted reports of child abuse and refused to act on the complaints of either an adult or a child. The prevailing mind-set held that the family unit was more important than the individual child, and officials were, therefore, reluctant to interfere and possibly break up a family. Most newspapers, emphasizing only the news that was fit to print, refused to deal with such sordid cases in their news columns.

As a former police reporter for a metropolitan newspaper, I can report at some length on this. While working as a reporter, I ran across one of the most horrible incidences of child sexual abuse that I have ever seen: A man had raped his one-year-old daughter, damaging her genital area so severely that the surgeons spent several hours suturing her and trying to repair the injury. Angered and upset, I wrote a story about it, only to have it returned by the city editor with the approval of the publisher. They said that they could understand my distress and anger but such matters should not appear in a family newspaper.

Finally, the issue has come to public attention, but like most such issues, there has been an overreaction, as if this would compensate for past neglect. Unfortunately, in the United States it is nearly impossible to do serious research into the topic; it is even difficult to study individuals who order pornography depicting adult-child sex, as possession of such pictures is against the law. Research into the topic has not kept up with the real need of the public to know, and the result has been the appearance of a lot of self-proclaimed experts who know little about sexuality or about children. Definitions remains ambiguous, charges can be devastating, and few want to really explore the topic for fear of being accused of being a child abuser. In a sense, research on the topic is similar to that on masturbation at the turn of the century. Americans knew little about it, and what little they knew was probably erroneous.

HOMOSEXUALITY

If child abuse has been a difficult subject to study, this is no longer true of lesbianism. The second wave of feminism again has had a liberating effect. Anyone who examines the studies of homosexuality in the past is immediately struck about how little attention has been paid to lesbianism. If nothing else, this emphasis was another indicator of the male domination of the study of sex. Though there were important early studies of lesbianism, particularly by Davis, most of the research concentrated on male homosexuality.

If anything, the disparity between studies on homosexuality as compared to lesbianism increased with the gay liberation movement. Though there were always cooperative efforts between gays and lesbians, ultimately lesbianism owed much more to the consciousness raising of the feminist movement than it did to male homosexuals. But perhaps this is putting the emphasis wrong, as lesbians themselves were among the leaders in the feminist movement, and in studying and researching women, lesbians were better able to understand themselves. Much of the lesbian scholarship, in fact, has been devoted to establishing a lesbian identity, its relationship to both lesbian and heterosexual communities, and to its survival. This work has been dominated by humanistically oriented scholars and the study of history and literature has played a major role in it. Lesbians have adopted varying definitions, some of which go beyond the traditional view of sexuality. This is particularly the case for those who define themselves as *political lesbians*, an identity that grew out of a feminine consciousness and implies a conscious rejection of the patriarchy, of traditional roles for women, and of limitations placed on women's control of their own lives. It has become a conscious embrace by women of women, as their primary emotional, erotic, and spiritual attachments.

Though lesbianism has benefited and been part of the homophile movement, it has also been separate and distinct, as evidenced by the founding of the Daughters of Bilitis in 1955 and the publication of the *Ladder*. Though the early studies on lesbianism, such as those of Davis, were done by women and some of these were not published until later,[91] most of the quantitative studies have combined lesbianism with homosexuality and been done by male and female researchers.[92] Many of the lesbian studies have concentrated on female friendship patterns,[93] recovering and identifying lesbians in the past, establishing a self-identity,[94] child custody, and women's health.[95] Though many of the lesbians who have children had them before they identified themselves as lesbians, there are also a number of female couples who have either adopted children or turned to artificial insemination, sometimes self-insemination using such traditional household items as turkey basters.[96]

Though lesbianism is part of the women's movement it is also very much part of the movement for gay power and has played a significant role in bringing about changes in attitudes toward homosexuality. One of the major factors in such changes was the findings of sex researchers, particularly of Kinsey, who brought home to homosexuals of both sexes that they were not alone and that though they might be a minority in most matters they were the same as others.

A particularly influential study in this regard was by the psychologist Evelyn Hooker, who had become acquainted with a gay male neighbor in the late 1940s. This made her somewhat curious about male homosexuality, and as she investigated, she found that few clinicians had ever studied homosexuals outside of mental health facilities or prisons. She undertook an eight-year longitudinal study of some carefully chosen thirty homosexual men whom she matched with thirty heterosexual men in age, education, and IQ. All sixty took the Rorschach test, the Thematic Apperception Test (TAT), and the Make-A-Picture-Story test (MAPS), which are projective tests then commonly used as diagnostic aids. Two testing experts evaluated the results blind; that is, they did not know the subject or his sexual orientation. Neither judge did better than chance in telling homosexuals from heterosexuals. General adjustment scores were the same for both groups. Through these and other measures, Hooker concluded that homosexuals were "very ordinary individuals, indistinguishable, except in sexual pattern, from ordinary individuals who are heterosexual." She argued that homosexuality as a clinical entity did not exist and that the forms of homosexuality were as varied as those of heterosexuality.[97]

Hooker's research, though on a small scale, was important for demonstrating that homosexual men did not have any psychopathology that she could measure and thus they could not be considered ill. It does not necessarily follow, however, that homosexual men were in every way other than sex indistinguishable from heterosexual men. Both groups in her sample represented highly selected individuals who were included because of their good adjustment. Hooker, to her credit, always emphasized the tentative conclusions of her studies and emphasized that there was not one homosexuality but a variety of homosexualities, an assumption now taken for granted. Still, she, as had Kinsey, openly challenged the psychiatric establishment's views on this, which tended to hold otherwise. Moreover, Hooker was not alone. Others joined the fray. Michael Schofield, for example, argued that homosexuals as a group probably differed from heterosexuals only in their object choice.[98]

The most comprehensive study yet made was the large-scale study by Bell, Weinberg, and Hammersmith on the lives of a sizable sample of homosexual and heterosexual men and women in the San Francisco Bay Area (see chapter 8). Among other findings, the three investigators reported that homosexual men and women were more likely to report poor relationships with their fathers than were the heterosexual members of the study group. It was not clear, however, whether the poor relationships were due to their gender nonconformity or whether parental rejection was itself part of the causal sequence for homosexuality. The most common element in the childhoods of both lesbians and homosexual men was gender nonconformity. Many of the boys had developed a homosexual pattern in their early years, although there was no evidence that this was due to a lack of opportunity for heterosexual interaction.[99]

Such studies increasingly brought homosexuality and lesbianism to public attention and also led homosexuals and lesbians to become more public themselves. Much of the organized homosexual movement of the early part of the twentieth century as exemplified by Magnus Hirschfeld's political campaigns had disappeared with the rise of the Nazis and the massive disruption resulting from World War II. The Dutch groups remained particularly important, and for a time the International Committee for Sexual Equality, founded in 1951, had served an important role as a clearinghouse for the exchange of information and opinion. Even this slight effort led to a denunciation of the group by the American writer R. E. L. Masters, who claimed it was "by far the most powerful body in the history of homosexual organization and may control to an extent of which few even dream the policies and organizational activities of homosexual groups throughout the world."[100] This kind of sensationalism sold books in the early 1960s, even though it had no correlation with reality. Still a number of organized homosexual groups and publications reappeared in Europe during the late 1940s and early 1950s.

In the United States, organization was much slower. The most important of the early groups was the Mattachine Society founded by Harry Hay in Los Angeles in 1950. Originally a secret organization modeled on the cell structure of the Communist Party, it spread rapidly. In a 1953 convention, the original Mattachine Society was restructured, and the reorganized society moved to San Francisco. It gradually went national, although in name only, as many of the groups that called themselves Mattachine had no connection with the organizing body. Out of the original society came the magazine *One*, which was the dominant publication in the gay community for two decades. Later, the *Mattachine Review* also made an appearance. These publications were symbolic of an increasingly public facade for the gay community and were victors in some of the early struggles. *One*, for example, had difficulties with the U.S. Post Office, which tried to prevent its circulation on the grounds that it was obscene simply because of its subject matter. Taking the issue to court, it won the right to go through the mails in 1958.[101] Giving further encouragement to organizations of homosexuals was Donald Webster Cory's (a pen name for Edward Sagarin) book about being homosexual in America; it was the first semiobjective survey of the homosexual lifestyle in the United States.[102] Among other things, the book's success made publishers aware of a potential homosexual book-buying public, which previously had been ignored.

Neither *One* nor the *Mattachine Review* had been the first gay magazine to be published in Los Angeles. That honor belongs to *Vice Versa*, some nine issues of which were published in 1947 and 1948. Edited by lesbians, it aimed to reach the lesbian community. Although it soon disappeared, many of the people associated with it were instrumental in the publication of *One* and later the *Ladder*.

Homosexual organizations grew rapidly in the 1950s and 1960s.[103] They also became increasingly politicized. In general, however, homosexual activities were ignored by the popular media, and the word *homosexual*, for example, could not be listed as the name of an organization in telephone directories, let alone in the columns of the daily newspapers.

Much of this changed with the so-called Stonewall Rebellion in June 1969 in New York City, an extensively covered event that marked the public emergence of homosexuality onto the pages of the family newspaper. Stonewall Inn was a dimly lit dance bar in Greenwich Village, which served as a haven for cross-dressing prostitutes (street queens) as well as for lesbian cross-dressers who were usually denied entrance to most other bars. Stonewall and similar bars at the lower end of the social scale in the gay community served as a strong magnet for police raids, because the customers were among the least powerful and most visible in the gay and lesbian world and were also most likely to be engaged in open solicitation.

During one of the police raids (shortly after midnight on June 27), one of the lesbian cross-dressers struggled with the police, and her struggle galvanized not only the drag queens but the watchful crowed outside. The crowd, if only because the bar was in a predominantly gay neighborhood, also included a large number of gays and lesbians. Roused by the struggles that ensued, the crowd began to taunt the police. Soon the arresting officers found themselves threatened by a hailstorm of cobblestones and bottles. The result was three days and nights of confrontation between the police and the street queens and their allies. The brief struggle became immortalized in gay legend as the Stonewall Rebellion, and the media, too, sensationalized its significance. The incident brought to public consciousness the existence of homosexuals and lesbians on a large scale, and it emphasized to the gay and lesbian community itself the importance of making their presence known. The result was gay liberation, in which homosexuals, following the civil rights movement and the feminist movement, now sought the kind of government recognition and protection that the other groups had.

There had been various factors outside of the gay community that made gay issues timely. A number of groups and organizations had challenged the traditional criminal and sickness model of homosexuality and lesbianism. In 1957, the Wolfenden Commission, a British Parliamentary commission, had urged the decriminalization of sexual activities between consenting adults.[104] In 1964 in Los Angeles and in 1965 on the national level, the ACLU had advocated a change in laws dealing with homosexuality.[105] The American Friends Service Committee had argued that the quality of sexual relations was more important than the kind of sexual activity.[106] The American Law Institute had also urged that sexual behavior between consenting adults be decriminalized. Even the U.S. government got into the act through a task-force on homosexuality set up by the National Institute of Mental Health, which was chaired Hooker. It also urged that sexual behavior between consenting adults be decriminalized,[107] a stand that not only caused a delay in publication of the report but a withdrawal of the government from any sponsored research that had direct bearing on the issue.

The psychiatrist Thomas Szasz emphasized to the public as well as to his colleagues how traditional ideas of sickness and health were tools of social repressiveness at the worst and narrow conventionalism at best.[108]

Perhaps the inevitable result took place in 1974 when the American Psychiatric Association voted to remove homosexuality from its catalog of mental illnesses, declaring it to be instead a "sexual orientation disturbance." Ultimately, it abandoned even this language.[109] In this action, they had been preceded by the American Psychological Association. Critics wondered whether the vote of a majority of psychiatrists could suddenly remove homosexuality from the category of illness. Some claimed the action was similar to one declaring that pneumonia was no longer an illness. The change, however, was basically a reflection of research findings that had undermined traditional psychiatric assumptions.

Still, the change was also political, in part the result of pressure on the psychiatric community, from both within and without. Not all psychiatrists accepted the decision of their nomenclature committee, and at their insistence, a referendum on the issue was held that affirmed the decision of nomenclature committee. Weinberg and Colin Williams, in their 1974 study of homosexuality, complained that psychoanalysts, by their emphasis on cure, had hindered theoretical progress and prevented a better understanding of variations in sexual behavior. This resulted in methodological deficiencies in their studies:

First, the samples used have been extremely small. This in itself need not always be a serious defect, even if it does limit more complex analysis of the data. A much more important problem is that such samples are usually made up of persons who are patients of the clinicians doing research and cannot provide much knowledge about homosexuals in toto. While a representative sample of homosexuals may be impossible to achieve, certainly less biased groups can be obtained. . . . Another major defect of such studies has been that control groups are rarely used. Comparison groups are crucial if, for example, one is concerned with determining the degree to which homosexuals are maladjusted (instead of claiming it by fiat). A heterosexual control group is essential to answer this question as well as etiological questions. Finally most studies of homosexuality have been culture bound.[110]

With the action of the psychiatrists, homosexuality had in essence been removed from the medical model, and the door was opened for new kinds of research. It also officially marked the end of the dominance of Freudian psychoanalysis on sexual research, since much of the decision to change the American Psychiatric Association definition had been based on research that psychiatrists had little to do with.

The 1970s and succeeding years saw a literal explosion in studies on homosexuality and lesbianism. Special journals such as the *Journal of Homosexuality* were established that gave an outlet for publication not only to the scientist and scholar but also to the dedicated lay person. University presses such as that of Columbia established special series to emphasize the importance of studies in homosexuality and lesbianism. Every discipline from anthropology and art to zoology has added its own insights. Moreover, insights gained from research into homosexuality, such as that by David McWhirter and Andrew M. Mattison's into gay couples, could as easily be applied to heterosexual couples.[111] As Denis Altman claimed, there had been a homosexualization of America and an Americanization of the homosexual.[112]

SUMMARY

In both feminist studies and homosexuality, social constructionism has carried the day. David Greenberg, in his massive study of the construction of beliefs about homosexuality, however, offered some words of caution, which he said held for all forms of sexual expression. He emphasized that there is always the possibility that any legal or societal interpretation put on a behavior can have unanticipated social consequences. The nineteenth-century inverts who argued that homosexuality was innate did not anticipate what the degeneracy theorists would later make of their claim. In its time, the medical model was an emancipating one, although it was later seen as repressive. Greenberg also asserted that sexologists and others should always keep this in mind.[113] In short, there is still much that is not known about human sexuality, whether of males or females, heterosexuals or homosexuals. Furthermore, the feminist and gay perspectives, though they have offered us new insights, do not offer us the final and ultimate truth, something that is never possible in science anyway.

10

PROBLEMS OF AN EMERGING SCIENCE

As sexology entered the twentieth century, it had become essentially another aspect of medical research. As such, research was based on patient populations and was aimed at helping physicians diagnose and treat individuals who consulted them. The course of the twentieth century saw the field broaden its horizons to include large numbers of nonphysician researchers who were not so much interested in establishing new diagnostic categories and treatment modalities as exploring new frontiers. Data were gathered not only from patients but also from carefully chosen statistical samples. Even many of the physicians who continued to do research in the field, such as Ellis, Hirschfeld, Bloch, Dickinson, and Masters, tended to abandon the diagnostic categories and to turn to data-gathering methods of the social sciences and even humanities. Biochemists, geneticists, physiologists, and endocrinologists all established a strong presence in the field, helping strengthen the scientific knowledge about sex.

The very complexity of the subject, however, worked against the dominance of research by any one discipline. While biology in its various specialties is essential to understanding sexual functioning, so is a knowledge of the social, cultural, and psychological aspects of sexuality. This requires the expertise of anthropologists, historians, psychologists, sociologists, literary specialists, art historians, musicologists, and others. If help or treatment is sought by an individual, a number of professions can become involved, not only the physicians who dominated the field in the nineteenth century but nurses, social workers, psychologists, social psychologists, and various kinds of counselors and/or therapists might also be called on. Because many aspects of sex behavior are involved with the legal system, lawyers and law enforcement officials also have an interest. So do educational professionals because one of the burning issues of the day, as it has been from the beginning of the century, is what should children be taught about sex and who should teach it.

With the influx of new researchers from a greater variety of disciplines and professions, the study of sexuality finally seemed to achieve respectability, at least of a sort. No one individual or even group of individuals, however, gained the dominance that Kinsey had in the 1950s or Masters and Johnson had in the 1960s. Even Money, who had pioneered gender research (see chapter 8), represented only one voice in a rapidly expanding field. Increasingly, individuals from a number of disciplines contributed to new findings, and at the same time, sex researchers in general became more narrowly focused. The wide-ranging generalists such as Kinsey, Money, and Masters and Johnson were less likely to appear as the knowledge about sexuality grew.

A major change in sex research has been the increasing contributions to sexology from previously silent disciplines such as history and sociology and professions such as law and nursing. The agenda for research also seemed to be less influenced by those who regarded themselves as sex researchers and more by outside forces. Though outside forces have always been important in sex research from the time of Ulrichs and his studies on homosexuality, during the much of the twentieth century it was a small group of professionals who dominated and even defined for the public what was known. Ellis and Kinsey are two of the more outstanding examples.

As the barriers to public discourse on sexuality were gradually removed, however, all sorts of people emerged as sex experts. They differed from the nineteenth-century writers on sex, whose ideas were outlined earlier, by their emphasis on the joys and pleasures of sex instead of the dangers with which their predecessors had been concerned. They also differed in their source of authority. The nineteenth-century writers for the most part were driven by religious morality, which they interpreted as being hostile to sex.

Writers in the last part of the twentieth century claimed science as their authority, although they also were more interested in propagandizing than in research findings. Much of the publicity about new developments in human sexuality was not generated by research results but rather by the rhetoric of members of some of the special-interest groups. Large segments of the interested public found it difficult to distinguish between legitimate research and wishful thinking. Unfortunately, this confusion was not limited to general public but often proved to be a problem even for those who regarded themselves as sexologists.

This was because the entry of new disciplines and professions, while broadening the scope of sex research, also increased the potential for ghettoization both by topic and discipline or profession. How does a biologist, for example, evaluate what is good history, and how does a specialist in eighteenth-century sex literature judge what is valid psychological research? In fact, how does a person from one specialized area even learn about the literature from another, especially because many researchers publish in their own specialized journals for others doing research on the same narrow subject matter? This is a problem in many other areas than sexology, but the issue is more serious in sexology simply because as inhibitions about public discussion about sex have been reduced, demand for information and perhaps even titillation has increased. Publishers, editors, talk show hosts, and columnists all try to respond to this interest, and the result is massive amounts of what can only be called "pop sex"; a mishmash of real data mixed with fantasy is being disseminated to feed the demands of a voracious public appetite.

Still, if even the expert has trouble keeping abreast of developments, how can the public tell junk from basic research? Moreover, in spite of the newfound respectability of the sex researcher, the researcher still has to exercise caution, because the potential for controversy remains very great in many areas of human sexuality. Moreover, many of the pop experts claim the same credentials as experts, and much of what is written is based on so-called case studies of clients, in which individual cases are used to generate universal examples. Anyone can, and many do, claim to be sex experts without any special knowledge base. Though state licensing laws generally have limited the term *sex therapist* to the licensed professionals in the regulated fields, most of them still lack basic expertise in sexual matters.

Although the American Association of Sex Educators, Counselors, and Therapists (AASECT) tried to dominate treatment modalities and educational programs through its certification program, such an attempt was doomed to failure, if only because so many of those engaged in sex therapy and education were already certified by their own professional associations and licensed by the state. Thus they saw no need to pay extra or train longer for AASECT certification. Furthermore, the history of a profession policing itself, as AASECT attempted and the American Medical Association (AMA) claimed to do, has a long history of failure in the United States.

AASECT's failure as a certifying agent for more than a select few proved a serious blow to the professionalization of sex therapy. Still, it also allowed organized sexology continued to develop more as a science than a profession, with a variety of disciplines entering the field and establishing expertise. It is the methodology used in the study that is important in determining the significance of the work, not the person or professional who wrote it up. This has been both its strength and its weakness, since as the educational system now exists, it is exceedingly difficult for someone to start out as a sexologist. Rather individuals seem to stumble into the field as they expand their horizon beyond their own discipline or add to their professional knowledge base.

SEXOLOGICAL EDUCATION

In fact, it remains almost impossible to enter into the field through a normal graduate school program. Though two American universities—New York University and the University of Pennsylvania—have graduate programs in human sexuality, both are in the school of education and are aimed at those seeking to be sex educators, not sex researchers or sex therapists. Neither program is particularly research oriented, and one of them has only one full-time faculty member in the area, though it has awarded degrees to scores of graduates. Several other universities allow graduate students to concentrate on certain aspects of sexual behavior within a particular discipline, and these institutions often encourage coursework in other disciplines. These programs, however, are individually tailored and available only to the extremely persistent.

The one other alternative has been the Institute for Advanced Study of Human Sexuality in San Francisco, which grew out of training programs established by the National Sex Forum. The forum itself began in 1968 as an outreach service of the Methodist-oriented Glide Foundation. The foundation set up a special program at the Glide Memorial Church in downtown San Francisco to reach out to gays and lesbians as well as to various street people. The program was under the direction of Ted McIlvenna, who was both an effective missionary and an entrepreneur. It was under his direction that the institute was formally incorporated as a graduate program in 1976 under the laws of California. It is a private, nonprofit school that caters to those who already have degrees in other professions. It offers the interested individual the opportunity to learn more about the developments in sexuality. Pomeroy served as dean until his retirement in the late 1980s. Though the institute awarded doctorates, its program was directed toward therapy and was not particularly research oriented. Few of its doctoral dissertations ever led to published articles.[1]

Beginning in the 1980s, several universities developed special coordinating committees in gay and lesbian studies that emphasized the interdisciplinary nature of such a discipline, but these programs have not yet progressed beyond their concentration on just one aspect of sexuality. Similar programs also developed in gender studies, usually as an offshoot of women's studies. Traditionally, many universities, particularly the land-grant ones, have provided departments or courses in marriage and family, which increasingly have included serious study of sexuality. So far, however, no university has taken the next step and established a graduate program in research-oriented sexology.

In essence, this means that sex research is not institutionalized in any major U.S. university or college, although it is in the Université de Québec. Sex, however, is widely taught on college and university campuses by professors from a variety of disciplines, and most institutions offer at least one course on the topic as part of a general education requirement or option. In most cases, the professor teaching the course does not regard it as a specialty but simply a departmental requirement that someone has to offer. Still, one result of such courses has been the appearance of numerous textbooks, the best of which summarize the current research findings in the area of human sexuality.[2] Anyone interested in these results would do well to read one of the textbooks. In spite of the popularity of such courses, rarely are faculty promotions based on one's expertise in sexology.

The greatest problem caused by the lack of an institutional base for sex research, however, is that the individual researcher is on his or her own, and without easy access to interdisciplinary and interprofessional cooperation with colleagues in the same field. Some of the worst errors in the field have occurred when individuals, experts in their own fields, tried to incorporate information from other specialties without being aware of the pitfalls of so doing. About the only way such interdisciplinary cooperation can come about is through the contacts made in such organizations as the Society for the Scientific Study of Sex (SSSS). The SSSS has made special efforts to maintain its interdisciplinary and interprofessional orientation while still emphasizing the importance of methodology in sex research. This is why the model of the San Francisco–based institute offers the best potential for the future of sexology, although such a model needs to go beyond the therapy and must be part of a full-fledged university rather than a stand-alone institute. Individuals who are experts in a particular discipline certainly can contribute to the growth of the sexual knowledge base. Their contribution would be much more valuable if they could exchange ideas and interact with researchers in other disciplines and professions. In this way all sexologists can fill in the gaps in some of their own knowledge. They can then become sexologists in the real sense of the term, even though their own research will undoubtedly continue to reflect their own discipline or profession.

VIEWS OF THERAPY AND SEXOLOGY

One of the questions continuing to trouble sexologists is what kind of approach they should take in their sex studies. Should it be a purely dispassionate study and examination of sexuality or should the sexologists be more involved in the politics of sexuality and with the people they study? This issue effects not only the researchers but the therapists as well, since within the therapy field there is a division between what some have called the humanistic therapists and the scientific ones.[3] Masters, Johnson, and Kaplan are said to be the leaders among the scientific-oriented therapists; whereas Hartman, Fithian, and Pomeroy are the leaders of the more humanistic ones. In actuality, the differences primarily involve treatment modalities, but even these are not as great as some have claimed.

One of the differences between the two viewpoints is said to be over the question as to whether the therapists should use such things as guided imagery, body image exercises, massage, and sexual exercises. Hartman and Fithian, for example, emphasized that therapists should engage in touch between clients and therapists and believed that superficial touch from one or both therapists in a responsible, professional situation was often necessary to get clients comfortable.[4] The more scientific therapists thought the touching could be carried too far and generally looked down on massage involving the therapist and the client.

One of the difficulties with the humanist-oriented therapists, in general, is that few of them published more than anecdotal accounts. Hartman and Fithian, for example, who made hundreds of observations on their clients, did not publish statistical records of their success rate, because they felt results could not be measured in short-term success but were dependent on attitudinal changes. Masters and Johnson would agree with the general tenor of this argument about success rate, but they thought it was important to emphasize short-term results as well.

Another difference between the two is in the use of surrogates who act as partners in treating sexual dysfunction in patients who lack their own partners. Although Masters and Johnson originally had used surrogates, they stopped because of a lawsuit brought against them by a husband who claimed that his wife was working as a surrogate. They ultimately felt that they did not need them. Hartman and Fithian continued to use surrogates, as did others, especially after the issue of recruitment of surrogates was, at least partly, solved by the foundation of the International Professional Surrogates Association (IPSA) in Los Angeles in 1973. The group established its own training program, code of ethics, and referral service, which gave the therapist a source to which he or she could turn. The issue, however, still remains controversial among therapists, with the more humanistic ones among the strongest advocates.

Another issue that separated the humanistic from the more scientific, at least for a time, was the use of the Sexual Attitude-Reassessment (SAR) workshops developed by the National Sex Forum. This entailed watching sexually explicit movies running simultaneously on a number of screens and portraying every variety of sexual activity to desensitize the would-be professional to various sexual problems. The assumption was that the therapists could then better handle the sexual problems of their clients and researchers could make their research presentations in a matter-of-fact way.[5] In popular jargon such demonstrations were often called *fuckeramas* and as such were initially criticized by Masters and Johnson. Criticism brought about modification, what some have called an intellectualization, of the process, and the modified SAR became widely accepted as a necessary first step to becoming a sexologist or therapist. Often an SAR session is used for clients, although here some differences of opinion among the experts on their effectiveness remain.[6]

Though the difference was put in terms of the humanist-oriented versus the scientific-oriented therapist, underlying the debate was a concern over what was regarded as respectable and what kind of professional image the sex therapist and sexologist should project. Sex research and sex therapy were regarded by the public at large, as well as by many of those engaged in research and therapy, as not quite respectable. Some seemed to feel that the more the topic could be intellectualized, the better chance it had, if not of becoming fully respectable, at least of having its research accepted. Kinsey, for example, tried to avoid any kind of advocacy position on the grounds that scientists were dispassionate observers. He even refused to join together with other sexologists for fear they as a group might take stands on issues that he felt would undercut his scientific impartiality.

Kinsey's attempt at scientific detachment, however, did not ultimately save him from a congressional investigation. Though matters have improved greatly since Kinsey, there is still a kind of taint attached to those who engaged in sex research or therapy, in spite of the growing respectability of the field. I, for example, have been investigated by the FBI and labeled a security risk, in part because of my studies of human sexuality.[7] Calderone, as noted earlier, was called a Communist.

Individuals have failed to get tenure, because their research into sexual topics has been dismissed as not scientific. Moreover, some areas of sexual behavior have been almost impossible to research. Several of the scholars who attempted to investigate adult-child sexual relations found themselves out of a job, had their careers threatened, and in a few cases, served jail terms. Lawyers, such as Lawrence Stanley, were even threatened with disbarment because they defended individuals allegedly involved in child pornography.

Most sex researchers and therapists become used to the giggles, the laughs, and the putdowns as the nature of their research comes to be known to their colleagues, friends, or wider audiences. In the past, it has taken determination to succeed in the field, and in the process, two conflicting stances about sexology seemed to evolve. At one end is what can only be called a missionary attitude: Sex is good, and let me convince you. In a sense, this was the attitude of Hirschfeld, and it is very influential among the more humanistic therapists. Others emphasize the dispassionate attitude and scientific detachment: Sex is an important topic to investigate and I am doing it scientifically. This was the attitude of Krafft-Ebing and Kinsey, and it underlies that of the more scientific researchers.

Most researchers and therapists end up somewhere in between, even those who are labeled by themselves or others as being in the humanist or scientific camps. Many of the people who end up in the sex field, as do many in the various helping professions, probably initially were attracted as a way to find out information about themselves or their loved ones. Certainly, as has been indicated, both feminists and homosexuals have played a disproportionate role in the study of sex and gender issues. Though this does not ultimately prevent them from arriving at objective conclusions, it raises a problem that cannot be ignored. The SSSS, for example, faced this problem in terms of membership of individuals who come from sexually stigmatized minority groups. Such individuals who can be, and often are, researchers in their own right serve a valuable purpose to other researchers, since they give insights to sexual issues available only to the insider and can help the researchers gain access to sample populations. Many obviously also have a mission. Nevertheless, the society decided to admit such individuals as regular members. Like any other members, they could submit abstracts or articles to the peer review for eventual presentation or publication.

The inclusion of members of such minority groups has given the SSSS an unusual appearance for a professional society, one that still embarrasses some SSSS members who want to be more professional, that is, eliminate such individuals. For other members, however, it is what makes the society worthwhile. The issue then is one between inclusion and exclusion. It is a problem faced by every sexological group, and one that has no easy answers. Even if a sexological group adopted a rigid exclusionary policy, admitting only those who are married and have children, there is no guarantee that all members would follow an acceptable heterosexual lifestyle pattern. Moreover, a policy of having a variety of members is less likely to isolate the potential subjects from the researcher; instead it emphasizes the collaborative effort of both researcher and subject to understand sexual behavior.

SEXOLOGY FADS

There are what might be called fashionable topics or areas in sexology, just as there are in other research and clinical areas. Sex therapy had its greatest growth years in the 1970s, after which there was both a decline and a reorientation. The decline was in the kind of therapy pioneered by Masters and Johnson that dealt with anorgasmic women, impotent men, and other sexual dysfunctions that exist among couples. Solutions became more complicated, perhaps because the easier problems had been solved and the persistent problems required medical intervention or long-term therapy. Many of the individuals who had specialized in sex therapy turned to more general therapy, although they still proclaimed sex therapy as one of their specialties.

The change was also assisted by the increasingly reluctance of insurance companies to reimburse individual sex therapists. Health maintenance groups, the growing trend in health care, were likely to include a sex therapist on their staffs, but the standards for acceptance into the groups became the same as the other professionals. This usually meant that the sex therapist was a member of one of the other registered professions. Even then, however, sex therapy was not offered on a long-term basis, but rather the therapist was encouraged to make a diagnosis and treat with medication or at most a few counseling sessions. This emphasis on quick treatment encouraged the healer to find a diagnostic category into which the client could be placed.

This led to two somewhat contradictory trends. Often traditional terms like *frigidity* or *impotence* were dressed up in new, less judgmental terminology that was also more professional, such as *anaorgasmic* and *erectile dysfunction*. In the past, many of those seeking help had what had been classified as perversions, pathologies, disorders, or otherwise deviant, new clients found themselves classed as having a paraphilia. This term, first used by Stekel, was picked up and popularized by Money and was ultimately adopted by the American Psychiatric Association in the third edition of their *Diagnostic and Statistical Manual*. Perhaps the term is less judgmental, although to call something abnormal or disordered love, even if it is done in Greek, still implies a judgmental stance to me. It is, however, a term that the ordinary client might not understand.

Money, in fact, has devoted much of his time in recent years to finding new terms for old, familiar diagnoses, perhaps in the process isolating the client from other individuals who share many of the same general factors, if not exactly the same manifestation. Whether the term *gynemimesis* (one who engages in female impersonation) is any more accurate because it is Greek than the plain English description is highly debatable, but it certainly makes it more difficult for the lay person to know what is going on. Whether *coprophilia* (being smeared with or ingesting feces) is radically different in its sources of origin from *urophilia* (being urinated upon or swallowing urine), or whether a leather fetishist differs from a rubber fetishist in base causes is unclear.[8] Putting many of these terms into Greek, however, seems an effort to emphasize that sexology is a science and that perhaps we know more than we do. This again returns to the question of how sexologists should look on themselves. Should they attempt to use the jargon of science and emphasize their statistical and anatomical knowledge or should they speak in languages that their clients understand?

At the same time that the criticism of jargon appears, there is a counter-claim that sex researchers are not specific enough or fail to distinguish between the pleasurable and the harmful. For example, much of the research into sadomasochism deals with what might be called theatrical sado-masochism. That is, it is activities engaged in by individuals who willingly choose to do so and for which the level of pain is essentially controlled by the masochist.[9] It is often viewed more as a social behavior than a pathology, and in the process the term has lost meaning. Weinberg, Williams, and Moser have argued there is no such personality type as a sadomasochist but rather it is an activity, and the focus should be more on the role than the person.[10]

Others, however, argue that there are real people out there, some of whom appeared as guards in Nazi concentration camps, who are sadists and who enjoy inflicting pain on others, particularly to those unable to fight back. There are also masochists, people who enjoy suffering. In fact, the history of Christian martyrs demonstrates the existence of thousands of such individuals. Should there be a distinction between the theatrical sadomasochism engaged in by many who today join S/M clubs or groups and the other, more traditional type? Are new terms needed? Are these real pathologies or are they simply socially constructed? Probably, sexologists need to speak of sado-masochisms and be much more specific than they so far have been.

Whatever they are called, the growing existence of social groups of sado-masochists, of cross-dressers, and of other previously stigmatized groups emphasizes the changing attitudes toward sex. In 1976, Robinson pro-claimed the modernization of sex, by which he meant a growth in sexual enthusiasm and a broadening of the range of legitimate sexual behavior. He claimed that this was the result of the research and publications of Havelock

Ellis, Kinsey, Masters, and Johnson. He worried, however, that in the process of modernization, the romanticism associated with sex was being threatened, because the sexual experience was being separated from its elaborate emotional associations. He feared that Americans were separated from their romantic past, gladly ridding themselves of that repression and embracing a deromanticized future, in which the greater freedom would lead to a emotional emptiness.[11]

Robinson was correct in arguing for a growing public consciousness of sexuality. The potentialities of sexual pleasures have led to such books as Alex Comfort's *The Joy of Sex* and its successor *More Joy*[12] and to an outpouring of books on erotic art. Certainly, the public seems to have an insatiable need to discuss sex, and both radio and television talk shows seem to be dominated by sexual issues. Ruth Westheimer has become a public personality, called by some the greatest sex therapist, because of the success of her television show. But others, better known for their research than their television personas, have also tapped into the public eagerness for sex information. Reinisch, the director of the Kinsey Institute, turned her expertise to writing a nationally syndicated column on sex problems, and the advice columnists from Dear Abby to Ann Landers to Joyce Brothers published in the same newspapers devote a considerable amount of their space to discussion of sexual issues.

Perhaps the best indicator of how far the concept of sex as pleasure has permeated society has been the growing concern with sexually disenfranchised people: the mentally and physically disabled, the incarcerated, and the elderly. The challenge to the assumption that the sole purpose and function of human sexuality is or should be procreation has led to an examination of the denial of sexuality to such subgroups. It is emphasized by various researchers that the ability to engage in sexual activities tends to enhance self-worth.[13] Only those individuals most opposed to any expression of sexuality would disagree with such a general statement.

The problem, as it is with so much in the sex field, is how far public policy should go. If all that those in charge of institutionalized individuals had to do would be to give their residents opportunities for privacy, only a minority might object. Masturbation has come to be recognized as an important form of sexuality, and it is comparatively easy to have staff members tiptoe out of the room when they see such practices or even when they see a husband and a wife having sex together. But issues become more complicated when the sexual activity involves two persons of the same sex or individuals who are not married to each other. Still, administrators can accept this if they are certain their clients still have enough mental capacity to make such decisions. But what if the clients are judged not to have such capacities? Complicating the issue even further is when the client's disability involves the lack of control over bowel and bladder or the inability to move freely about. What kind of special help should the attendant give in positioning or turning the person or even helping him or her to masturbate? Many individuals are helped in their masturbatory fantasies if they have access to sexually explicit materials, films, books, and illustrations. Should institutions or attendants engage in the distribution of pornography? What if a person wants and needs a sexual partner but is unable to find one? Should sexual surrogates be hired?

Inevitably, administrators or caretakers often find it easiest to not open the Pandora's box of sexuality beyond giving a modicum of privacy to those individuals under their care. They are not only frightened by what the public reaction might be if they went further but also realize that they themselves are unable to handle sexual questions. Complicating the issue further is the reemergence of sexually transmitted diseases as a major public health problem.

AIDS AND HIV

The greatest challenge is AIDS (acquired immunodeficiency syndrome), which hit the United States with a vengeance in the 1980s. If the growth in sex-for-pleasure books and a growing willingness to discuss sexual issues emphasizes changing public attitudes toward sexuality, the anxiety and fear raised by AIDS emphasizes the difficulty government has in coping in any positive way with sexual problem areas.[14]

When the first cases of AIDS were reported in 1981, epidemiologists at the Centers for Disease Control in Atlanta immediately began tracking the disease backward in time as well as forward. They then determined that the first cases of AIDS in the United States probably occurred in 1977, although that date is continually being pushed further back.[15]

By early 1981, AIDS had been reported in fifteen states, the District of Columbia, and two foreign countries, but the total remained low, 158 men and 1 woman. Perhaps because so many of those diagnosed as having the disease were known to be homosexuals, there was a reluctance to intervene too drastically. Still, the potential danger of AIDS to spread beyond the gay community was clear by 1983, when 3,000 cases of AIDS had been reported in adults from forty-two states, the District of Columbia, and Puerto Rico, and the disease had been recognized in twenty other countries.[16]

During 1983–84, a number of symptoms became known as AIDS-related complex (ARC), and these included chronic generalized lymphadenopathy (swollen glands), extreme fatigue, weight loss, fever, chronic diarrhea, mild immune system abnormalities, decreased levels of blood platelets, and fungal infection in the mouth. Some ARC patients appeared to remain stable indefinitely; others developed the various symptoms associated with AIDS. The retrovirus that causes AIDS was identified in 1983 in France, where it was called lymphadenopathy-associated virus and in the United States in 1984 where it was named human T-lymphotropic virus. Eventually both countries accepted the term human immunodeficiency virus (HIV) advocated by the International Committee for the Taxonomy of Viruses.[17]

Researchers found that before HIV can reproduce, it must make DNA from its RNA. Thus it may remain dormant for some time, until some factor stimulates it to begin reproducing itself. As it makes copies of itself, HIV destroys the ability of the body's T4 helper lymphocytes to stimulate the immune system to fight diseases. Individuals are diagnosed as having AIDS when they develop one of the opportunistic infections or diseases associated with AIDS—Kaposi's sarcoma, pneumocystis carinii pneumonia, or cytomegalovirus infections—and test positive for HIV. One of the puzzles confronting AIDS researchers and physicians is that individuals can test positive for AIDS infection, but still not show the symptoms of these opportunistic infections. Exposure to HIV is determined by a blood test known as an enzyme-linked immunosorbent assay (ELISA), but the risk of subsequently transmitting or contracting AIDS is not clear. Cases are now known of individuals who have tested positive for HIV for more than ten years but who have not yet come down with AIDS.

The number of deaths from AIDS grew rapidly in the 1980s, especially among the young adult males. By 1986, it accounted for at least 15 percent of adult male deaths in San Francisco and at least 4.4 percent of such deaths in New York City. As of this writing, it is the major killer of both young men and young women. For sex researchers, the growth and spread of AIDS emphasizes the danger of government and foundation neglect of serious sex research. For example, one of the principal difficulties in dealing with AIDS, even after the cause of the disease had been diagnosed and some treatment options became available, was the lack of reliable information about the sex habits of Americans. The best study was that of Kinsey done in 1947, which, as indicated earlier, by no means used a representative sample.

Though there had been other quantitative studies of sex habits, none has matched the scale of the Kinsey studies. Morton Hunt's study sponsored by *Playboy*[18] was one of the better ones, but as various polls have shown, including two by Gallup,[19] answers depend on the way the questions are asked. Some of the more specialized polls, such as J. D. LeMater and P. MacCorquodale's on premarital sexuality,[20] are helpful perhaps because they did concentrate on specific sexual behavior. The same is true of studies of married couples, such as that by P. W. Blumstein and P. Schwartz.[21]

There have been a number of broad-scale surveys carried out, for example, by *Redbook*,[22] *Cosmopolitan*,[23] *Psychology Today*,[24] and *Playboy*[25] based on volunteer responses from readers. Shere Hite independently gathered data from a variety of volunteer informants.[26] These studies and others like them[27] might be regarded as volunteer case studies and as such might give valuable insights into problem areas, but their value is limited, despite the fact that many of them involve thousands of participants and attract a lot of media attention. Generally, they represent only those people who choose to offer information about their sex lives (volunteers are usually more liberal minded about sex matters than nonvolunteers), and the analysis applied to such information is often simplistic.

Diamond, a well-known sex researcher from the University of Hawaii, urged would-be users of such data to be guided by three rules about the facts contained in them. The first rule is that readers need to be aware that "facts are always accompanied by attitudes or emotions." This simply means that data collected by a religious fundamentalist group may differ quite radically from data collected by a nonreligious or less doctrinaire organization. The same set of facts, in different hands, can be used to prove different points, and certain facts may be suppressed if they are considered morally or politically unacceptable. Diamond's second rule is that even though "researchers talk about populations and trends," their data are gathered from individuals. The researchers' or interviewers' own sexual experiences seriously color their acceptance and interpretation of data collected. Most important is the third rule: "One must always distinguish between what is and what might or should be." Culture, history, law, and religion possess an inertia that resists change, and readers should be skeptical about reports of any dramatic break with tradition unless there is widespread documentation of change.[28]

Part of the difficulty is that scientifically valid sample studies are very costly. The one major post-Kinsey study supported by an agency of the U.S. government was, interestingly enough, conducted by the Kinsey Institute itself. In the late 1960s, the director of the National Institute of Mental Health (NIMH), Stanley Yolles, established the NIMH Taskforce on Homosexuality, which was chaired by Hooker (see chapter 9). Several studies were authorized, including a national survey not so much of sexual experiences, but of the respondents' moral judgments about sexual activities.

The study quickly became plagued with difficulty. Though the questions were constructed with care, and the interviewers received special training, the part of the questionnaire that dealt with respondents' sexual experiences was self-administered, and there was no check on validity. The actual survey was carried out by a contract with the National Opinion Research Center (NORC) at the University of Chicago. The survey proved more costly than anticipated, and when NIMH funds ran out, there was dispute about how NORC was to be paid. Moreover, those researchers affiliated with the Kinsey Institute itself became involved in serious internal disagreements.

The basic finding that the moral attitudes of Americans toward sex had not changed in the 1960s was an important antidote to those who said there had been a global shift in moral attitudes. The difficulty, however, is that moral attitudes do not necessarily correspond to practice. The study, in effect, failed to distinguish between what is and what should be. Moreover, such changes that were occurring, took place among high school and college students and young adults who were not generally included in the study. Because it was the attitudes of these people that were determining the conduct of large sections of the population and that would become increasingly influential, the study was misleading. To complicate the issue further, the internal feuds kept the study from being published until 1989, at which time it had little more than historical value.[29]

With the departure of Yolles from NIMH, the section devoted to sexual studies was closed down, and the U. S. government again tried to ignore sexuality. Enterprising publishers, seeing that books on sexuality often sold well, quickly labeled any number of publications as sex surveys. The result was often not only unscientific but naive surveys such as the so-called *Janus Report*, which ignored much of the methodology developed for survey research over the past thirty years.[30]

One consequence of the AIDS epidemic, however, was to force the U.S. government to once again pay attention of sexual issues. The National Institute of Child Health and Human Development, knowledgeable of the fact that increasing numbers of children were being born HIV positive, sponsored a study conducted by the Batelle Memorial Institute into the sexual behavior of men between twenty and thirty-nine years old in the United States; the study specifically focused on the use of prophylactics. The researchers found that once a week was the median number of times a male had sex, that 7.3 was the median number of sexual partners per man, that more white than black men either performed oral sex or received it, that more whites than blacks had anal sex, and that blacks were more likely than whites to use condoms (mostly for contraceptive purposes).

The study also noted that for those engaged in one-night stands, bisexuals and homosexuals were more likely to use condoms than married men having sex with their spouses. The major controversial finding was the comparatively low number of men reporting same-sex contacts, 2.3 percent over the previous five years, a figure that might have been influenced by the way the poll was taken.[31] This finding was almost immediately contradicted by a Louis Harris poll of 739 men, which found that 4.4 percent of the men surveyed had sex with another male within the past five years. Still, the Batelle data were not out of line with what contemporary British and French studies found, namely 1.4 percent of the men in those national samples had engaged in same-sex activities within the past five years.[32]

Finally recognizing the need for more data, various governmental agencies began to draw up plans for a national sex survey in the early 1990s. A preliminary sampling was completed, and a government contract was signed with the National Opinion Research Center at the University of Chicago to do the study. Perhaps because the publicity about the study had been greater than that for the Batelle one, there was soon controversy, and before it could get under way, Senator Jesse Helms intervened to prevent funding. This further emphasizes the political connotations still inherent in sex research.

Edward Laumann (who is now the provost of the University of Chicago) was to head up the study and set out to raise funds to continue the study. He raised $1.7 million from private foundations. The study is presently being readied for publication, and the results should be the definitive study for late-twentieth-century American sexual practices. Similar efforts to study sex habits in the United Kingdom were vetoed by Prime Minister Margaret Thatcher, but the Wellcome Foundation furnished money to do the British studies. Only in France was the government itself able to complete the studies. In sum, in spite of all the talk about sexuality and the importance of sex research, governmental agencies, fearful of hostile public reaction, still are forced to move with great care and caution in this area.

Undoubtedly, it was the realization of the government's sensitivity in this area that many groups concerned with AIDS research felt the need to organize as political pressure groups. In fact, if AIDS did nothing else for sex research, it emphasized just how important it is. To do any kind of epidemiological long-range planning, it is important to know not only what percentage of the population is homosexual, bisexual, and heterosexual but how many different sexual partners people have in their life spans or in any one year. This has to be measured by age cohorts, and this means it is important to determine the age at which sexual activity begins, the number of sexual acts engaged in, and the age of the partners. Since sex behavioral patterns are influenced by educational levels, social and economic class, belief systems, and any number of variables, sex surveys must of necessity be far more comprehensive than simply counting the number of sex contacts.

Preliminary data from the University of Chicago sample indicate that about 4.5 percent of the population has had a same-sex contact at least once in their lifetime, but whether one contact is indicative of a lifetime of homosexual activity is doubtful, and the data on this have not yet been released. The average number of sexual partners for both homosexuals and heterosexuals seems to be rounding out between 6 and 7, but the lack of breakdown by age cohorts make some of these data presently meaningless.

Not only was the government slow to react to the AIDS crisis but so also were some of the organized groups of sex professionals. The dominant voices in AASECT, for example, responded to AIDS by emphasizing abstinence. SSSS was much more active, and some of its members early engaged in AIDS-related research. Gays and lesbians in the SSSS were active in pushing for changes in sex practices among gay men, including decreasing the number of sexual partners, using condoms, and reducing such potentially dangerous practices as fist fucking (inserting one's fist into a partner's anus). SIECUS for its part also emphasized the necessity of public education about the dangers of AIDS and the necessity to reach teenagers and subteenagers.

Educational campaigns about the need for safe sex, at least among gays, turned out to be far more effective than any campaign of abstinence. One way to measure effectiveness, without a mass survey, is by examining other sexually transmitted diseases over the same time period. The decrease in gonorrhea among men that has taken place since the AIDS-oriented safer sex campaign began is regarded to be the result of the adoption of safer sex practices by homosexuals. In 1983, F. N. Judson reported a decline in gonorrhea among gay men, and no change among heterosexual men and women, emphasizing that the gay response was almost immediate.[33]

OTHER SEXUALLY TRANSMITTED DISEASES

AIDS, however, is just one of the sexually transmitted diseases (STDs), and several others are epidemic. Regarded as particularly dangerous is chlamydia, which now ranks as the number-one disease on college campuses and in Planned Parenthood clinics. It is potentially dangerous because one of its effects is infertility. It is easy to diagnose and treat when the person comes for treatment, but the issue with this disease and with others STDs is the lack of prevention programs. There has been no long-term effort by any government agency, at least to my knowledge, to develop vaccines for any sexually transmitted diseases except AIDS. The various government agencies act as if they were in the middle of the nineteenth century, when it was not permissible to talk about sex.

Some of this reluctance might have been the result of complacency toward STDs that grew out of the high midcentury cure rate with the used of antibiotics and sulfonamides, though this cannot fully explain the hesitation to develop such vaccines. Penicillin, as indicated earlier, was shown to be effective in curing syphilis, and both the sulfonamides and penicillin proved effective against gonorrhea. By 1957, the new wonder drugs pushed the number of primary and secondary cases of syphilis to an all-time low of 3.9 per 100,000, and gonorrhea cases had fallen to 127.4 per 100,000. The sequelae of primary syphilis had largely disappeared or was drastically reduced, and admissions to state mental hospitals for tertiary syphilis dropped dramatically.[34]

This drop was not long lasting, as new resistant strains of the syphilis developed, and the number of cases began to rise after 1980. Cases of gonorrhea began to escalate after 1960, and what had been regarded as comparatively minor STDs, such as herpes and venereal warts, became widespread. Chlamydia, which had not even been diagnosed in 1940, reached epidemic proportions in the 1980s. Venereal warts are known to cause cancer of the cervix. This means that the prognosis for good health for women with an active sexual life has deteriorated radically in the past twenty years, even more than that for men (AIDS is excluded from this statement), yet the government has done almost nothing in the way of educational campaigns concerning preventative measures. Some of the diseases could potentially be dealt with by vaccines, but little money has gone into finding such vaccines. Much can also be done by massive public health–oriented preventative drives but the government has been reluctant to sponsor such efforts. The result has been disastrous and a public health nightmare. Thus, despite new findings about sexuality and the evidence of public interest in sex, effective action in dealing with sex problems remains subject to all kinds of political pressure, a pressure so intense that it is easier for the bureaucrat and politician to do nothing than to be involved in controversy.

Undoubtedly, the difficulty is that research into sexually transmitted disease is looked on as sex research, and except for the period when the Rockefeller Foundation was sponsoring it and the brief reign of Yolles, sex research has never been very well funded—until the AIDS epidemic forced some rethinking. It is true that funding was available for areas of what might be called social ills associated with homosexuality or lesbianism—for example, lesbianism and alcoholism; homosexuality and the use of social services; and in the 1980s, almost anything concerned with the criminal justice system, including child abuse—but this still means that basic issues were ignored.

Some types of sexologists did receive money from specific disciplinary granting agencies for sex-oriented studies. Money, for example, had his research supported for most of his career by grant money, because he was able to tie in his sex research with other projects that the agencies were interested in. Lack of funding, however, has served as an effective deterrent to research into human sexuality by research teams dependent on grant money—the standard method of doing research in science. This has left the field to individuals who are already well established, who can get money for other projects and bootleg some off onto sex research, who have access to a private foundation, or who can support their own research. The individual willing to do this has to be highly motivated. Such a policy certainly has prevented the establishment of graduate research that is so common in most academic disciplines.

Perhaps a personal story would be pertinent here. In late 1966, NIMH was under the direction of Yolles, and I was invited to apply for an NIMH grant to study homosexuality. I was unable to do so, because I was overseas and had already been awarded a grant for another project that was to begin on my return to the United States. I was told, however, to apply as soon as I could, because my kind of expertise was much needed. By 1969, when I could apply to the NIMH, the funds in the agency had disappeared. The agency did issue a report on homosexuality a few years later, but it was very much abbreviated.

Fortunately, the Erickson Educational Foundation established by Reed Erickson came to my rescue and allowed me to continue research over the following six years. Erickson had also supported Money and several other researchers. The foundation, however, was small and the demands on the resources were great. By the middle of the 1970s, Erickson began to redirect the funding, although those who had grants were allowed to finish them off. No other private foundation has since stepped in to fund sexually oriented studies. It was possible, for a time, to get money from the National Endowment for the Humanities for sex-oriented research—as long as it was not too controversial. Much of these funds were cut off in the last years of the Reagan administration and during the Bush presidency.

The situation is not entirely hopeless, as evidenced by the University of Chicago's success in getting funds to carry out their sex survey, but as of this writing, no foundation specifically lists sex as one of the areas it is likely to fund and no government agency does. Sometimes, however, it is possible to write a grant so it can be funded under the rubric of feminist, women's, or criminal justice studies, all more likely sources of funding.

SEXOLOGY AND SCIENCE

This discussion emphasizes the difficulties under which sex research has been conducted. Serious research has, however, been carried out. The question is whether sex research can be called science. The answer is yes, but with qualifications. The popular image of the scientist is one of a dispassionate observer, cautiously carrying out experiments, continually verifying data, and then carefully publishing the conclusions in a refereed journal. This is an ideal, and occasionally it does happen that way. The image ignores the fact that the scientist or investigator is a flesh-and-blood creature and that it is a basic fact of human existence as social beings that individuals find it extraordinarily difficult to step outside their own convictions and act as detached observers. This was particularly true in the early history of sex research, since researchers had to contend with the idea that sexual activity in itself was sinful and so was researching it.

Perhaps this is why, as challenges to the religious model were made by the emerging scientists in the seventeenth and eighteenth centuries, those investigating sex tended to give a scientific imprimatur to the dangers of unbridled sex, as Tissot did. This was what the Christian Church taught, and though researchers such as Tissot believed they were putting sex into a scientific perspective, their basic Christian assumptions simply remained unexamined.

The combination of sex, religious values, and observational science, used by Tissot to justify the status quo, has led some to argue that science itself was simply a weapon of the establishment wielded to maintain control of human conduct. There certainly is some merit in such arguments, but they also represent a rather nineteenth-century view of science. This view was best exemplified by the French philosopher Auguste Comte (1798–1857), the founder of sociology, who proclaimed that science would replace theological and metaphysical knowledge and would, in addition, afford knowledge about how to construct a far better society.

Postmodernists have led a full-scale assault on this view. One of the leaders of the attack in the sex field was the French philosopher-historian Foucault (see chapter 9). Foucault held that the eighteenth century saw a growth of administrative bodies designed to manage life's activities, or to use his word, the state exercised growing control of *bio power*. According to Foucault, control of sexuality was the key to the control and regulation of people, the "means of access to the life of the body and the life of the species."[35] Though Foucault and his disciples ignore the fact that government long before the eighteenth century had been an agency of sex regulation, it is true that it did take a more dominant role as the influence of the Church declined. This does not mean that the government relied on science for its decisions, only that it selected from the scientific data that which it wanted to use. Government did so to give it a basis for intervention, particularly in those areas that the Church previously had regarded as sin.[36] Certain that through their selective use of science they had a new and more verifiable form of truth than religious doctrine, state officials quickly extended their control, prisoners of the unexamined assumptions of their own time.

There is, however, a different way of interpreting the data from that of the postmodernists, in which everything seems to be relative to the point of view of the observer. Presuppositions do play an important role in science, but it is impossible to have a scientific "view from nowhere." To say one's perspective is unaffected by any assumptions and that it is the one correct view of the world is to play God, or do what Donna Haraway called the "God trick," and no scientist should do this.[37] Moreover, science does not work this way. What is important in science is that presuppositions or theories change in response to the data, although such changes are not easy. In fact there is a general tendency to stick with the same old assumptions, fitting new data into these assumptions until they became so modified and complicated that a new explanation seems more valid. Thus science did not mount a full-fledged attack on Church assumptions about the dangers of sexuality but accepted such assumptions without examining them, since they had no reason to challenge them.

The classic illustration (far removed from the field of sexology) of the acceptance of continuing assumptions and the difficulty in bringing about change is the shift from the theory of a geocentric universe to a heliocentric one. The long dominance of the geocentric theory demonstrates just how strong a hold presuppositions often have on us and how important it is to recognize them. Though the Greeks had briefly toyed with the idea that the sun might really be the center of the universe, it was the geocentric view adopted by Aristotle that dominated Western thought. The issue was not a matter simply of observation, because observations were not accurate enough to prove one view more valid than the other, but it was the assumptions and explanations tied in with the geocentric view that gave it dominance. An arrow or an object tossed into the air fell to the ground because the earth, a material thing, was at the center and every material thing wanted to come home. Fire rose because it aimed toward the sphere of fire, which was above that of earth, water, and air. Seemingly every occurrence could be drawn into such a theory.

In the thirteenth century, St. Thomas Aquinas went further than Aristotle and tied the geocentric theory to Christian theology and made it the cornerstone of Christian faith. In Aquinas's time, developments in impetus physics, astronomical observations, and other kinds of data forced an ever more complicated geocentric theory far beyond the simple universe that Aristotle had conceived. Copernicus, in the sixteenth century, intrigued by his discovery that the Greeks had for a time played with the idea of a heliocentric universe, wondered how they could have done so, and his treatise on the possibility was published posthumously in 1543. This new theory, using much of the same data as the geocentric theory, continued to circulate, even though it was rejected by the scientists of the time. The growing observations and experiments, however, forced increasingly complex spheres and counterspheres of motion to be assumed, and the geocentric theory steadily grew more elaborate.

Still, to unseat the geocentric theory, it took a new instrument of observation, the telescope, to show that there were many occurrences in the universe that could not be explained by the geocentric viewpoint, for example the existence of Jupiter's moons and the phases of Venus. Galileo, who first made these observations, said that the earth revolved around the sun instead of the sun revolving around the earth. He was forced to recant his theories because of the Inquisition. Ultimately, however, the data for assuming a heliocentric theory grew stronger, and when Isaac Newton developed his theory of gravity to explain in a new way why objects fell to the earth, the heliocentric theory came to be accepted as the more satisfactory one.

Because the Vatican had canonized Aquinas and based a theology on the geocentric theory, Catholics who adopted the new theory were in difficulty with their Church until the twentieth century was almost over. Eventually, of course, the sun was shown to be part of the Milky Way Galaxy, and though it might be the center of our solar system, it is not now regarded as a particularly important star in terms of the entire universe.[38]

The historian and philosopher of science Thomas Kuhn explained that while the development of science has been incremental, there have also been periodic shifts in insights, which are necessary to establish a new paradigm. These shifts result when the data, no matter how they were interpreted or squeezed within the existing paradigm, become increasingly unable to explain various anomalies, and this eventually leads to a new kind of commitment, or a shift in paradigm to a new theory.[39] The new theory, however, does not throw out the old data that have been empirically verified, but rather they are used to verify the new theory. It is this willingness of science to eventually follow up where the data lead that is the key to scientific growth. Though this view of science has been used by many postmodernists to denigrate scientific truths and to argue that everything is relative to direction from which the data are approached, this was not the way Kuhn envisioned it, nor is it my view.

Still, as Karl Popper pointed out in the 1950s and 1960s, it is not possible conclusively to verify events by sensory experience. This led Popper to introduce the idea of science as vulnerable knowledge—knowledge lacking in the absolute certainty. Science must, therefore, be based on the principle of falsification rather than verification. That is, scientists should subject any conjecture to the severest imaginable empirical tests, in the hope that if the theory is false the tests will reveal it. Theories thus tested should be held until confronted by a hypothesis, itself critically tested, that gives evidence of a repeatable violation.[40]

In short, science is a method of verification, the best we have in fact, but we cannot be absolutely certain that what we believe today will not be challenged tomorrow. Instead, science gives us a way of achieving what might be called "objective knowledge," because it is not an individual product but rather the product of a community of scientists negotiating with each other. Results of data gathering and observations are published and judged by one's peers, and this means that objectivity in science are those views that seem to be agreed on by the scientific community at any one point in time, but these views can change with new data or new assumptions.

This discussion, however, has reviewed the ideas of historians of science in the last half of the twentieth century. These were not the ideas of those writing about issues in sexuality in the nineteenth or early twentieth century. One of the difficulties is that even the best of the early researchers believed that their observations were value free and were impartial, objective views, and they were willing to accept that public policy be made on the basis of their findings. The issue was further complicated by the fact that much of nineteenth-century sex research was not carried out by dispassionate scientists but by those who were either interested in reinforcing the status quo or those who regarded themselves as reformers. In either case, they were quick to establish new theories to back their stances in accord with Comtean science.

Looking back on that period, we find that the postmodernists have a valid point in emphasizing that many of those involved in sex research built new theoretical bases or, in sociological terms, social constructions of how society should be, based on their belief that science proved their point of view. Some of these system makers proved every bit as dogmatic as their theological predecessors had been. They were, however, looking at the data from the prisms of the past, unwilling to allow new light to appear.

The development of a sexual science has posed unique problems to traditional Western culture in terms of what kind of limits society should put on sexual behavior. It has challenged traditional Western assumptions, for example that sexual activity is a major cause of illness or cause of criminal activity. It has cast enough doubt on the idea that sex is sinful that the theologians have begun to reexamine traditional religious assumptions. The mainstream of current religious thinking accepts sex as a natural component of being human, which in itself is a revolution.

SEXOLOGY AND THE LAW

Although many American laws remain in place that were founded on assumptions that have been challenged by sex researchers, the law itself is changing. From the perspective of a historian, there have been radical changes in official attitudes, and as emphasized earlier, not only has sex has come to be recognized as pleasurable but it is now accepted that sexual pleasure is independent of procreation and that sexual pleasure can take many forms. As these new ideas have been incorporated into the old paradigm, many people hold contradictory ideas and are trying somehow to hold onto the traditional while accepting the new. Churches and synagogues have been busy reinterpreting what their attitudes should be while trying to keep within their religious traditions. Others have insisted that it is not a matter for discussion or debate, although even the most conservative churches in this respect have changed in spite of themselves, or at least their members have. Certainly, only a minority continue to believe the nineteenth-century views about the dangers of masturbation or that menstruation is a nervous reaction.

It is not so much what new findings do or do not demonstrate, however, but how these are interpreted by those in power, since sex remains a political issue. The continuance of outmoded beliefs is demonstrated in the struggle over admitting homosexuals into the U.S. Armed Forces. Though the conservatives who fear homosexuality opposed this policy because they do not want to spread homosexuality by seemingly recognizing it as simply one form of sexual attraction, by exempting homosexuals from military service (as the United States tried to do during the Vietnam War), they in effect made it official policy that only heterosexuals would be killed in any war. In point of fact, the exclusion of homosexuals is almost solely a political stance by the conservatives, both in and out of the service, because the military has always admitted homosexuals, as long as they did not proclaim that they were or were not too obvious. In fact, one of the reasons for the growth of homosexual organizations in post–World War II America was that the induction of millions of men into service had allowed homosexuals an opportunity to meet and contact others and realize that they were not alone in their feelings.

Sexual findings seem to point to a different paradigm about sexual beliefs from what had existed before; it is still not clear what use should be made of research's results. Once sex is accepted as a pleasurable experience, and it certainly has been in much of the Western world, it would seem very difficult to limit it to the confines of marriage or even to heterosexuality, since the very belief that sex can be pleasurable leads to the conclusion that pleasure is found in different ways by different individuals as a result of biological, cultural, psychological, and social factors.

It is probably for this reason that self-identified conservative, family-oriented groups proclaim their opposition to sex research. They fail to realize that research or scientific data can point out the direction but, in the end, public policy is a political decision. For example, much of the American public seems to agree that any kind of consenting sex could be tolerated, but putting the word *consent* in there immediately sets limits on what can or cannot take place, since some people in terms of age or mental development cannot really consent. Sex researchers can make their conclusions public, but those who define the members of the nonconsenting groups have so far ignored what researchers have found. Most states use a simple definition of age, but is this the best way? It is evident that teenagers are having sex even though in many states they are not old enough to drive a car or purchase cigarettes or alcohol.

Unfortunately, we still know little about sexuality in children and adolescents, and laws now on the books prevent anyone from doing any kind of serious research. More data are now available about sexual activity that does not involve another person, such as engaging in masturbation, reading erotic books, watching erotic movies, and cross-dressing in private. Most research indicates that such activities are not harmful but are widespread; most sex researchers tend to urge that laws designed to prohibit such activities be eliminated. It is not what the researchers find, however, that governs public policy, nor should it.

Moreover, not all sexual activity that is pleasurable to a given individual can be done without an audience. The voyeur or the exhibitionist, for example, need others who are unwilling participants. Though research might find such activity comparatively harmless, the act of voyeurism is an invasion of the privacy of others, which for many in the sex field is more important than the individual's desire to exhibit himself or herself. American society, in particular, and Western society, in general, have increasingly been willing to tolerate voyeurs who pay to see sex shows; perhaps we can now charge exhibitionists to put on a show. In a sense, we pay exhibitionists when we pay to see strip tease artists or nude models. The concept just needs to be extended. Such decisions again, however, are political decisions, since conflicting rights are involved and the ultimate decision is not one that sex researchers can make.

Similarly, prostitution usually involves the voluntary coming together and exchange of money or services. Sex research indicates that this could be a decision left to the individuals involved, but the problem is that it is often a public activity. Research has indicated that zoning certain areas as sex areas tends to increase crime and that neighborhoods that involuntarily become center of prostitution suffer. Perhaps the answer here is more discrete solicitation, exclusion of such activities from residential areas, and even open advertising. Most sex research argues against the licensing of prostitutes, since this is so stigmatizing to a group of women who usually are only temporarily in the business. This, however, should not prevent the authorities from regarding certain sexual activities as business enterprises from massage parlors to peep shows to stores that specialize in sexually related supplies and materials, including pornography.

Sex research can certainly help ease fears about the unknown, but how sexual activities should be regulated still remains a political question, one for which there is not one correct answer. This, however, is true of scientific investigation on many other issues facing society. Environmental scientists can indicate problems, but solutions are political. Even matters of military defense such as the so-called Star Wars defense plan are not so much matters of scientific truth as political decisions. This simply is what happens when government is involved. Information and data are available, but governments tend to select that which suits them at a particular time.

Moreover, the data about sex do not always agree. In fact, the real problem with sexology today, as has already been emphasized, is not just coping with the stigma associated with the field in the past, but the ever growing public demand for information about sexual matters. Almost anyone can write a sex book that at least one person will buy and believe, even if it has no scientific backing at all. Remainder tables are full of books that offer new remedies for sexual inadequacy; that discuss secret sexual information; that reveal the cures of a sex therapist; and that are aimed at heterosexual, homosexual, and many other kinds of sexual subgroups. Many, if not most, of these books are not worth the paper they are printed on, but sex, especially advice books, seems to sell. The problem is not determining the good from the mediocre but determining which have valid information and which are simply fluff or wishful thinking. Unfortunately, sexology has not yet found a way to focus on the better books and point out the fallacies of the worst. Many sex books on the market today deal with satanism and sex or other such issues that sex researchers long ago demonstrated had little factual support. Still, they sell.

SUMMARY

This leads to the ultimate question of whether sexology is a science. It certainly has a subject matter deserving of scientific investigation, and it has a strong database of information, much of it tested by traditional scientific methods. Because it differs from many traditional sciences as a result of its interdisciplinary nature and the influences of various biological, social, psychological, cultural, and historical factors on it, sexology demands more of its researchers. It lacks funding and institutional support, but it has a hard core of dedicated researchers determine to overcome these obstacles. Not all aspects of sex research, however, are on the same level, and this means that it is still on its way to becoming a science and is not yet a full-fledged discipline. Certainly, its biological wing meets all criteria for a science, while the social and behavioral part is just beginning to develop. But they are every bit as scientific as research in other areas of these disciplines.

The next decade will probably be crucial in what happens to sexology: Will it strive to remain an interdisciplinary endeavor or will it simply split up into specific professional fragments, each going its own way? In spite of lack of organization and in spite of lack of funding, sex researchers have held together throughout much of the twentieth century, and these ties were strengthened in the 1960s as scientific and professional societies with a sexology focus emerged again. Perhaps what held these researchers together was the stigma attached to the research; they went to SSSS meetings or international gatherings to get moral support for what they were doing. If this was the case in the past, it is not clear it will be so in the future. Moreover, in this age of specialization, sex research is also becoming more specialized. Subgroups are appearing, as are more specialized journals. There is still an effort to cross disciplinary and professional lines, but it becomes increasingly difficult as the literature grows and the number of professionals involved increase. In sum, as sexology becomes more scientific and demanding in testing its research findings, it finds itself changing. Perhaps this is the cost of becoming a science. The once-beleaguered group of researchers achieved respectability and success, and in the process, they found that their specialty itself had grown up and changed.

In conclusion, it should be emphasized that sexology, like most other areas of human investigation, has depended on answering questions that have been asked, and these vary in part with societal needs and demands at any given time. This book touches only the surface of the research of the past twenty years, and many names and studies simply could not be included because of space limitations. For failing to mention them and their significant studies, I apologize to my colleagues and urge them to continue their research. There are still a lot of things we need to know and much more we would like to know, if enough money, time, and professionals can be found. Sexology as a science has made important contributions to society, but how we ultimately use our newfound knowledge of sexuality remains a societal decision. I hope that books such as this will help provide the background for making those decisions and help society realize that sexology has made real contributions to humanity's well-being.

Notes

Introduction

1. For a lengthy discussion of this see Vern L. Bullough, *Sexual Variance in Society and History* (Chicago: University of Chicago Press, 1976).
2. Paul Robinson, *The Modernization of Sex* (New York: Harper & Row, 1970).
3. Edward M. Brecher, *The Sex Researchers* (Boston: Little, Brown, 1969).
4. It is important to emphasize that Darwin's views on sex changed over the course of his research. See Richard F. Michod and Bruce R. Levin, *The Evolution of Sex* (Sunderland, Mass.: Sinauer Associates, 1988), 10–11. Darwin's more developed view, which is summarized by Michod and Levin, appeared in Charles Darwin, *The Effects of Cross and Self Fertilization in the Vegetable Kingdom* (London: Murray, 1876), chap. 1. The individual who explicitly stated that sex existed for the good of the species was August Weismann, *Essays upon Heredity and Kindred Biological Problems*, trans. E. B. Poulton, S. Shonland, and A. E. Shipley, 2d ed., 2 vols. (Oxford, UK: Clarendon, 1891), chap. 1.
5. Cesare Lombroso, *Criminal Man* (reprint, Montclair, N.J.: Patterson Smith, 1972).

Chapter 1: Sex Research and Assumptions

1. Vern L. Bullough, "An Early American Sex Manual, or Aristotle Who?" *Early American Literature* 7 (1973): 236–47.
2. Aristotle, *History of Animals*, v, 6 (541b), trans. D'Arcy Wentworth Thompson,

in *The Works of Aristotle*, vol. 4, ed. J. A. Smith and W. D. Ross (Oxford, UK: Clarendon Press, 1910).

3.   Much of the material in this section has long been known and is not the result of my research. For greater detail, see Herbert Wendt, *The Sex Life of the Animals*, trans. Richard Winston and Clara Winston (New York: Simon & Schuster, 1912). See also various histories of biology. For the individuals concerned see the biographical sketches in *The Dictionary of Scientific Biography*, 16 vols., (New York: Scribner's, 1970–80).

4.   See Vern L. Bullough, Brenda Shelton, and Sarah Slavin, *The Subordinated Sex* (Athens: University of Georgia Press, 1988).

5.   Aristotle, *Generation of Animals*, 729A, 25–34, trans. A. L. Peck (London: Heinemann, 1953).

6.   For some of these see Joseph Needham, *A History of Embryology* (New York: Abelard-Schuman, 1959), 43–44.

7.   Hippocrates, *On Intercourse and Pregnancy*, a translation of *On Semen and on the Development of the Child*, by T. U. H. Ellinger (New York: Schuman, 1952), chap. 1, pp. 21ff.

8.   Avicenna, *Canon of Medicine*, trans. O. Cameron Gruner (London: Luzak, 1930), Book I, section 126, p. 23.

9.   Albertus Magnus, *De animalibus libri XXVI*, vols. 15 and 16, ed. Herman Studler (Munster: Beitrage zur Geschichte des Mittelalters, 1916–20), lib. IX, tract 2, cap. 3; lib. XV, tract 2, caps. 4–11, pp. 1026 ff.

10. St. Thomas Aquinas, "De conceptione Christi quod activum principium," in pt. 3 of *Summa Theologica* (New York: Benzinger, 1947), Part III, question 32, "De conceptione Christi quod activum principium," iv.

11. Plato, *Timaeus*, 91C, trans. and ed. R. G. Bury (London: Heinemann, 1961).

12. Soranus, *Gynecology*, I, 3, viii, trans. Owsei Temkin (Baltimore, Md.: Johns Hopkins University Press, 1956), p. 9.

13. For more on this topic, see Ilza Veith, *Hysteria: The History of a Disease* (Chicago; University of Chicago Press, 1965).

14. Arthur W. Meyer, *An Analysis of the De Generatione Animalium of William Harvey* (Stanford, Calif.: Stanford University Press, 1936); Howard B. Adelman, *Embryological Treatises of Hieronymus Fabricus of Aquapendente* (Ithaca, N.Y.: Cornell University Press, 1967), 113–21; and Elizabeth B. Gasking, *Investigations into Generation, 1651–1828* (Baltimore, Md.: Johns Hopkins Press, 1967), 16–36.

15. For a popular account of this see Clifford Dobell, *Antony van Leeuwenhoek and His "Little Animals"* (New York: Russell & Russell, 1958).

16. Quoted in John Farley, *Gametes and Spores: Ideas about Sexual Reproduction 1750–1914* (Baltimore, Md.: Johns Hopkins University Press, 1982), 11.

17. See Joseph Needham, *A History of Embryology* (New York: Abelard-Schuman, 1959), and F. J. Cole, *Early Theories of Sexual Generation* (Oxford, UK: Clarendon, 1930). See especially Farley, *Gametes and Spores*.

18. Farley, *Gametes and Spores*, 3.

19. For details, see Vern L. Bullough, *Sexual Variance in Society and History* (Chicago: University of Chicago Press, 1976); and Bullough, Shelton, and Slavin, *Subordinated Sex*.

20. Quoted in S. A. D. Tissot, *Onanism: Or a Treatise upon the Disorders of Masturbation*, trans. A. Hume (London: Pridden, 1766), 15. I have been unable to find the exact same quotation in Boerhaave's works. See Hermann Boerhaave, *Institutione medicae* (Leiden: Lugduni Batavorum, 1708, and other editions).

21. Tissot, *Onanism*, 17.

22. See John Brown, *The Elements of Medicine*, 2 vols., rev. Thomas Beddoes (Portsmouth, N.H.: Treadwell, 1803); the quote is from pp. viii–ix. See also Théophile de Bordeu, *Recherches sur les maladies chroniques* (Paris, 1775). For a brief discussion of some of these ideas, see Lester S. King, *The Medical World of the Eighteenth Century* (Chicago: University of Chicago Press, 1958), 143–7.

23. Tissot, *Onanism*. Tissot wrote in Latin but his work was rapidly translated into various languages.

24. Philippe Ricord, *Traité pratique des maladies vénérienne* (Paris: Rouvier & Bouvier, 1838), and other editions. See also W. E. Pusey, *The History and Epidemiology of Syphilis* (Springfield, Ill.: Thomas, 1933); and Theodor Rosebury, *Microbes and Morals* (New York: Viking, 1971).

25. Benjamin Rush, *Medical Inquiries and Observations upon the Diseases of the Mind* (Philadelphia: Kimber & Richardson, 1812), 347.

26. Sylvester Graham, *A Lecture on Epidemic Diseases Generally, and Particularly the Spasmodic Cholera* (Boston: Campbell, 1838), 5–7.

27. Ibid. The book went through ten editions between 1834 and 1848 and was translated into several foreign languages. He also wrote other books: Sylvester Graham, *A Lecture to Young Men, on Chastity, Intended Also for the Serious Consideration of Parents and Guardians*, 10th ed. (Boston: Pierce, 1848), which was originally published in 1834. For a discussion of Graham, see Stephen Willner Nissenbaum, *Sex, Diet and Debility in Jacksonian America: Sylvester Graham and Health Reform* (Westport, Conn.: Greenwood, 1980).

28. There is a vast and growing literature on the sexual beliefs in Victorian America and Europe. In addition to the works already cited, see Charles E. Rosenberg, "Sexuality, Class and Role in 19th-Century America," *American Quarterly* 25 (1973): 131–53; Carroll Smith Rosenberg and Charles Rosenberg, "The Female Animal: Medical and Biological Views of Women and Her Role in Nineteenth-Century America," *Journal of American History* 60 (1973): 332–56; G. J. Barker-Benfield, *The Horrors of the Half Known Life: Male Attitudes Toward Women and Sexuality in Nineteenth-Century America* (New York: Harper & Row, 1976); and John D'Emilio and Estelle B. Freedman, *Intimate Matters: A History of Sexuality in America* (New York: Harper & Row, 1988). Less scholarly but not less valuable is John Money, *The Destroying Angel* (Buffalo, N.Y.: Prometheus, 1985). There are many more and I apologize to my historian colleagues for not citing their works on this topic.

29. J. H. Kellogg, *Plain Facts for Old and Young* (Burlington, Iowa: Senger, 1882), 332–44.

30. George M. Beard, *Sexual Neurasthenia: Its Hygiene, Causes, Symptoms, and Treatment with a Chapter on Diet for the Nervous*, ed. A. D. Rockwell (New York: Treat, 1884), 58, 134–207. See also George M. Beard, "Neurasthenia or

Nervous Exhaustion, *Boston Medical Journal* 3 (1869): 217; and George M. Beard, *American Nervousness: Its Causes and Consequences* (New York: Putnam's, 1881). For discussions, see Charles E. Rosenberg, "The Place of George M. Beard in Nineteenth Century Psychiatry," *Bulletin of the History of Medicine* 26 (1962): 245–59; Philip P. Weiner, "G. M. Beard and Freud on 'American Nervousness,'" *Journal of the History of Ideas* 17 (1956): 269–74; and John S. Haller and Robin M. Haller, *The Physician and Sexuality in Victorian America* (Urbana: University of Illinois Press, 1974), 5–43.

31. See Vern L. Bullough and Martha Voght, "Homosexuality and Its Confusion with the 'Secret Sin' in Pre Freudian America," *Journal of the History of Medicine* 38 (1973): 283–8.

32. Henry Thomas Kitchener, *Letters on Marriage* (London: Chapple, 1812), 1:24.

33. Samuel Solomon, *A Guide to Health*, 64th ed. (n.p., n.d.), 189–93.

34. L. T. Nichols, *Esoteric-Anthropology* (London: Nichols, 1853), 84.

35. W. J. Hunter, *Manhood Wrecked and Rescued* (New York: Health Culture, 1900), 118.

36. Kitchener, *Letters on Marriage*, 1:26.

37. Ibid., 1:49.

38. Allen W. Hagenbach, "Masturbation as a Cause of Insanity," *Journal of Nervous and Mental Diseases* 6 (1879): 603–12.

39. Alfred Hitchcock, "Insanity and Death from Masturbation," *Boston Medical and Surgical Journal* 36 (1842): 283–6.

40. A. Jacobi, "On Masturbation and Hysteria in Young Children," *American Journal of Obstetrics* 8 (1876): 595–6.

41. Hagenbach, "Masturbation as a Cause of Insanity."

42. Joseph W. Howe, *Excessive Venery, Masturbation, and Continence* (New York: Treat, 1899), 419–27.

43. James Foster Scott, *The Sexual Instinct* (New York: Treat, 1899), 419–27.

44. G. Stanley Hall, *Adolescence* (New York: Appleton, 1904), 1:435, 445.

45. L. F. Bergeret, *The Preventive Obstacle or Conjugal Onanism* (New York: Turner & Mignard, 1897), 125.

46. James Foster Scott, *Heredity and Morals* (New York: Treat, 1899), 434.

47. Kitchener, *Letters on Marriage*, 2:247–58.

48. Solomon, *Guide to Health*, 213.

49. Hunter, *Manhood Wrecked and Rescued*, 126.

50. William Alcott, *The Physiology of Marriage* (Boston: Dinsmor, 1866), 118.

51. Dio Lewis, *Chastity or Our Secret Sins* (New York: Maclean, 1875), 111.

52. E. P. Miller, *Abuses of the Sexual Function* (New York: Gray & Green, 1867), 32–33.

53. Edward Dixon, *Treatises on the Diseases of the Sexual System* (New York: Rorback, 1855).

54. Alcott, *Physiology of Marriage*, 98.

55. Kitchener, *Letters on Marriage*, 2:318.

56. Lewis, *Chastity*, 117.

57. Nichols, *Esoteric-Anthropology*, 98.

58. Ibid., 155.

59. Alexander J. C. Skene, *Treatises on the Diseases of Women* (New York: Appleton, 1889), 929–30.

60. William Acton, *The Function and Disorders of the Reproductive Organs in Childhood, Youth, Adult Age, and Advanced Life Considered in Their Physiological, Social, and Moral Relations*, 3d American ed. (London: J. A. Churchill, 1871), 135–40.

61. E. B. Duffy, *Relations of the Sexes* (New York: Wood & Holbrook, 1876), 219.

62. Bergeret, *Preventative Obstacle*, 105.

63. Nichols, *Esoteric-Anthropology*.

64. *Satan in Society* (Cincinnati: Vent, 1871), 168.

65. Elizabeth Osgood Goodrich Willard, *Sexology as the Philosophy of Life* (Chicago: Walsh, 1867), 306–8.

66. For an analysis of this see Vern L. Bullough, "Technology for the Prevention of 'Les Maladies Produits par la Masturbation,'" *Technology and Culture* 28 (October 1987): 828–32.

67. John Power, *Essays on the Female Economy* (London: Burgess & Hill, 1831); G. F. Girdwood, "Theory of Menstruation," *Lancet* 1 (1842–3): i, 825–30; and J. Bennet, "On Healthy and Morbid Menstruation," *Lancet* 1 (1852): i, 35, 65, 215, 328, 353.

68. M. M. Smith, "Menstruation and Some of Its Effects upon the Normal Mentalization of Woman," *Memphis Medical Monthly* 16 (1896): 393–9; and Frederick Fluhman, *Menstrual Disorders, Diagnosis and Treatment* (Philadelphia: Saunders, 1939).

69. E. F. W. Pflüger, *Ueber die Eierstöcke der Sügethiere und des Menschen* (Leipzig: Engelmann, 1863).

70. Edward H. Clarke, *Sex in Education; or a Fair Chance for Girls* (Boston: Osgood, 1873), 37–38.

71. Ibid., 156–7.

72. For further amplification of his views see Vern L. Bullough and Martha L. Voght, "Women, Menstruation, and Nineteenth-Century Medicine," *Bulletin of the History of Medicine* 47 (1973): 66–82.

73. John Goodman, "The Menstrual Cycle," *Transactions of the American Gynecological Society* 2 (1877): 650–62; and John Goodman, "The Cyclical Theory of Menstruation," *American Journal of Obstetrics* 11 (1878): 673–94.

74. George W. Englemann, "The American Girl of Today: The Influence of Modern Education on Functional Development," *Transactions of the American Gynecological Society* 25 (1900): 8–45.

75. Board of Regents, *Annual Report for the Year Ending, September 30, 1877* (Madison: University of Wisconsin, 1877), 45.

76. For a discussion, see Jill Harsin, *Policing Prostitution in Nineteenth Century Paris* (Princeton, N.J.: Princeton University Press, 1985); for an overall picture see Vern L. Bullough and Bonnie Bullough, *Women and Prostitution* (Buffalo, N.Y.: Prometheus, 1988).

77. For a discussion of this, see Peter Gay, *Education of the Senses*, vol. 1 of *The Bourgeois Experience: Victoria to Freud* (New York: Oxford University Press, 1984).

78. Ibid., 99–101.

79. A. J. B. Parent-Duchâtelet, *De la prostitution dans la ville de Paris*, 2 vols., 2d
     ed. rev. (Paris: Baillière, 1837); it was originally published in 1836.

80. Among the investigations published were Herman Joseph Löwenstein, *De men-
     tis aberrationibus ex partium sexualium conditione abnormi oriundus* (1823);
     Joseph Haeussler, *Ueber die Beziehungen des Sexualsystemes zur Psyche* (1826);
     and Heinrich Kaan, *Psychopathia Sexualis* (1844). For details, see Ral Seidel,
     "Sexologie als positive Wissenschaft und sozialer Anspruch" (Ph.D. diss., Uni-
     versity of Munich, 1961); and Iwan Bloch, *Das Sexualleben unserer Zeit in Sein
     Beziehungen zur modernen kultur* (Berlin: Louis Marcus, 1908).

81. Johann Ludwig Casper, *A Handbook of the Practice of Forensic Medicine, Based
     upon Personal Experience*, 3d ed., trans. George William Balfour (London: New
     Sydenham Society, 1863), 3:330–46.

82. See, for example, Johan Ludwig Casper, *Handbuch der gerichtlichen Medizin*,
     rev. and aug. Carl Liman (Berlin: Hirschwald, 1889).

83. This idea was initially developed by Mary McIntosh, "The Homosexual Role,"
     Social Problems 16 (1986): 182–92. Others have amplified the importance of this
     construct, for example, Jeffrey Weeks, "Discourse, Desire and Social Deviance:
     Some Problems in the History of Homosexuality," in *The Making of the Modern
     Homosexual*, ed. Kenneth Plummer (Totowa, N.J.: Barnes & Noble, 1981),
     76–111; and David Greenberg, *The Construction of Homosexuality* (Chicago:
     University of Chicago Press, 1989).

84. James D. Steakley, *The Homosexual Emancipation Movement in Germany* (New
     York: Arno, 1975), 10.

85. Hubert C. Kennedy, *Ulrichs: The Life and Work of Karl Heinrich Ulrichs, Pio-
     neer of the Modern Gay Movement* (Boston: Alyson, 1988), 9.

CHAPTER 2: HOMOSEXUALITY AND OTHER FACTORS

1.   Much of this is based on Hubert C. Kennedy, *Ulrichs: The Life and Work of
     Karl Heinrich Ulrichs, Pioneer of the Modern Gay Movement* (Boston: Alyson,
     1988), 1–45.

2.   The twelve titles were published collectively as Karl Heinrich Ulrichs, ed.,
     *Forschungen über das Rätsel der mannmännlichen Liebe (Researches into the
     Riddle of Love Between Men)* (12 vols. in 1, reprint, New York: Arno Press, 1975).
     An English edition, *The Riddle of "Man-Manly" Love*, translated by Michael
     Lombardi-Nash, exists in 2 volumes (Buffalo, N.Y.: Prometheus Books, 1994).

3.   Quoted from Plato, *Symposium*, 3:304, in *The Dialogues of Plato*, trans. Ben-
     jamin Jowett (4 vols. in 1, New York: Random House, 1937).

4.   The legend is recounted in Hesiod, *Theogony*, 126–93, ed. M. L. West (Oxford,
     UK: Clarendon, 1966).

5.   "Vier Brief von Karl Heinrich Ulrichs [Numa Numantius] an seine Ver-
     wandten," ed. with an introduction by Magnus Hirschfeld, *Jahrbuch für sex-
     uelle Zwischenstufen* 1 (1899), 63.

6.   Kennedy, *Ulrichs*, 107; see also Albert Moll, *Perversions of the Sex Instinct: A
     Study of Sexual Inversion*, trans. Maurice Popkin (Newark, N.J.: Julian, 1931), 43.

7. James D. Steakley, *The Homosexual Emancipation Movement in Germany* (New York: Arno, 1975), 5.

8. Ulrichs, *Forschungen*, 3:57.

9. Hubert C. Kennedy, "The 'Third Sex' Theory of Karl Heinrich Ulrichs," *Journal of Homosexuality* 6 (1980–81): 107–8.

10. Carl Westphal, "Die Konträre Sexualempfindung," *Archiv für Psychiatrie und Nervenkrankheiten* 2 (1869): 73–108.

11. For a discussion of Kertbeny and the various terms, see Manfred Herzer, "Kertbeny and the Nameless Love," *Journal of Homosexuality* 12 (1985): 1–26; J. C. Feray, "Une histoire critique du mot homosexualité," *Arcadie* 325 (1981): 11–21; J. C. Feray, "Une histoire critique du mot homosexualité, II," *Arcadie* 326 (1981): 115–24; J. C. Feray, "Une histoire critique du mot homosexualité, III," *Arcadie* 327 (1981): 171–81, 328 (1981): 246–58; and Manfred Herzer, "Ein Brief von Kertbeny in Hannover an Ulrichs in Würzburg," *Capri* 1 (1987): 25–35.

12. For the pamphlets see [Károly Mária Kertbeny], *Das Gemeinschädliche des § 143 des preussischen Strafgesetzbuchs vom 14 April 1851 und daher seine notwendige Tilgun als § 152 im Entwurfe eines Strafgesetzbuches für den Norddeutschen Bund* (Leipzig: Serbe, 1869); and [Károly Mária Kertbeny], *§ 143 des Preussischen strafgesetzbuches vom 14 April 1851 und seine Aufrechterhaltung als § 152 im Entwurfe eines Strafgesetzbuches für den Norddeutschen Bund* (Leipzig: Serbe, 1869). The pamphlets were reprinted in *Jahrbuch für sexuelle Zwischenstufen* 6 (1905), i–iv, 3–66. For a discussion see Herzer, "Kertbeny and the Nameless Love."

13. Quoted in Herzer, "Kertbeny and the Nameless Love," 11.

14. There is a brief biographical sketch in Richard von Krafft-Ebing, *Textbook of Insanity*, trans. Charles Gilbert Chaddock (1876; reprint, Philadelphia: Davis, 1904). See also *Wiener Klinische Wochenschrift* 16 (1903): 21–22.

15. Edward M. Brecher, *The Sex Researchers* (Boston: Little, Brown, 1969), 56–59.

16. Richard von Krafft-Ebing, *Psychopathia Sexualis with Especial Reference to Contrary Sexual Instinct: A Medical Legal Study*, 12th ed., trans. F. J. Rebman (1906; reprint, Brooklyn, N.Y.: Physicians & Surgeons, 1933), 470, n. 1.

17. Richard von Krafft-Ebing, "Neue studien auf dem Gebiet der Homosexualität," *Jahrbuch für Sexuelle Zwischenstufen* 3 (1901): 1–36. For additional information see Albert Caraco, *Supplément à la Psychopathia Sexualis* (Lausanne: Edition L'Âge d'Homme, 1983); and Klaus Pacharzina and Karin Albrecht-Désirat, "Die Last der Ärzte," in *Der unterdrückte Sexus*, ed. J. Hohmann (Lollar: Achenbach, 1977), 97–113.

18. Charles Darwin, *The Descent of Man, and Selections in Relation to Sex*, 2 vols. (London: Murray, 1871), 402.

19. Richard von Krafft-Ebing, *Psychopathia Sexualis with Especial Reference to Contrary Sexual Instinct: A Medical Legal Study*, 7th enlarged, rev. ed., trans. Charles Gilbert Chaddock (Philadelphia: Davis, 1894), 1. All editions of this work, as far as I have been able to consult them, contained this statement.

20. Ibid., 5.

21. In each edition of *Psychopathia Sexualis*, Krafft-Ebing added new cases, for example, there were only 192 in the 7th edition.

22. Alfred Binet, "Le fétichisme dans l'amour, étude de psychologie morbide," *Revue Philosophique* 24 (1887): 143.

23. Marquis de Sade, *The Complete Justine*, trans. Richard Seaver and Austry Wainhouse (1791; reprint, New York: Grove, 1965); and Marquis de Sade, *Juliette*, trans. Richard Seaver and Austry Wainhouse (1797; reprint, New York: Grove, 1965). During his confinement to an insane asylum during the last part of his life, de Sade wrote *Les Journées de Sodome*, which remained unpublished until, in the twentieth century, Iwan Bloch published it in both French and German.

24. Leopold von Sacher-Masoch, *Venus in Pelz* (reprint, Munich: Verlag, 1967).

25. "Editorial," *British Medical Journal*, 1 (June 24, 1893): 1325–6.

26. Binet, "La fétichisme dans l'amour."

27. For a discussion of this see Frank J. Sulloway, *Freud, Biologist of the Mind: Beyond the Psychoanalytic Legend* (New York: Basic, 1979), 286–7.

28. Albert von Schrenck-Notzing, "Un cas d'inversion sexuelle amélioré par la suggestion hypnotique," in *Premier Congrès International de l'Hypnotisme Expérimental et Thérapeutique: Comptes rendus*, ed. Edgar Bérillon (Paris: Octave Doin, 1889), 319–22.

29. Albert von Schrenck-Notzing, *Therapeutic Suggestions in Psychopathia Sexualis (Pathological Manifestations of the Sexual Sense), with Especial Reference to Contrary Sexual Instinct*, trans. Charles Gilbert Chaddock (Philadelphia: Davis, 1895).

30. Sulloway, *Freud*, 287.

31. See Albert Moll, *Die Konträre Sexualempfindung* (Berlin: Fischer's Medicinische Buchhandlung, 1891), 185. This was translated as *Perversions of the Sex Instinct: A Study of Sexual Inversion Based on Clinical Data and Official Documents*, trans. Maurice Popkin (Newark, N.J.: Julian Press, 1931).

32. Havelock Ellis, *Psychology of Sex: A Manual for Students* (New York: Emerson, 1933),105.

33. Sulloway, *Freud*, 290–3.

34. James G. Kiernan, "Sexual Perversion and the Whitechapel Murders," *The Medical Standard* 4 (November 1888): 129–30.

35. Frank Lydston, "Sexual Perversion, Satyriasis and Nymphomania," *Medical and Surgical Reporter* 61 (1889): 253–8, 281–5. This was developed more fully in G. Frank Lydston, *The Diseases of Society* (Philadelphia: Lippincott, 1904).

36. Cesare Lombroso, *Criminal Man: According to the Classification of Cesare Lombroso, Briefly Summarized by His Daughter Gina Lombroso-Ferrero*, with an introduction by Cesare Lombroso (New York: Putnam's, 1911).

37. Quoted in Sulloway, *Freud*, 296.

38. Moll, *Die Konträre Sexualempfindung*. For a biographical sketch, see Heinz Goerke, ed., *Berliner Ärzte selbstzeugnisse* (Berlin: Verlag, 1965), 236–63. See also Albert Moll, *Ein Leben als Arzt der Seele: Erininnerungen* (Dresden: Verlag, 1936).

39. *Die Konträre Sexualempfindung*, 167, n. 1.

40. Albert Moll, *Untersuchungen über die Libido Sexualis* (Berlin: Fischer's Medicinische Buchhandlung, 1897), 10. Much of this was translated into English as

Libido Sexualis: Studies in the Psychosexual Laws of Love Verified by Clinical Case Histories, trans. by David Berger (New York: American Ethnological Press, 1933).

41. Ibid., 44.

42. Ibid., 46–47.

43. Max Dessoir, "Zur Psychologie der Vita sexualis," *Allgemeine Zeitschrfit für Psychiatrie* 50 (1894): 941–75.

44. Moll, *Untersuchungen,* 421–5.

45. Ibid., 326–8, 500. See also Albert Moll, *Das Sexualleben des Kindes* (Berlin: Walther, 1909); and Sulloway, *Freud,* 290–1, 303–5.

46. This change is noticed in Moll, *Untersuchungen;* and particularly in Albert Moll, *Handbuch der Sexualwissenschaften* (Berlin: Voge, 1912). In Moll, *Ein Leben als Arzt der Seele,* he stated that most homosexuality was acquired by improper sexual experiences.

47. Albert Eulenburg, *Sexual Neuropathie* (Leipzig: Spohr, 1895); Albert Eulenburg, ed., *Real-Encyclopädie der gesammten Heilkunde, medicinisch-chirurgisches Handworterbuch für praktische Arzte,* 26 vols. (Vienna: Urban & Schwarzenberg, 1894–1901).

48. Hermann Rohleder, *Vorlesungen ueber Sexualtrieb und Sexualleben des Menschen* (Berlin: Verlag, 1901), iii.

49. Albert Eulenburg, *Algolagnia: Sadism and Masochism,* trans. Harold Kent (New York: New Era, 1934).

50. Schrenck-Notzing, *Therapeutic Suggestions,* ix–x. The German original, which I have not seen, was entitled *Die Suggestionstherapie bei krankhaften Erscheinungen des Geschlechtsinnes* (Stuttgart: Enke, 1892).

51. Richard von Krafft-Ebing, *Alterations of Sexual Life, After the "Psychopathia Sexualis,"* trans. Arthur V. Burbury (London: Staples, 1959), 10–12.

52. For further amplification of this see Vern L. Bullough, *Sexual Variance in Society and History* (Chicago: University of Chicago Press, 1976).

53. Quoted in David Hothersall, *History of Psychology* (Philadelphia: Temple University Press, 1984), 221.

54. See Peter Gay, *Education of the Senses* (New York: Oxford University Press, 1984), 316.

55. Karl Pearson, *The Ethics of Freethought: A Selection of Essays and Lectures* (London: Unwin, 1888), 371.

56. Francis Galton, *Memories of My Life* (New York: Dutton, 1908).

57. C. P. Blacker, *Eugenics: Galton and After* (London: Duckworth, 1952).

58. By the early twenties professional geneticists had become alarmed by the misuse of science by the eugenicists, most of whom were ignorant of the latest advanced in genetics. See Kenneth M. Ludmerer, "American Geneticists and the Eugenics Movement: 1905–1935," *Journal of the History of Biology* 2 (1969): 337–65. See also Kenneth M. Ludmerer, *Genetics and American Society* (Baltimore, Md.: Johns Hopkins University Press, 1972); Mark Haller, *Eugenics: Hereditarian Attitudes in American Thought* (New Brunswick, N.J.: Rutgers University Press, 1963); Donald K. Pickens, *Eugenics and the Progressives* (Nashville: University of Tennessee Press, 1968); D. J. Kevles, *In the Name of*

Eugenics: Genetics and the Use of Human Heredity (New York: Knopf, 1985); and Philip R. Reilly, *The Surgical Solution: A History of Involuntary Sterilization in the United States* (Baltimore, Md.: Johns Hopkins University Press, 1991).

59. For some of these works see Lothrop Stoddard, *The Rising Tide of Color* (New York: Scribner's, 1922); Lothrop Stoddar, *Revolt against Civilization* (New York: Scribner's, 1923); and Alfred P. Schulz, *Race or Mongrel* (Boston: Page, 1908). Such views continued in a somewhat more gentlemanly form well into the twentieth century. See Carleton Putnam, *Race and Reason* (Washington, D.C.: Public Affairs Press, 1961).

60. There are a vast number of books and articles dealing with censorship. One classic work that goes over the changes in law is Norman St. John-Stevas, *Obscenity and the Law* (London: Secker & Warburg, 1956), 66–67.

61. Ibid., 71, 126–7.

62. There are literally bookcases of studies about Oscar Wilde. The verbatim transcripts of the trial were published as *The Three Trials of Oscar Wilde*, compiled with an introduction by H. Montgomery Hyde (New York: University Books, 1956). In Lord Alfred Douglas, *Autobiography* (London: Secker, 1929), Douglas attempted to portray himself as an innocent victim, but the evidence is against him. See the book by Douglas's nephew: Marquess of Queensberry, *Oscar Wilde and the Black Douglas* (London: Hutchinson, 1949). An interesting account is by Wilde's son, whose name was changed as a result of the trial: Vyvyan Holland, *Son of Oscar Wilde* (London: Hart-Davis, 1954). See also Rupert Hart-Davis, ed., *The Letters of Oscar Wilde* (London: Hart-Davis, 1962); and Rupert Hart-Davis, ed., *More Letters of Oscar Wilde* (New York: Oxford University Press, 1985).

63. For quote, see Fawn M. Brodie, *The Devil Drives: A Life of Sir Richard Burton* (New York: Norton, 1967), 291.

64. There are numerous biographies of Burton plus a library of his writings. Besides the Brodie biography *The Devil Drives*, Byron Farwell's *Burton: A Biography of Sir Richard Francis Burton* (New York: Viking, 1963) was most helpful.

65. Richard Burton, ed. and trans., *A Plain and Literal Translation of the Arabian Nights' Entertainments, Now Entitled The Book of the Thousand Nights and a Night. With Introduction, Explanatory Notes on the Manners and Customs of Moslem Men and a Terminal Essay Upon the History of the Nights*, 10 vols. (London: Kama Shastra Society, 1885). This has often been reprinted. See also Richard Burton, *Supplemental Nights to the Book of the Thousand Nights and a Night. With Notes Anthropological and Explanatory by Richard F. Burton* (London: Kama Shastra Society, 1886–88).

66. *Kama Sutra of Vatsyana*, trans. Richard Burton (London: Kama Shastra Society, 1883); *Ananga Ranga; (Stage of the Bodiless One) or, The Hindu Art of Love*, trans. Richard Burton (London: Kama Shastra Society, 1885); *The Perfumed Garde of the Cheikh Nefzaoui: A Manual of Arabian Erotology*, trans. Richard Burton (London: Kama Shastra Society, 1886); and *Priapeia or the Sportive Epigrams of Divers Poets on Priapus: the Latin Text now for the First*

Time Englished in Verse and Prose, trans. Richard Burton (London: Kama Shastra Society 1890). There were several others in this series.

67. Hermann Heinrich Ploss, *Das Weib in der Natur-und Völkerkunde*, 2 vols. (Leipzig; Grieben, 1885). Ploss died shortly after the publication of his book, and after the first edition of fifteen hundred copies sold out, a gynecologist colleague of Ploss, Maximilian Bartels, took over revision and republication. Bartels found thousands of jotted-on scraps of paper on subjects, such as prostitution, that were not covered in the first edition. The second edition, with Bartels as second author, became much more comprehensive and aimed more toward the general public. It deals not only with gynecology but the ages and stages of women from birth to death. Illustrations were also added. The third and final edition was edited by Maximilian and Paul Bartels, but this was never published in German because the plates were destroyed by the Nazis. It was, however, translated into English by Eric John Dingwall (London: Heinemann, 1935), with some additions.

68. Paula Weideger, *History's Mistress: A New Interpretation of a Nineteenth Century Ethnographic Classic* (New York: Penguin, 1985), 31.

69. Iwan Bloch, *Beiträge zur Aetiologie der Psychopathia Sexualis* (Dresden: Dohrn, 1903), 2:192–206. This was also translated into English, although Bloch was not always well served by his translators. Bloch, *Anthropological Studies in the Strange Sexual Practices of All Race in All Ages, Ancient and Modern, Oriental and Occidental, Primitive and Civilized*, vol. 1, trans. Keene Wallis (New York: Anthropological Press, 1933); vol. 2, trans. Ernst Vogel (New York: Anthropological Press, 1935).

70. Ibid., 2:363–5. Bloch's explanation about the need for varied sexual stimuli as a source of sexual "aberrations" had previously been suggested by Alfred Hoche, "Zur Frage der forensischen Beurtheilung sexueller Vergehen," *Neurologisches Centralblatt* 15 (1896): 57–68.

71. Bloch, *Beiträge*, 2:363–5.

72. Iwan Bloch, *Das Sexualleben unserer Zeit in seinen Beziehungen zur modernen Kultur* (Berlin: Marcus, 1906). The sixth German edition was translated by M. Eden Paul as *The Sexual Life of Our Time in Its Relations to Modern Civilization* (London: Rebman, 1910).

73. Iwan Bloch, *Die Prostitution*, 2 vols. (Berlin: Marcus, 1912–25). The second volume is incomplete and what exists of it was finished by others.

74. Iwan Bloch, *Der Ursprung der Syphilis*, 2 vols. (Jena: Fisher, 1901–11).

75. Ral Seidel, "Sexologie als Positive Wissenschaft und Sozialer" (Anspruch: (Ph.D. diss., University of Munich, 1961), 42. Bloch had actually spelled out the concept of sexualwissenschaft six years before he invented the term in his study of the Marquis de Sade. See Eugen Dühren [Iwan Bloch], *Der Marquis de Sade und seine Zeit* (Berlin: Verlag, 1900), 1–19. Popularized English translations of the 1900 book usually omit Bloch's discussion about the scientific approach to sex.

76. Sigmund Freud, *Three Essays on the Theory of Sexuality*, in vol. 7 of *The Standard Edition of the Complete Psychological Works of Sigmund Freud*, trans. and

ed. under the direction of James Strachey (London: Hogarth, 1953), 7:139, n. 2.

77. Friedrich S. Krauss, *Das Geschlechtsleben in Glauben, Sitte, Baruch und Gewohnheitsrecht der Japaner* (Leipzig: Verlag, 1911).

78. F. Karsch-Haack, *Das Gleigeschlechtliche Leben der Naturvölker* (Munich: Verlag, 1911).

79. Pisanus Fraxi, *Bibliography of Prohibited Books*, with an introduction by G. Legman (reprint, New York: Brussell, 1962). *My Secret Life*, 11 vols., with an introduction by G. Legman (reprint, New York: Grove Press, 1962).

80. For full citations of these criticisms, see Norman St. John-Stevas, *Obscenity and the Law* (London: Secker & Warburg, 1956), 54–5, 57–8.

81. Emphasizing (perhaps overemphasizing) this contribution was Erwin J. Haeberle, "The Jewish Contribution to the Development of Sexology," *Journal of Sex Research* 18 (1982): 305–23. The connection of the rise of modern sexology with Judaism has a mixed meaning, because the Nazis regarded the work of the German-Jewish physicians in the understanding of sex only as a further sign of Jewish decadence. Even non-Nazis in Germany were hostile. See Otto Weininger, *Sex and Character* (London: Heinemann, 1908), 303, 309.

82. Enoch Heinrich Kisch, *Das Geschlectsleben des Weibes* (Berlin: Urban & Schwarzenberg, 1907). For biographical data see *Neue deutsche Biographie* (Berlin: Duncker & Humblot, 1953–8), 11:680.

83. *Neue deutsche Biographie*, 4:683.

84. Max Dessoir, *Vom Jenseits der Seele: Die Geheimuissenschaften in kritische Betrachtung* (Stuttgart: Ferdinand Enke Verlag, 1931).

85. *Neue deutsche Biographie*, 3:617.

86. Moll, *Ein Leben als Arzt der Seele*.

87. Eulenburg, *Algolagnia*, 163.

88. Max Marcuse, "Wandlungen des Fortpflanzungsgedankens und Willens," *Abhandlungen aus dem Gebiete der Sexualforschung* 1 (1918): 29.

89. Quoted in Dennis B. Klein, *Jewish Origins of the Psychoanalytic Movement* (New York: Praeger, 1918), 138.

90. H. G. Adler, *The Jews in Germany: From the Enlightenment to National Socialism* (Notre Dame, Ind.: University of Notre Dame Press, 1969), 78.

CHAPTER 3: HIRSCHFELD, ELLIS, AND FREUD

1. For a fictionalized treatment of some of this, see Roger Peyrefitte, *The Exile of Capri* (London: Secker & Warburg, 1961); see also William Manchester, *The Arms of Krupp* (Boston: Little, Brown, 1968).

2. Ernest Jones, *Sigmund Freud, Life and Work* (London: Hogarth, 1957), 2:95.

3. R. W. Clark, *Freud, the Man and the Cause* (London: Cape, 1980), 219.

4. Th. Ramien, *Sappho und Socrates, Wie erklärt sich die Liebe der Mannër und Frauen zu Personen des eigenen Geschlechts?* (Leipzig: Spohr, 1896).

5. Albert Moll, *Die Konträre Sexualempfindung* (Berlin: Fisher's, 1891).

6. I have not seen the pamphlet, which was titled *The Case of Wilde and the Prob-*

lem of Homosexuality, but it is quoted at length in Charlotte Wolff, *Magnus Hirschfeld: A Portrait of a Pioneer in Sexology* (London: Quartet, 1986), 37–40.

7. Benedict Friedlander, *Die Renaissance des Eros Uranios* (Schmargendorf-Berlin: Verlag Renaissance, 1904).

8. Hans Blüher, *Die drei Grunformen der Sexual Inversion (Homosexualität)* (Leipzig: Spohr, 1913); and Hans Blüher, *Die Role der Erotik in der männlichen Geselschaft,* 2 vols. (Jena: Diedrichs, 1917–19).

9. James D. Steakley, *The Homosexual Emancipation Movement in Germany* (New York: Arno, 1975); and John Lauritsen and David Thorstad, *The Early Homosexual Rights Movement, 1863–1935* (New York: Times Change, 1974).

10. Steakley, *Homosexual Emancipation Movement,* 38–40; and Manfred Herzer, "Politik und Wissenschaft beim Magnus Hirschfeld," in *Rebellion gegen das Valiumzeitalter über legungen zur Gesundheitsbewegung,* ed. Stefan Lundt (Berlin: Dokumentation des Gesundheitsages, 1981), 81. Moll grew increasingly hostile to Hirschfeld as a person and to his theories, and in Albert Moll, *Handbuch der Sexualwissenschaften* (Leipzig: Vogel, 1926), revised ed., part 1, 766–772; part 2, 850, he argued against inborn homosexuality.

11. A journal with the same name, this time with Bloch and Eulenburg as editors, began to be published in 1914, and though it survived a number of tribulations resulting from the difficulties of publication during the war and the hard economy of the 1920s, it succumbed in 1929 with volume seventeen, at which time Max Marcuse was editor.

12. Magnus Hirschfeld, *Die Transvestiten, Eine Untersuchung über den erotischen Verkleidungstrieb* (Berlin: Pulvermacher, 1910); this was translated by Michael Lombardi-Nash as *The Transvestites: An Investigation of the Erotic Drive to Cross Dress,* trans. Michael Lombardi-Nash (Buffalo, N.Y.: Prometheus, 1991).

13. Magnus Hirschfeld, *Die Homosexualität des Mannes un des Weibes* (Berlin: Marcus, 1914). Hirschfeld summarized his ideas in English in Magnus Hirschfeld, "Homosexuality," in *Encyclopedia Sexualis,* ed. Victor Robinson (New York: Dingwall-Rock, 1936), 321–34.

14. Magnus Hirschfeld, *Naturgesetze der Liebe* (Berlin: Pulvermacher, 1912). There was considerable research going on in this area, although much of it was by French scientists. Claude Bernard (1813–78) had shown the metabolic relationship of glycogen to diabetes miletus and the importance of pancreatic juice for digestion. Charles-Édouard Brown-Séquard (1818–94), his successor, had shown the importance of the adrenal glands and testicular organs in the chemical process.

15. Wolff, *Magnus Hirschfeld,* 129.

16. Lawrence Birken, *Consuming Desire: Sexual Desire and the Emergence of a Culture of Abundance* (Ithaca, N.Y.: Cornell University Press, 1988), 88–89.

17. Ibid., 89.

18. Hermann Rohleder, *Die Masturbation, Eine Monographie für Ärzte, Pädagogen und gebildet Eltern,* 2d ed. (Berlin: Fischer 1902).

19. Quoted in Norman Haire, ed., *Sexual Anomalies and Perversions: Physical and Psychological Development Diagnosis and Treatment. A Summary of the Works*

of the Late Professor Dr. Magnus Hirschfeld, 2d ed. (London: Encyclopedic, 1966), 124.

20. Magnus Hirschfeld, *Geschlechtliche Entwicklungsstörungen,* vol. 1 of *Sexualpathologie* (Bonn: Marcus & Weber, 1916), 179.

21. Magnus Hirschfeld, *Sexualpathologie,* 3 vols. (Bonn: Marcus & Weber, 1916, 1918, 1921).

22. This is based on a statement he made in "Magnus Hirschfeld," in *Encyclopaedia Sexualis,* ed. Victor Robinson (New York: Dingwall-Rock, 1936), 317–21. Though his major works were not translated during his lifetime, several of his lesser ones now have been: Magnus Hirschfeld, *Sexual Anomalies* (reprint, New York: Emerson, 1948); Magnus Hirschfeld, *Sexual Pathology,* trans. Jerome Gibbs (reprint, New York: Emerson, 1945); and Magnus Hirschfeld, *Men and Women: The World Journey of a Sexologist* (New York: Putnam's, 1935).

23. Summary contents of the various congresses are given in Wolff, *Magnus Hirschfeld;* and Norman Haire, "World League for Sexual Reform," in *Encyclopedia Sexualis,,* ed. Victor Robinson (New York: Dingwall-Rock, 1936), 811–14. More detailed accounts are available: Magnus Hirschfeld, "Zur I. Internationalen Tagung für Sexualreform auf sexualwissenschaft lichen Grundlage," *Jahrbuch für Sexuelle Zwischenstufen* 21 (1921): 99–105; Bertha Riese and J. R. Leunbach, eds., *Proceedings of the Second Congress of the World League for Sexual Reform* (Copenhagen: Levin & Munksgaard, 1928); Norman Haire, ed., *Proceedings of the Third Congress for Sexual Reform* (London: Kegan Paul, 1930); H. Steiner, ed., *Sexualnot und Sexualreform* (Vienna: Elbemühl, 1931); and Joseph Weikopf, "Der Brünner Sexualkongress," *Sexus* 1 (1933): 26–33.

24. Albert Moll, *Verhandlungen auf dem Internationalen Kongress für Sexualforschung* (Berlin: Marcus & Weber, 1928).

25. For Moll's criticism see Albert Moll, ed., *Handbuch der Sexualwissenschaften* (Leipzig: Verlag Vogel, 1912) and especially in the 2d ed. (1926), and Albert Moll, "Der 'reaktionare' Kongress für Sexualforschung," *Zeitschrfit für Sexualwissenschaft* 13, no. 10 (1927): 321–31.

26. Quoted in Wolff, *Magnus Hirschfeld,* 247, from the introduction by Norman Haire to the English edition of Hirschfeld's *Sex in Human Relations.*

27. Ellis had the language advantage for English-speaking readers, and so his works are much more easily available than any other of the early sex researchers, except Freud. Six volumes of *Studies in the Psychology of Sex* had appeared by 1910, after which he proceeded to revise them. The first volume was published in German and listed J. A. Symonds as a coauthor; it appeared in English without Symonds's name: Havelock Ellis, *Sexual Inversion* (London: Watford University Press, 1897). The publication of the series was then transferred to the United States over the imprint of F. A. Davis of Philadelphia. The order and the names of the volumes and editions were revised and the first six American volumes are as follows: Havelock Ellis, *The Evolution of Modesty; The Phenomena of Sexual Periodicity and Auto-eroticism,* vol. 1 of *Studies in the Psychology of Sex* (New York: Davis, 1900); Havelock Ellis, *Sexual Inversion,* vol. 2 of *Studies in the Psychology of Sex* (New York: Davis, 1901); Havelock Ellis, *Analysis of the Sexual Impulse,* vol. 3 of *Studies in the Psychology of Sex* (New York: Davis,

1903); Havelock Ellis, *Sexual Selection in Man*, vol. 4 of *Studies in the Psychology of Sex* (New York: Davis, 1905); Havelock Ellis, *Erotic Symbolism. The Mechanism of Detumescence. The Psychic State in Pregnancy*, vol. 5 of *Studies in the Psychology of Sex* (New York: Davis, 1906); and Havelock Ellis, *Sex in Relation to Society*, vol. 6 of *Studies in the Psychology of Sex* (New York: Davis, 1910). There were several reprintings and a seventh volume was added in 1928: Havelock Ellis, *Eonism and Other Supplementary Studies*, vol. 7 of *Studies in the Psychology of Sex* (Philadelphia: Davis, 1928).

28. Havelock Ellis, *My Life* (Boston: Houghton Mifflin, 1939). There are numerous biographies or accounts of Ellis, including Isaac Goldberg, *Havelock Ellis: A Biographical and Critical Survey* (New York: Simon & Schuster, 1926); Houston Peterson, *Havelock Ellis: Philosopher of Love* (Boston: Houghton Mifflin, 1928); Arthur Calder Marshall, *The Sage of Sex* (New York: Putnam's, 1949); John Stewart Collis, *An Artist of Life* (London: Cassell, 1959); Rose F. Ishill, *Havelock Ellis* (Berkeley Heights, N.J.: Oriole Press, 1959); and Vincent Brome, *Havelock Ellis: Philosopher of Sex* (London: Routledge & Kegan Paul, 1979).

29. Quoted in Brome, *Havelock Ellis*, 27.

30. [George Drysdale], *Physical, Sexual and Natural Religion, by a Student of Medicine* (London: Edward Truelove, 1854). Later editions were published under the title of *The Elements of Social Sciences* and it was under this title that Ellis saw it. It was extremely popular and went through some thirty-five English editions by 1905 and was translated into ten European languages.

31. Havelock Ellis, general preface to *Evolution of Modesty*, iii–vi.

32. Françoise Delisle, *Friendship Odyssey* (London: Heinemann, 1946).

33. Havelock Ellis, *Man and Woman* (London: Scott, 1894). This went through numerous editions and a new condensed edition appeared in 1926 (Boston: Houghton Mifflin, 1926).

34. John Addington Symonds, *A Study in Greek Ethics* (London: 1883); this was written in 1873 and was published in a limited edition of only ten copies. John Addington Symonds, *A Problem in Modern Ethics* (London: 1891) was also originally issued in a limited edition of ten copies. Both have been widely reprinted, sometimes singly, and sometimes combined into one volume. Symonds also wrote his autobiography but this was never published. It was used as a basis for Phyllis Grosskurth, *The Woeful Victorian: A Biography of John Addington Symonds* (New York: Holt, Rinehart & Winston, 1864).

35. Grosskurth, *John Addington Symonds*, 284–5.

36. At least this is what Ellis wrote to Symonds on January 3, 1893.

37. Ellis, *Sexual Inversion*, 355–6.

38. Ibid., 316–7.

39. Ibid., 83.

40. Ibid., 244–57.

41. Ibid., 258.

42. Ibid., 257.

43. Ellis, "Auto-eroticism," *Evolution of Modesty*, 161.

44. Ibid., 162.

45. Paul Robinson, *The Modernization of Sex* (New York: Harper & Row, 1970), 13.

46. For an example of Freud's reaction to Ellis, see the letter of Sigmund Freud to Wilhelm Fliess dated January 3, 1899, in Sigmund Freud, *The Origins of Psycho-Analysis; Letters to Wilhelm Fleiss, Drafts and Notes: 1887–1902*, ed. Maurice Bonaparte, Anna Freud, and Ernest Friss, trans. Eric Mosbacher and James Strachey (New York: Basic, 1954).

47. Ellis, *Analysis of the Sexual Impulse*, 63–65.

48. Ibid., 66–68.

49. Ibid., 59.

50. Ellis, *Erotic Symbolism*.

51. Havelock Ellis,"The Sexual Impulse in Women," in *Analysis of the Sexual Impulse*, vol. 3 of *Studies in the Psychology of Sex* (New York: Davis, 1903).

52. Ibid., 256.

53. Ellis, *Eonism and Other Supplementary Studies*.

54. Sigmund Freud and Josef Breuer, *Studies in Hysteria*, vol. 2 of *Complete Psychological Works of Sigmund Freud*, ed. and trans. by James Strachey et al. (London: Hogarth, 1955).

55. Lancelot Law Whyte, *The Unconscious before Freud* (New York: Basic, 1960).

56. Frank J. Sulloway, *Freud, Biologist of the Mind: Beyond the Psychoanalytic Legend* (New York: Basic, 1979), 185.

57. Sigmund Freud, "'Civilized' Sexual Morality and Modern Nervousness," in *Collected Papers*, trans. Joan Riviere (New York: Basic, 1959), 2:76–99.

58. This appears in Sigmund Freud, *Three Essays on the Theory of Sexuality*, in *Standard Edition of the Complete Psychological Works of Sigmund Freud* (London: Hogarth, 1953), 7:125–43.

59. Sulloway, *Freud*, 98.

60. Ibid., 211; and also Sigmund Freud, *The Origins of Psycho-Analysis*, 303–4.

61. Iwan Bloch, *The Sexual Life of Our Time in Its Relation to Modern Civilization*, trans. M. Eden Paul (London: Rebman, 1910), 756.

62. See, for more detail, Sulloway, *Freud*, 319.

63. Sigmund Freud, "The Psychogenesis of a Case of Homosexuality in a Woman," in *Collected Papers*, authorized translation under supervision of Joan Riviere (New York: Basic, 1959), 2:202–31.

64. See Timothy F. Murphy, "Freud Reconsidered: Bisexuality, Homosexuality, and Moral Judgment," *Journal of Homosexuality* 9, no. 2–3 (1983–4): 65–77.

65. This appears in a footnote which was not in the 1910 article of Sigmund Freud, "Leonardo da Vinci and a Memory of His Childhood," It was added in 1919 and appears in *Complete Psychological Works of Sigmund Freud* (London: Hogarth, 1955), 11:99, n. 2.

66. Sulloway, *Freud*, 183; and Sigmund Freud, "'A Child Is Being Beaten': A Contribution to the Origin of Sexual Perversions," in *Standard Edition of the Complete Psychological Works of Sigmund Freud* (London: Hogarth, 1953–74), 17:200–1.

67. For much more detail see Kenneth Lewes, *The Psychoanalytic Theory of Homosexuality* (New York: Simon & Schuster, 1988); and Murphy, "Freud Reconsidered," 65–77. See also Warren Johnson, "Freudian Concepts," in *Encyclopedia of Homosexuality*, ed. Wayne R. Dynes (New York: Garland, 1990), 1:434.

68. "Historical Notes: 'A Letter from Freud,'" *American Journal of Psychiatry* 107 (April 1955): 786–7. The letter was uncovered by Kinsey.
69. Kate Millett, *Sexual Politics* (Garden City, N.Y.: Doubleday, 1970).
70. Sigmund Freud, "An Outline of Psychoanalysis," trans. Joan Riviere, in *Standard Edition of the Complete Psychological Works of Sigmund Freud* (London: Hogarth, 1964), 23:23–24; and Sigmund Freud, "Civilization and Its Discontents," in vol. 21 of *Standard Edition of the Complete Psychological Works of Sigmund Freud* (London: Hogarth, 1961).
71. Sigmund Freud, "The Taboo of Virginity," in *Collected Papers*, trans. Joan Riviere (New York: Basic Books, 1959), 4:116; and Sigmund Freud, "Some Psychical Consequences of the Anatomical Distinction between the Sexes," in *Collected Papers*, ed. and translated under supervision of James Strachey (New York: Basic, 1959), 5:186–97.
72. See, for example, William N. Stephens, *The Family in the Cross Cultural Perspective* (New York: Holt, Rinehart, & Winston, 1953), 246; and especially, J. D. Unwin, *Sexual Regulations and Human Behavior* (London: Williams & Norgate, 1933), ix–x, 85, 87, 108; and J. D. Unwin, *Sex and Culture* (London: Oxford University Press, 1934).

CHAPTER 4: THE AMERICAN EXPERIENCE

1.   For discussion of some of these see Vern L. Bullough, *Sexual Variance in Society and History* (New York: Wiley Interscience, 1976), 587–8.
2.   G. Frank Lydston, *Lecture on Sexual Perversion, Satyriasis and Nymphomania* (Chicago: Philadelphia Medical and Surgical Reporter). This was a reprint from the *Philadelphia Medical and Surgical Reporter*.
3.   Denslow Lewis, *The Gynecologic Consideration of the Sexual Act: And an Appendix with an Account of Denslow Lewis*, ed. Marc H. Hollender (Weston, Mass.: MTSL Press, 1970). The quotes come from the introduction by Hollender. See also John C. Burnham, "The Progressive Era Revolution in American Attitudes Toward Sex," *Journal of American History* 59 (1973): 885–908.
4.   C. H. Hughes, "Postscript to Paper on 'Erotopathia,'" *The Alienist and Neurologist* 14 (October 1893): 731–2.
5.   "Review of *Sexual Inversion*," *American Journal of Insanity* 59 (1902): 182.
6.   William Noyes, Reviewer, *Psychological Review* 4 (1897), 447.
7.   Randolph Winslow, "Report of an Epidemic of Gonorrhea Contracted from Rectal Coition," *Medical News* 49 (August 14, 1886): 180–2.
8.   G. Alder Blumer, "A Case of Perverted Sexual Instinct," *American Journal of Insanity* 39 (1882): 22–35.
9.   Lydston, *Lecture*.
10. J. Richardson Parke, *Human Sexuality: Medico-Literary Treatise* (Philadelphia: Professional, 1906), 251.
11. John Burnham, "The Physicians' Discovery of a Deviate Community in America," *Medical Aspects of Human Sexuality* (1973), and quoted in Vern L. Bullough, *Sexual Variance*, 590.

12. Allan M'Lane Hamilton, "The Civil Responsibility of Sexual Perverts," *American Journal of Insanity* 52 (1895–6): 503–11.

13. A. B. Holder, "The Bote: Description of a Peculiar Sexual Perversion Found among North American Indians," *New York Medical Journal* 1 (1889): 623–5.

14. Will Roscoe, *The Zuni-Man Woman* (Bloomington: University of Indiana Press, 1991).

15. Mark Thomas Connelly, *The Response to Prostitution in the Progressive Era* (Chapel Hill: University of North Carolina Press, 1980), 8–9.

16. Ibid., 8.

17. This concept is based on the ideas of Burnham, "Progressive Era Revolution."

18. See Keith Thomas, "The Double Standard," *Journal of the History of Ideas* 20 (1959): 195–216.

19. For a discussion see David Pivar, *Purity Crusade, Sexual Morality, and Social Control, 1868–1900* (Westport, Conn.: Greenwood, 1973).

20. For some of this literature see Vern L. Bullough, Barret Elcano, Margaret Deacon, and Bonnie Bullough, *Bibliography of Prostitution* (New York: Garland, 1977); and for the continuing fascination with it see Vern L. Bullough and Lilli Sentz, *Prostitution: An Annotated Bibliography* (New York: Garland, 1992).

21. [William Rathbone Greg], "The Great Sin of Great Cities," *Lancet* (January 20, 1855). It was also published as a pamphlet under the same title.

22. W. E. H. Lecky, *History of European Morals* (reprint, New York: Braziller, 1955), 2:283.

23. For a more lengthy discussion of this plus references see Vern L. Bullough, Brenda Shelton, and Sarah Slavin, *The Subordinated Sex* (Athens: University of Georgia Press, 1988), 275–312. See also Vern Bullough and Bonnie Bullough, *Women and Prostitution* (Buffalo, N.Y.: Prometheus, 1987): 232–328.

24. Allan Nevins and Milton Halsey Thomas, eds., *The Diary of George Templeton Strong* (New York: Columbia University Press, 1952), 1:318.

25. Havelock Ellis, *Sex in Relation to Society*, vol. 6 of *Studies in the Psychology of Sex* (Philadelphia: Davis, 1929), 288–9.

26. See, for example, L. Duncan Bulkley, *Syphilis in the Innocent (Syphilis Insontium), Clinically and Historically Considered with a Plan for the Legal Control of the Disease* (New York: Bailey & Fairchild, 1894). For a detailed overall study see Allan M. Brandt, *No Magic Bullet: A Social History of Venereal Disease in the United States Since 1880* (New York: Oxford University Press, 1985).

27. For a discussion of this see Owsei Temkin, "Therapeutic Trends and the Treatment of Syphilis before 1900," *Bulletin of the History of Medicine* 39 (July–August 1955): 309–16. For a more general discussion see Brandt, *No Magic Bullet*, 12–7.

28. Quoted in Leonard J. Goldwater, *Mercury: A History of Quicksilver* (Baltimore, Md.: Johns Hopkins University Press, 1972), 215.

29. An excellent reference work on this topic is Brandt, *No Magic Bullet*, 12–13.

30. Martha Marquardt, *Paul Ehrlich* (New York: 1951); M. P. Earles, "Salvarsan and the Concept of Chemotherapy," *Pharmaceutical Journal* 204 (April 18, 1970): 340–2; and Isador Rosen and Nathan Sobel, "Fifty Years' Progress in the Treatment of Syphilis," *New York State Medical Journal* 50 (November 15,

1950): 1694–6. See also Theodore Rosebury, *Microbes and Morals* (New York: Viking, 1971).

31. There are many accounts of these campaigns. For a general overall survey see Bullough and Bullough, *Women and Prostitution*, 259–90. For a more specialized study of the American scene see David Pivar, *Purity Crusade.*

32. There is a vast literature on these various alternatives. See, for example, Martin Henbry Blatt, *Free Love and Anarchism* (Urbana: University of Illinois Press, 1989); Lawrence Foster, *Women, Family and Utopia* (Syracuse, N.Y.: University of Syracuse Press, 1991); Louis J. Kern, *An Ordered Love: Sex Roles and Sexuality in Victorian Utopias—The Shakers, the Mormons, and the Oneida Community* (Chapel Hill: University of North Carolina Press, 1981); Raymond Lee Muncy, *Sex and Marriage in Utopian Communities* (Bloomington: University of Indiana Press, 1973); Stephen Nissenbaum, *Sex, Diet, and Debility in Jacksonian America: Sylvester Graham and Health Reform* (Westport, Conn.: Greenwood, 1980); and Taylor Stoehr, *Free Love in America: A Documentary History* (New York: AMS, 1979). For discussions of free love see Hal D. Sears, *The Sex Radicals: Free Love in High Victorian America* (Lawrence, Kans.: Regents Press, 1977); and John C. Spurlock, *Free Love: Marriage and Middle-Class Radicalism in America, 1825–1860* (New York: New York University Press, 1988). Even some of the less sexually oriented utopian communities had a variety of sexual policies: Everett Webber, *Escape to Utopia: The Communal Movement in America* (New York: Hastings House, 1959); and Edith Roelker Curtis, *A Season in Utopia* (New York: Nelson, 1961). A good summary is in John d'Emilio and Estelle B. Freedman, *Intimate Matters: A History of Sexuality in America* (New York: Harper & Row, 1988).

33. Victoria Woodhull, "The Elixir of Life: or Why do We Die" (speech given at the tenth annual convention of the American Association of Spiritualists, Chicago, September 18, 1873). The speech was printed in tract form, a few copies of which have survived, including one in the Vern and Bonnie Bullough Collection at the California State University at Northridge.

34. A good overview is Paul S. Boyer, *Purity in Print* (New York: Scribner's, 1968).

35. D'Emilio and Freedman, *Intimate Matters*, 160.

36. Alfred Fournier, *Syphilis and Marriage*, trans. Prince Albert Morrow (New York: Appleton, 1881); and Prince A. Morrow, *Venereal Memoranda: A Manual for the Student and Practitioners* (New York: Wood, 1885).

37. Burnham, "Progressive Era Revolution," 893.

38. Fournier, *Syphilis and Marriage*; and Prince Morrow, *Social Diseases and Marriage: Social Prophylaxis* (New York: Lea Brothers, 1904).

39. Bryan Strong, "Ideas of the Early Sex Education Movement in America, 1890–1920," *History of Education Quarterly* 12 (1972): 129–61.

40. Burnham, "Progressive Era Revolution"; Brandt, *No Magic Bullet*, 38; Connelly, *Response to Prostitution*; Ruth Rosen, *The Lost Sisterhood: Prostitution in America, 1900–1918* (Baltimore, Md.: Johns Hopkins University Press, 1982); and James Gardner, "Microbes and Morality: The Social Hygiene Crusade in New York City, 1891–1917" (Ph.D. diss., Indiana University, 1973).

41. D'Emilio and Freedman, *Sexual Intimacy*, 207.

42. For a brief discussion of the social hygiene movement see Charles Walter

Clarke, *Taboo: The Story of the Pioneers of Social Hygiene* (Washington, D.C.: Public Affairs, 1961), 82–84.

43. For the development of the rubber condom see Vern L. Bullough, "A Brief Note on Rubber Technology: The Diaphragm and the Condom," *Technology and Culture* 22 (January 1981): 104–11.

44. W. P. J. Mensinga, *Über facultative Sterilität*, 2 vols., 2d ed. (Leipzig: Heuser, 1884).

45. Edgar Bliss Foote, *Medical Common Sense* (New York: Published by the author, 1863). The book was copyrighted in 1862, and this might be the date of his invention for which he said he had applied for a patent. There is no evidence it was granted and no record in the patent office.

46. A good discussion of the chemical contraceptives from the point of view of the 1930s is Cecil I. B. Voge, *The Chemistry and Physics of Contraceptives* (London: Cape, 1933).

47. Brandt, *No Magic Bullet*, 96.

48. Quoted by Brandt, ibid., 96.

49. Brandt, *No Magic Bullet*, 96–121.

50. The findings were found in volume ten of her unpublished works by a Stanford historian, who wrote it up: Carl Degler, "What Ought to Be and What Was: Women's Sexuality in the Nineteenth Century," *American Historical Review* 79 (December 1974): 1467–90. The complete survey was published as Clelia Duel Mosher, *The Mosher Survey: Sexual Attitudes of Forty-five Victorian Women*, ed. James Mahood and Kristine Wenburg (New York: Arno, 1980).

51. Katheryn Allamon Jacob, "The Mosher Report," *American Heritage* 23, no. 4 (June–July, 1981): 56–65.

52. He wrote up his use of models in Robert Latou Dickinson, "The Application of Sculpture to Practical Teaching in Obstetrics," *American Journal of Obstetrics and Gynecology* 40 (1940): 662–70.

53. Among his writings was James C. Cameron, Edward P. Davis, Richard C. Norris, Robert L. Dickinson, eds., *American Text Book of Obstetrics*, 2d ed. (1895; reprint, Philadelphia: Saunders, 1902). Many of his most important publications came after his retirement from practice, including Robert L. Dickinson, *Birth Atlas* (New York: Maternity Center Association, 1940); Robert Latou Dickinson, *Control of Conception* (Baltimore, Md.: Williams & Wilkins, 1931); and Robert Latou Dickinson, *Human Sex Anatomy* (Baltimore, Md.: Williams & Wilkins, 1949).

54. Robert Latou Dickinson, "A Program for American Gynecology: Presidential Address," *American Journal of Obstetrics and Gynecology* 1 (1920): 2–10.

55. R. L. Dickinson and L. Beam, *A Thousand Marriages* (Baltimore, Md.: Williams & Wilkins, 1931); and R. L. Dickinson and L. Beam, *The Single Woman* (Baltimore, Md.: Williams & Wilkins, 1934).

56. Dickinson and Beam, *Thousand Marriages*, 420.

57. Felix Roubaud, *Traité de l'impuissance et de la stérilité chez l'homme et chez la femme*, 2d ed. (Paris: Baillière, 1876).

58. Dickinson and Beam, *Thousand Marriages*, chap. 7, p. 93, figs. 55, 67, 142, 145, 146.

59. Dickinson, *Human Sex Anatomy*.

60. Ernst P. Boas and Ernst F. Goldschmidt, *The Heart Rate* (Springfield, Ill.: Thomas, 1932). This is summarized and graphed in Dickinson, *Human Sex Anatomy*, 86, 126–7.

61. Max J. Exner, "Sex Education in the Colleges and Universities," *Journal of the Society for Sanitary and Moral Prophylaxis* 6 (October 1915): 131–3.

62. M. J. Exner, "Prostitution in Its Relation to the Army on the Mexican Border," *Social Hygiene* 3 (April 1917): 202–11. I am indebted to Brandt, *No Magic Bullet*, for the references to Exner.

63. M. J. Exner, *Problems and Principles of Sex Education* (New York: Association Press, 1915). I was unable to consult the original but did see a reprint of the pamphlet that noted that it was unchanged from the original. It was published by the Association Press in 1922, indicating that the pamphlet continued to circulate.

64. Kinsey, for example, felt that the incidence of masturbation was so low that it represented a failure to obtain the facts. Alfred C. Kinsey, Wardell B. Pomeroy, and Clyde E. Martin, *Sexual Behavior in the Human Male* (Philadelphia: Saunders, 1948), 499. He also arrived at lower incidences of premarital sexual intercourse than Kinsey did. Ibid., 552.

65. Quoted in Peter Collier and David Horowitz, *The Rockefellers: An American Dynasty* (New York: Holt, Rinehart & Winston, 1976); see also Raymond B. Fosdick, *John D. Rockefeller, Jr.: A Portrait* (New York: Harper, 1956).

66. Rosen, *Lost Sisterhood*, 124–7; Roland Richard Wagner, "Virtue Against Vice: A Study of Moral Reformers and Prostitution in the Progressive Era" (Ph.D. diss., University of Wisconsin, 1971); and "The Rockefeller Grand Jury Report," *McClure's* 35 (1910): 471–3.

67. For an account of her work at Bedford Hills see Estelle B. Freedman, *Their Sisters' Keepers* (Ann Arbor: University of Michigan Press, 1981), 134. See also Eugenia C. Lekkerker, *Reformatories for Women in the United States* (The Hague: Wolters, 1931), 105. Davis herself also wrote about this aspect in Katharine Bement Davis, "The Fresh Air Treatment for Moral Disease," *Proceedings of the Annual Congress of the National Prison Association of the United States* (1905), and Katharine Bement Davis, "Outdoor Work for Women Prisoners," *Proceedings of the National Conference of Charities and Corrections* (1909).

68. There is a brief biography of her: Ellen Fitzpatrick, in her introduction to *Katharine Bement Davis, Early Twentieth-Century American Women, and the Study of Sex Behavior*, ed. Ellen Fitzpatrick (New York: Garland, 1987). I have relied on this for some data.

69. For a firsthand account of the laboratory, see Katharine Bement Davis, introduction to *The Mentality of the Criminal Woman*, by Jean Weidensall (Baltimore, Md.: Warwick & York, 1916), ix–xiv. Weidensall was the director of the lab, and before it closed in 1918 it had undertaken the study of 761 women at six institutions. No one or two outstanding causes of prostitution were discovered.

70. George Kneeland, *Commercialized Prostitution in New York City* (New York: Century, 1913).

71. Abraham Flexner, *Prostitution in Europe* (New York: Century, 1913); Raymond B. Fosdick, *European Police Systems* (New York: Century, 1915); Raymond B. Fosdick, *American Police Systems* (New York: Century, 1921); and H. B. Woolston, *Prostitution in the United States* (New York: Century, 1921).

72. Edward L. Bernays, *Biography of an Idea: Memoirs of a Public Relations Counsel* (New York: Simon & Schuster, 1965).

73. Katharine Bement Davis to John D. Rockefeller Jr., April 23, 1920, Rockefeller Family Archives, Rockefeller Boards, Bureau of Social Hygiene, record group 2, Rockefeller Foundation Archives, Pocantico Hills, North Tarrytown, New York.

74. John D. Rockefeller Jr. to Katharine Bement Davis, October 16, 1920, Rockefeller Family Archives, Rockefeller Boards, record group 3, Rockefeller Foundation Archives.

75. Katharine Bement Davis, *Factors in the Sex Life of Twenty-two Hundred Women* (New York: Harper, 1929).

76. Ibid., xvi.

77. Ibid., 152–3.

78. Ibid., 15–21.

79. Ibid., 62–94.

80. See S. D. Aberle and G. W. Corner, *Twenty-five Years of Sex Research: History of the National Research Council for Research in Problems of Sex. 1922–47* (Philadelphia: Saunders, 1953). See also Vern L. Bullough, "Katharine Bement Davis, Sex Research, and the Rockefeller Foundation," *Bulletin of the History of Medicine* 61 (1988): 74–89; and Vern L. Bullough, "The Rockefellers and Sex Research," *Journal of Sex Research* 21 (1985): 113–25.

81. Request for an Appropriation of $20,000 to the Bureau of Social Hygiene to Be Used in Promoting the Working out of a Plan for Research in the Field of Sex, Rockefeller Family Archives, Rockefeller Boards, record group 1, Rockefeller Foundation Archives. For earlier stages of the proposal see M. J. Exner, M.D., to John D. Rockefeller Jr., June 7, 1921, Rockefeller Family Archives, Rockefeller Boards, record group 2, Rockefeller Family Archives; and John D. Rockefeller to Katharine Bement Davis, June 23, 1921, Bureau of Social Hygiene, Minutes, ser. 1, box 3, Rockefeller Foundation Archives.

82. John D. Rockefeller, Jr., to Katharine Bement Davis, June 23, 1921, Bureau of Social Hygiene, Minutes, ser. 1, box 3, Rockefeller Foundation Archives. For earlier Flexner responses see Simon Flexner to John D. Rockefeller Jr., June 13, 1921, Rockefeller Family Archives, Rockefeller Boards, record group 2, Rockefeller Foundation Archives. There are two Flexner letters for that date, one written before Flexner received a letter from Exner, and one after. Both say the same thing, although the second is more cautious.

83. See Earl F. Zinn, "History, Purpose, and Policy of the National Research Council's Committee for Research on Sex Problems," *Mental Hygiene* 8 (1924): 94–105.

84. Ibid.; and Exhibit B, October 28, 1921, Conference on Sex Problems, Rockefeller Family Archives, Rockefeller Boards, record group 2, Rockefeller Foundation Archives.

85. Vernon Kellogg, Permanent Secretary, National Research Council, to John D. Rockefeller Jr., March 25, 1922, Rockefeller Family Archives, Rockefeller Boards, record group 2, Rockefeller Foundation Archives.
86. Gilbert V. Hamilton, *A Research in Marriage* (New York: Boni, 1929), 154–5.
87. Earl F. Zinn to G. V. Hamilton, May 28, 1928, Bureau of Social Hygiene, ser. 3, box 9, Rockefeller Foundation Archives.
88. This is what Kinsey said, in Kinsey, Pomeroy, and Martin, *Sexual Behavior in the Human Male*, 25–26.
89. Dickinson and Beam, *Thousand Marriages*, vii.

CHAPTER 5: ENDOCRINOLOGY RESEARCH AND CHANGING ATTITUDES

1. I have not been able to find this report in the Rockefeller Archives. See Katharine Bement Davis to Raymond B. Fosdick, October 16, 1925, Rockefeller Family Archives, Rockefeller Boards, record group 2, Rockefeller Foundation Archives, Pocantico Hills, North Tarrytown, New York. See also Sophie Bledsoe Aberle and George W. Corner, *Twenty-five Years of Sex Research: History of the National Research Council for Research in the Problems of Sex, 1922–1947* (Philadelphia: Saunders, 1953), 16.
2. Several researchers supported by the CRPS, however, did present papers at the Second International Congress of Sex Research held in London in 1930. This was a continuation of Moll's group and was a rival of Hirschfeld and Ellis's group. This was the last meeting of the group. See A. W. Greenwood, ed., *Proceedings of the Second International Congress for Sex Research* (London: Oliver & Boyd, 1931).
3. Yerkes' three-page guide has been reprinted: Aberle and Corner, *Twenty-five Years of Sex Research*, 102–4. It included six basic areas: (1) genetics of sex, (2) determination of sex, (3) sex development and differentiation, (4) the problem of sex interrelations, (5) sex functions, and (6) systematics of sex in plants and animals. Yerkes argued that category four, which he thought was essentially a human problem, should be included in the physiological, psychological, and sociological division of the committee, although he agreed that the potential for biological research existed as well. The category outlined in the greatest detail was the third one, which he believed could be divided into two general areas: the description, including histological studies, of the gonads at all ages and the problem of sex hormones, which was subdivided into a number of other groups. He also devoted a couple of paragraphs to sex function, the study of which involved altering the sex glands under experimental conditions (for example, by vasectomy or x-ray), and proposed a study of the causes of sterility.
4. For a brief account of his career see Ray L. Watterson, "Frank R. Lillie," in *Dictionary of Scientific Biography*, ed. Charles Coulston Gillespie (New York: Scribner's, 1973), 8:354–60.
5. Clarence E. McClung, "The Accessory Chromosome; Sex Determination," *Biology Bulletin* 3 (1902): 43–84.

6. F. R. Lillie, "The Theory of the Freemartin," *Science* 43 (1916): 611–3; F. R. Lillie, "The Free-Martin: A Study of the Action of Sex Hormones in the Foetal Life of Cattle," *Journal of Experimental Zoology* 23 (1917): 371–472.

7. See, for example, the "News and Comments" column, *Science* 258 (November 6, 1992): 880–2.

8. Aberle and Corner, *Twenty-five Years of Sex Research*, 29.

9. See in particular George W. Corner, *The Hormones in Human Reproduction*, 2d ed. (Princeton, N.J.: Princeton University Press, 1947); and for more detail, see Edgar Allen et al., *Sex and Internal Secretions*, 2d ed. (Baltimore, Md.: Williams & Wilkins, 1939).

10. Aristotle, "History of Animals," 631b19–632a32; trans. D'Arcy W. Thompson, in *The Complete Works of Aristotle*, Bollingen Series No. 71.2 (Princeton, N.J.: Princeton University Press, 1984); Aristotle, "Generation of Animals," 787b20–788a17; trans. A. Platt, in *The Complete Works of Aristotle*, and Aristotle, "Problems," 8946b19–894b34; in *The Complete Works of Aristotle*.

11. Arnold Adolph Berthold, "Transplantation der Hoden," *Bulletin of the History of Medicine* 16 (1944): 399–401. The original was published in 1849.

12. Emil Knauer, "Einige Versuch über Ovarientransplantation bei Kaninchen," *Zentralblatt fur Gynakologie* 20 (1896): 524–8.

13. Artur Biedl, *Innere Sekretion* (Vienna: Urban & Schwarzenburg, 1910). An English translation appeared in 1912, and the fourth edition (1922) includes an exhaustive bibliography.

14. Thomas Addison, *On the Constitutional and Local Effects of Disease of the Supra-Renal Capsules* (London: Highley, 1855); see also Thomas Addison, "Disease: Chronic Suparenal Insufficiency," *London Medical Gazette* 43 (1849): 517–8; it was an article that was later expanded into the book.

15. Much of this is recounted in Claude Bernard, *Leçons de physiologie expérimentale appliquée à le médecine*, 2 vols. (Paris: Baillière, 1855–6).

16. Charles-Édouard Brown-Séquard, "Expérience démonstrant la puissance dynamogénique chez l'homme d'un liquide extrait de testicules d'animaux," *Archives de Physiologie Normale et Patholoique*, ser. 1, 5 (1889): 651–8.

17. George Redmayne Murray, "Note on the Treatment of Myxedema by Hypodermic Injection of an Extract of Thyroid Gland of a Sheep," *British Medical Journal* 2 (1891): 796–7.

18. William Maddock Bayliss and Ernest Henry Starling, "The Mechanism of Pancreatic Secretion," *Journal of Physiology*, 28 (1902): 325–53. See also William Maddock Bayliss and Ernest Henry Starling, "Demonstration of the Existence of Secretin in the Duodenal Secretion," *Lancet* 1 (1902): 813.

19. Eugen Steinach, *Sex and Forty Years of Biological and Medical Experiments* (New York: Viking, 1930), 239–40.

20. Eugen Steinach, *Verjüngung Durch Experimentelle Neubelebung der Alternden Pubertätsdrüse*, (Berlin: J. Springer, 1920); see also Eugen Steinach, *Sex and Life* (New York: Viking Press, 1940).

21. For this and other examples see Stewart H. Holbrook, *The Golden Age of Quackery* (New York: Macmillan, 1959).

22. Gerald Carson, *The Roguish World of Doctor Brinkley* (New York: Holt, Rinehart & Winston, 1960).

23. The first report on the subject was made by Frederick G. Banting, Charles H. Best, and John J. R. Macleod at the American Physiological Society on December 28, 1921; this was published as Frederick G. Banting, Charles H. Best, and John J. R. Macleod, "The Internal Secretion of the Pancreas," *American Journal of Physiology* 59 (1922): 479. See also Frederick G. Banting and Charles H. Best, "The Internal Secretion of the Pancreas," *Journal of Laboratory and Clinical Medicine* 7 (1921–2): 251–66.

24. Edgar Allen and Edward Doisy, "An Ovarian Hormone," *Journal of the American Medical Association* 81 (1923): 819–21; and Edgar Allen and Edward Doisy, "The Induction of a Sexually Mature Condition in Immature Females by Injection of the Ovarian Follicular Hormone," *American Journal of Physiology* 69 (1924): 577–88.

25. Charles R. Stockard and George N. Papanicolaou, "The Existence of a Typical Oestrus Cycle in the Guinea-Pig, with a Study of Its Histological and Physiological Changes," *American Journal of Anatomy* 22 (1917): 225–83.

26. Selmar Aschheim and Bernhard Zondek, "Schwangerschaftsdiagnose aus dem Harn (durch Hormonnachweis)," *Klinische Wochenschrift* 7 (1928): 8–9.

27. Edward A. Doisy et al. "The Preparation of the Crystalline Ovarian Hormone from the Urine of Pregnant Women," *Journal of Biological Chemistry* 86 (1930): 499–509.

28. George W. Corner, *The Seven Ages of a Medical Scientist: An Autobiography* (Philadelphia: University of Pennsylvania, 1981), 235.

29. T. F. Gallagher and F. C. Koch, "The Testicular Hormone," *Journal of Biological Chemistry* 84 (1929): 495–500; C. R. Moore, T. F. Gallagher, and F. C. Koch, "The Effects of Extracts of Testis in Correcting the Castrated Condition in the Fowl and in the Mammal," *Endocrinology* 13 (1929): 367–74; and C. R. Moore and T. F. Gallagher, "On the Prevention of Castration Effects in Mammals by Testis Extract Injection," *American Journal of Physiology* 89 (1929): 388–94.

30. Adolf Friedrich Johann Butenandt, "Ueber die chemische Untersuchung der Sexualhormone," *Zeitschrift für angew Chem* 44 (1931): 905–8.

31. K. David, E. Dingemanse, J. Freud, and E. Laquer, "Über krystallinisches männliches Hormon aus Hoden (Testosteron), wirsamer als aus Harn oder aus Cholesterin beritetes Andosteron," *Hoppe-Seyler's Zeitschrift für physiologische Chemie* 233 (1935): 281–2.

32. Herbert M. Evans and Joseph A. Long, "The Effect of the Anterior Lobe Administered Intraperitoneally upon Growth Maturity and Oestrus Cycles of the Rat," *Anatomical Record* 21 (1921): 62–63.

33. Technically, it was not isolated until 1949. See Choh Hao Li, M. E. Simpson, and H. M. Evans, "Isolation of Pituitary Follicle-Stimulating Hormone," *Science* 109 (1949): 445–6.

34. Chao Hao Li, "Interstitial Cell Stimulating Hormone. II. Method of Preparation and Some Physico-Chemical Studies," *Endocrinology* 27 (1940): 803–8.

35. There is a vast and growing literature on this. Though the endnotes have cited

some of the key breakthroughs, one of the chief sources is the various editions of *Sex and Internal Secretions*. The first edition, *Sex and Internal Secretions*, ed. Edgar Allen (Baltimore, Md.: Williams & Wilkins, 1932), and much of the research in it was sponsored by the CRPS. Edgar Allen, *Sex and Internal Secretions*, 2d ed., eds. C. H. Danforth, and E. A. Doisy (Baltimore, Md.: Williams & Wilkins, 1939) updated the subject; and the last edition, *Sex and Internal Secretions*, 3d ed., ed. William C. Young, 2 vols. (Baltimore, Md.: Williams & Wilkins, 1961), is almost totally devoted to the biology of sex.

36. G. W. Barthelmez, *Histological Studies on the Menstruating Mucous Membrane of the Human Uterus*, Contributions to Embryology No. 142 (Washington, D.C.: Carnegie Institution, 1932).

37. This was the theory advanced by J. E. Markee, *Menstruation in Intraocular Endometrial Transplants in the Rhesus Monkey*, Contributions to Embryology No. 28 (Washington, D.C.: Carnegie Institution, 1940), 219–306; and J. E. Markee, "Morphologic and Endocrine Basis for Menstrual Bleeding," in *Progress in Gynecology*, ed. J. V. Meigs and S. H. Sturgis (New York: Grune & Stratton, 1946), 2:37–47.

38. There are a number of studies involved in this but the early ones were by F. L. Hisaw, "Development of the Grafian Follicle and Ovulation," *Physiological Review* 27 (1947): 95–119. See also Li, Simpson, and Evans, "Interstitial Cell Stimulating Hormone," 803–8; see Li, Simpson, and Evans, "Isolation of Pituitary Follicle-Stimulating Hormone."

39. George W. Corner and W. M. Allen, "Physiology of the Corpus Luteum," *American Journal of Physiology* 88 (1929): 326–46.

40. The change in temperature was noted early; W. Squire, "Puerperal Temperatures," *Transactions of the Obstetrical Society (London)* 9 (1868): 129. For its use in natural family planning, see J. Ferin, "Détermination de la période stérile prémenstruelle par la courbe thermique," *Bruxelles Medica* 27 (1947): 86–93.

41. J. Billings, *Natural Family Planning: The Ovulation Method* (Collegeville, Minn.: Liturgical, 1973); see also E. Billings and A. Westmore, *The Billings Method: Controlling Fertility without Drugs or Devices* (New York: Random House, 1980).

42. This was first discovered by A. Aschheim and B. Zondek, "Ei und Hormon," *Klinische Wochenschrift* 6 (1927): 1321; and A. Aschheim and B. Zondek, "Die Schwangerschaftsdiagnose aus dem Harn durch Hormonnachweis," *Klinische Wochenschrift* 7 (1928): 7, 8–9, 1404–11, 1453–7.

43. See Corner, *Seven Ages of a Medical Scientist*, 249–54; George Corner and William Myron Allen, "Physiology of the Corpus Luteum," *American Journal of Physiology* 88 (1929): 325–6. See also Chandler M. Brooks, Jerome L. Gilbert, Harold A. Level, and David R. Curtis, *Humors, Hormones and Neurosecretions* (New York: SUNY Press, 1962); John W. Everett, "The Mammalian Female Reproductive Cycle and Its Controlling Mechanisms," in *Sex and Internal Secretions*, 3d ed., 1: 497; Frederick L. Hisaw and Frederick L. Hisaw, Jr., "Action of Estrogen and Progesterone on the Reproductive Tract of Lower Primates," in *Sex and Internal Secretions*, 3d ed. (Baltimore, Md.: Williams & Wilkins, 1961), 556–89; and H. Maurice Goodman, *Basic Medical Endocrinology* (New York: Raven, 1988).

44. Sinclair Lewis, *Arrowsmith* (1925; reprint, New York: Grosset & Dunlap, 1945).

45. Paul De Kruif, "Jacques Loeb, the Mechanist," *Harper's Monthly Magazine* 146 (1922–3): 181–90. De Kruif describes his collaboration with Sinclair Lewis in his memoir, *Sweeping Wind: A Memoir* (New York, Harcourt, Brace and World, 1962). See also Mark Schorer, *Sinclair Lewis* (New York, 1961), 361–9; and Charles E. Rosenberg, "Martin Arrowsmith: The Scientist as Hero," *American Quarterly* 15 (Fall 1963): 447–58.

46. Jacques Loeb, "On the Nature of the Process of Fertilization and the Artificial Production of Normal Larvae (plutei) from the Unfertilized Eggs of the Sea Urchin," *American Journal of Physiology* 3 (1899): 135–8. For a fuller account of his views see Jacques Loeb, *The Mechanistic Conception of Life* (Chicago: University of Chicago Press, 1912).

47. See Donald Fleming, the introduction to Jacques Loeb, *The Mechanistic Conception of Life* (1912; reprint, Cambridge, Mass.: Harvard University Press, 1964), xxiii.

48. See G. Pincus and E. V. Enzmann, "Can Mammalian Eggs Undergo Normal Development *in Vitro*," *Proceedings of the National Academy of Sciences* 20 (1934): 121–2. See also, *New York Times* (May 13, 1934), sec. 8, p. 6, col. 3; and *Time* 23 (March 12, 1934): 57. Interestingly, it is now known that rabbit spermatozoa must undergo "seasoning" or capacitation in the female genital tract before they can activate the ovum, and it is believed that Pincus caused the rabbit eggs to develop through accidental parthenogenetic activation, since pathogenesis is easier to achieve in rabbits than *in vitro* fertilization. Live birth from *in vitro* fertilization was accomplished in 1959 by Min-Chueh Chang, Pincus's long-time co-worker. See M.-C. Chang, "Fertilization of Rabbit Ova in Vitro," *Nature* 184 (1959): 466–7. It was Chang who also discovered the necessity of the maturation of the sperm: M-C. Chang, "Fertilizing Capacity of Spermatozoa Deposited into the Fallopian Tubes," *Nature* 168 (1951): 697–8. This was also co-discovered and called capacitation by C. R. Austin in the same year: C. R. Austin, "Observation on the Penetration of the Sperm into the Mammalian Eggs," *Australian Journal of Scientific Research* 4 (1951): 581. See also Aldous Huxley, *Brave New World* (London: Chatto and Windus, 1932).

49. Gregory Pincus and E. V. Enzmann, "The Comparative Behavior of Mammalian Eggs *in Vivo* and *in Vitro*, II. The Activation of Tubal Eggs of the Rabbit," *Journal of Experimental Zoology* 73 (1936): 195–208. See also G. Pincus, *The Eggs of Mammals*, Experimental Biology Monographs (New York: Macmillan, 1936).

50. J. D. Ratcliff, "No Father to Guide Them, " *Collier's Magazine* (March 20, 1937): 19. See also *Newsweek* 7 (April 4, 1936): 4; *Time* 27 (April 6, 1936): 49–50; and *New York Times* (March 27, 1936).

51. *New York Times* (November 2, 1939), 18; "Rabbits without Fathers" [Editorial], *New York Times* (November 3, 1939), 20; *New York Times* (April 28, 1940), 8; and *New York Times* (April 30, 1941), 11. Articles that served as the source for the media were Gregory Pincus, "The Development of Fertilized and Artificially Activated Rabbit Eggs," *Journal of Experimental Zoology* 82 (1939):

85–120; Gregory Pincus, "The Breeding of Rabbits Produced by Recipients of Artificially Activated Ova," *Proceedings of the National Academy of Sciences* 25 (1939): 357–9; and Gregory Pincus and Herbert Shapiro, "Further Studies on the Activation of Rabbit Eggs," *Proceedings of the American Philosophical Society* 83 (1940): 163–5.

52. For this see Albert Q. Maisel, *The Hormone Quest* (New York: Random House, 1965), esp. chap. 4, pp. 59–81; see also for background to this section, James Reed, *From Private Vice to Public Virtue: The Birth Control Movement and American Society Since 1830* (New York: Basic, 1978), 317–33.

53. This theme is explored by Reed, *From Private Vice to Public Virtue*, 225.

54. Bonnie Bullough and George Rosen, *Preventive Medicine in the United States 1900–1990* (Canton, Mass.: Science History, 1992), 46–47. See also Allan M. Brandt, *No Magic Bullet: A Social History of Venereal Disease in the United States Since 1880* (New York: Oxford University Press, 1985), 122.

55. F. M. Thurston, *A Bibliography on Family Relationships* (New York: National Council of Parent Education, 1932).

56. C. B. Broderick, "To Arrive Where We Started: The Field of Family Studies in the 1930's," *Journal of Marriage and Family* 50 (1988): 569–84.

57. See, for example, Hal D. Sears, *The Sex Radicals: Free Love in High Victorian America* (Lawrence: Regents Press of Kansas, 1977); and Karen Lystra, *Searching the Heart: Women, Men, and Romantic Love in Nineteenth Century America* (New York: Oxford University Press, 1989).

58. Henrik Ibsen, *Ghosts* and *A Doll's House* in *The Complete Major Prose Plays of Henrik Ibsen*, trans. Rolf Fjelde (New York: Holt, Rinehart & Winston, 1967).

59. George Bernard Shaw, *Getting Married* in *The Doctor's Dilemma, Getting Married*, and *The Shewing-up of Blanco Posnet* (London: Constable, 1911).

60. Otto Weininger, *Sex and Character* (New York: Putnam's, 1906).

61. August Strindberg, *The Father* (1887), trans. Valburg Anderson (New York: Appleton-Century-Crofts, 1964); Strindberg, *Miss Julie* (1888) and the *Creditors* (1888), trans. Elizabeth Sprigge (Chicago: Aldine, 1962); Robert Herrick, *Together* (New York: Macmillan, 1908) and *One Woman's Life* (1913) (New York: AMS, 1964).

62. Ben Lindsay and Wainwright Evans, *The Companionate Marriage* (New York: Boni & Liveright, 1927).

63. For details of her life and activities see Ruth Hall, *Passionate Crusader: The Life of Marie Stopes* (New York: Harcourt Brace Jovanovich, 1977); I have relied on this for biographical data.

64. Marie Stopes, *Married Love* (London: Fifield, 1920); Marie Stopes, *Wise Parenthood* (London: Fifield, 1918); Marie Stopes, *Radiant Motherhood* (London, Putnam's, 1920); and Marie Stopes, *Enduring Passion* (London: Putnam's, 1928). All went through numerous editions. She wrote many other books and articles, including a number of poems and essays.

65. For a comprehensive discussion of this see Steven Seidman, *Romantic Longings: Love in America 1830–1980* (New York: Routledge, 1991).

66. Marie Carmichael Stopes, *Enduring Passion*, 4th ed. (New York: Putnam's, 1931), 21.

67. Margaret Sanger, *Happiness in Marriage* (New York: Blue Ribbon, 1926).

68. Hannah Stone and Abraham Stone, *A Marriage Manual* (New York: Simon & Schuster, 1939), 215.

69. Isabel E. Hutton, *The Sex Technique in Marriage*, 2d ed. (New York: Emerson, 1932), 107.

70. Stopes, *Enduring Passion*, 19.

71. Th. H. van de Velde, *Ideal Marriage: Its Physiology and Technique* (New York: Covici Friede, 1930), 2.

72. Ibid., 6.

73. I have based my information on van de Velde on the brief biography of him in Edward M. Brecher, *The Sex Researchers* (Boston: Little, Brown, 1969), 82–104. Brecher titled his chapter on van de Velde "He Taught a Generation How to Copulate," but it is important to emphasize that van de Velde was just one of the writers who did.

74. van de Velde, *Ideal Marriage*, 301.

75. Ibid., 211.

76. Ibid., 238–41.

77. See E. Haldeman-Julius, *The First Hundred Million* (New York: Simon & Schuster, 1928)

78. A good overview of the court decisions is Paul S. Boyer, *Purity in Print: Book Censorship in America* (New York: Scribner's, 1968).

79. For the English background see G. R. Scott, *Into Whose Hands* (London: Swan, 1961), 108–9.

80. Morris L. Ernst and Alan U. Schwartz, *Censorship: The Search for the Obscene* (New York: Macmillan, 1964), 72–79.

81. I have not seen the pamphlet, but it is extracted in ibid., 80–92.

82. Muriel Box, ed., *Birth Control and Libel: The Trial of Marie Stopes* (New York: Barnes, 1968).

83. Ernst and Schwartz, *Censorship*, 161–3. The case was *United States v. One Obscene Book Entitled "Married Love,"* April 6, 1931.

84. The cases is summarized in Ernst and Schwartz, *Censorship*, 93–107.

85. For information see James F. Cooper, *Technique of Contraception* (New York: Day-Nichols, 1928), 177–204.

86. "The Consumer and the Law," *Human Fertility* 8 (June 1943): 48–49; *Human Fertility* 9 (September 1944): 93–94.

87. Reed, *From Private Vice to Public Virtue*, 242.

88. Ibid., 168.

89. Ibid., 305.

90. See Norman B. Ryder, "The Emergence of a Modern Fertility Pattern: United States, 1917–66," *Fertility and Family Planning*, ed. S. J. Behrman, Leslie Corsa Jr., and Ronald Freedman (Ann Arbor: University of Michigan Press, 1969). For earlier studies see Ronald Freedman, Pascal K. Whelpton, and Arthur A. Campbell, *Family Planning, Sterility and Population Growth* (New York: McGraw-Hill, 1949); and Pascal K. Whelpton, Arthur A. Campbell, and John E. Patterson, *Fertility and Family Planning in the United States* (Princeton, N.J.: Princeton University Press, 1966).

CHAPTER 6: FROM FREUD TO BIOLOGY TO KINSEY

1. John C. Burnham, "The Influence of Psychoanalysis upon American Culture," *American Psychoanalysis: Origins and Development,* ed. Jacques M. Quen and Eric T. Carlson (New York: Brunner/Mazel, 1978), 60.

2. Nathan Hale, *Freud and the Americans: The Beginnings of Psychoanalysis in the United States, 1876–1917* (New York: Oxford University Press, 1971).

3. Floyd Dell, "Speaking of Psychoanalysis, the New Boon for Dinner Table Conversationalists," *Vanity Fair* 5 (December 1915): 53.

4. Quoted in Burnham, "Influence of Psychoanalysis," 61.

5. Seymour Fisher and Roger P. Greenberg, *The Scientific Credibility of Freud's Theories and Therapy* (New York: Basic, 1977), 285. See also Frank J. Sulloway, "Reassessing Freud's Case Histories: The Social Construction of Psychoanalysis," *Isis* 82 (1991): 245–73. Sulloway excluded Freud's ventures into psychobiography such as his study of Leonardo da Vinci from his case histories.

6. For example, see Jeffrey M. Masson, *The Assault on Truth: Freud's Suppression of the Seduction Theory* (New York: Farrar, Straus & Giroux, 1984).

7. Sigmund Freud, *The Complete Letters of Sigmund Freud to Wilhelm Fliess, 1887–1904,* trans. and ed. Jeffrey Mossaieff Masson (Cambridge, Mass: Belknap, 1985), 447.

8. Sulloway, "Reassessing Freud's Case Histories," 275.

9. Alfred Adler, *The Neurotic Constitution: Outlines of a Comparative Individualistic Psychology and Psychotherapy,* trans. Bernard Glueck and John E. Lind (New York: Moffat, Yard, 1917).

10. Alfred Adler, *Study of Organ Inferiority and Its Psychical Compensation,* trans. Smith Ely Jelliffe (New York: Nervous & Mental Disease Publishing, 1917). He later argued that secondary sexual characteristics of the opposite sex appear much more frequently among neurotics than among normal individuals and that it was this organic predisposition that led to feelings of individual personality. See Alfred Adler, "Der psychische Hermaphroditism im Leben und in der Neurose," *Fortschritt der Medizin* 38 (1910): 486–93.

11. Frank J. Sulloway, *Freud, Biologist of the Mind: Beyond the Psychoanalytic Legend* (New York: Basic, 1979), 430–1. See also Alfred Adler, *Understanding Human Nature,* trans. Walter Béram Wolfe (New York: Greenberg, 1927).

12. Wilhelm Stekel, *The Homosexual Neurosis,* trans. James Van Teslaar (New York: Emerson, 1940); Wilhelm Stekel, *Patterns of Psychosexual Infantilism* (New York: Grove, 1959); and Wilhelm Stekel, *Sexual Aberrations* (New York: Grove, 1963). For an interesting account see Wilhelm Stekel, *The Autobiography of William Stekel,* ed. Emil A. Gutheil (New York: Liveright, 1950).

13. Carl Gustav Jung, *The Collected Works of C. G. Jung,* vols. 7 and 9 (1966), ed. Gerhard Adler, Michael Fordham, and Herbert Read (Princeton, N.J.: Princeton University Press, vol. 7, 1966, vol. 9, 1968), parts 1–2. See also C. J. Jung, *Two Essays on Analytical Psychology* (New York: Pantheon, 1953).

14. Paul A. Robinson, *The Freudian Left* (New York: Harper & Row, 1969).

15. See Wilhelm Reich, *The Function of the Orgasm* (reprint, New York: World,

1971); Wilhelm Reich, *The Sexual Revolution* (New York: Farrar, Straus & Giroux, 1962).

16. *New York Times,* November 5, 1957, p. 31, col.4.

17. These ideas are developed in a series of books and articles. See Herbert Marcuse, *Reason and Revolution* (Boston: Beacon, 1960); Herbert Marcuse, *Eros and Civilization,* 2d ed. (Boston: Beacon, 1966); and Herbert Marcuse, "The Social Implications of Freudian 'Revisionism,'" *Dissent* 2, no. 3 (Summer 1955): 221–40.

18. There are a vast number of studies by Bissonette and his collaborators. See S. D. Aberle and G. W. Corner, *Twenty-five Years of Sex Research: History of the National Research Council for Research in Problems of Sex. 1922–47* (Philadelphia: Saunders, 1953), 144–7. Among the studies on mammals are T. H. Bissonette and A. G. Csech, "Modification of Mammalian Sexual Cycles. VII. Fertile matings of raccoons in December and in February induced by increasing daily periods of light," *Proceedings of Royal Society of London* 122 (1937): 246–54; T. H. Bissonette and A. G. Csech, "Modified Sexual Photoperiodicity in Cotton-Tail Rabbits," *Biology Bulletin* 77 (1939): 364–7; and T. H. Bissonette, "Sexual Photoperiodicity in Animals," *Journal of Heredity* 26 (1935): 284–6.

19. Much of this work was conducted under the direction of Herbert M. Evans (1882–1971). Among his most important articles was H. M. Evans and Katharine S. Bishop, "On the Existence of a Hitherto Unrecognized Dietary Factor Essential for Reproduction," *Science* 56 (1922): 650–1.

20. C. R. Moore and H. D. Chase, "Heat Application and Testicular Degeneration," *Anatomical Record* 26 (1923): 344; C. R. Moore and W. J. Quick, "A Comparison of Scrotal and Peritoneal Temperature," *Anatomical Record* 26 (1923): 344; and particularly C. R. Moore and W. J. Quick, "The Scrotum as a Temperature Regulator for the Testes," *American Journal of Physiology* 68 (1924): 70–9.

21. Robert Yerkes and Ada Watterson Yerkes, *The Great Apes: A Study of Anthropoid Life* (New Haven: Yale University Press, 1929). His writings are extensive but among pertinent studies are R. M. Yerkes and J. H. Elder, "The Sexual and Reproductive Cycles of the Chimpanzee," *Proceedings of the National Academy of Science* 22 (1936): 362–70; and R. M. Yerkes, "Sexual Behavior in the Chimpanzee," *Human Biology* 11 (1939): 78–111. Yerkes' papers are at the Yale University Medical School Library. There is a short biography of him: John C. Burnham, "Robert M. Yerkes," *Dictionary of Scientific Biography,* ed. Charles C. Gillespie (New York: Scribner's, 1976), 14:549–51.

22. See Frank A. Beach, "A Review of Physiological and Psychological Studies of Sex Behavior in Mammals," *Physiological Review* 27 (1947): 240–327.

23. Bronislaw Malinowski, *The Sexual Life of Savages in North-Western Melanesia: An Ethnographic Account of Courtship, Marriage and Family Life Among the Natives of the Trobriand Islands, British New Guinea* (New York: Harcourt, Brace, 1929).

24. H. R. Hays, *From Ape to Angel: An Informal History of Social Anthropology* (reprint, New York: Capricorn, 1964), 327.

25. Margaret Mead, *Coming of Age in Samoa* (New York: Morrow, 1928).

26. Much of this criticism of Mead was more or less subrosa until after her death. The major charges were leveled in the early 1980s by Derek Freeman and caused a major controversy in anthropology and led to a denunciation of Freeman at the 1983 meetings of the American Anthropological Association held in Chicago. See Melvin Ember, "Evidence and Science in Ethnography: Reflections on the Freeman-Mead Controversy," *American Anthropologist* 87 (1985): 906–9. The issue simmered until 1987, when Fa'apua'a, one of Mead's informants, was found. Then nearly ninety, she told how she and her girlfriend Fofoa (who died in 1936) had made up what they told Mead. See Larry Gartenstein, "Sex, Lies, Margaret Mead, and Samoa," *Geo* (June–August, 1991). For other papers see Derek Freeman, "Fa'apua'a, Famu and Margaret Mead," *American Anthropologist* (December 1989): 1017–22. Derek Freeman, "There's Tricks i'th' World," *Visual Anthropology Reviews* (Spring 1991) [the title is a quote from Hamlet]; and Derek Freeman, "Paradigms in Collision," *Academic Questions* (July 1992). At present there is growing agreement that Mead's statements about Samoa were based on misinformation and misunderstanding. See Lenora Foerstel and Angela Gilliam, eds., *Confronting the Margaret Mead Legacy: Scholarship, Empire, and the South Pacific* (Philadelphia: Temple University Press, 1992).

27. Margaret Mead, *Sex and Temperament in Three Primitive Societies* (New York: Morrow, 1935).

28. P. Mantegazza, *Anthropological Studies of Sexual Relations of Mankind* (New York: Anthropological Press, 1932).

29. Clellan S. Ford and Frank A. Beach, *Patterns of Sexual Behavior* (New York: Harper, 1951), 250.

30. Ibid., 257.

31. Ibid., 129–30.

32. Jane Belo, *Bali, Rangda and Barong*, American Ethnological Society Monograph No. 16 (New York: J. J. Augustin, 1949), 130.

33. Ford and Beach, *Patterns of Sexual Behavior*, 130.

34. For an overall survey of these data from a critical contemporary perspective see Vern L. Bullough, *Sexual Variance in Society and History* (Chicago: University of Chicago, 1976), 22–50.

35. A very early study of Greek sex life was that of M. H. E. Meier, in "Päderastie," *Encylopädie der Wissenschaften und Künsten*, ed. J. S. Ersch and J. G. Gruber (Leipzig: Bockhaus, 1837), vol. 9, sec. 3, pp. 149–88, and although this was known to the few specialists who consulted the encyclopedia, the subject was widely ignored in most studies of ancient Greece. One of the first works to publicize it was Hans Licht, *Sexual Life in Ancient Greece* (London: Routledge & Kegan Paul, 1932). Licht was the pseudonym for Paul Brandt, and the English edition is a kind of bowdlerized translation of Brandt's *Sittengeschichte Griechenlands*. Other studies such as [John Addington Symonds], A *Problem in Greek Ethics* (n.p., 1901) were not widely available. John Jay Chapman, *Lucian, Plato, and Greek Morals* (Boston: Houghton Mifflin, 1931), 120, emphasizes the shock that a classical scholar had in reconciling himself to the facts of Greek sex life.

36. Otto Kiefer, *Sexual Life in Ancient Rome* (London: Routledge & Kegan Paul, 1934); and John Jakob Meyer, *Sexual Life in Ancient India* (reprint, New York: Barnes & Noble, 1953).

37. Charles Forberg, *De figuris veneris* (reprint, New York: Medical Press of New York, 1963).

38. See for example, Mitchell S. Buck, trans., *The Priapeia: An Anthology of Poems on Priapus* (n.p., 1937).

39. *My Secret Life*, introduction by G. Legman (reprint, New York: Grove Press, 1966), 11: 2191–2.

40. Ibid., 2:206.

41. Dorothy Dunbar Bromley and Florence Haxton Britten, *Youth and Sex* (New York: Harper & Brothers, 1938). The questionnaires are reproduced in the appendix.

42. Lewis M. Terman, Paul Buttenweiser, Leonard W. Ferguson, Winifred B. Johnson, and Daniel P. Wilson, *Psychological Factors in Marital Happiness* (New York: McGraw-Hill, 1938).

43. See Carney Landis, Agnes T. Landis, and M. Marjorie Bolles, *Sex in Development* (New York: Hoeber, 1940); and Carney Landis, M. M. Bolles, and D. D'Esopo, "Psychological and Physical Concomittants of Adjustment in Marriage," *Human Biology* 12 (1940): 559–65.

44. See, for example, W. B. Johnson and Lewis M. Terman, "Personality Characteristics of Happily Married, Unhappily Married, and Divorced Persons," *Character and Personality* 3 (1935): 199–311; C. Landis and M. M. Bolles, "Psychosexual Immaturity," *Journal of Abnormal Social Psychology* 35 (1940): 449–52; and C. Landis, A. T. Landis, and M. M. Bolles, *Sex in Development* (New York: Hoeber, 1940). Conspicuous by its absence in the CRPS-supported projects is any study of homosexuality or other variant sexuality that so dominated much of the European research. For some of the problems, see Henry L. Minton, "Femininity in Men and Masculinity in Women: American Psychiatry and Psychology Portray Homosexuality in the 1930's," *Journal of Homosexuality* 13 (1986): 1–22.

45. C. C. Miles and Lewis M. Terman, "Sex Differences in the Association of Ideas," *American Journal of Psychology* 41 (1929): 165–206; Lewis M. Terman, "Sex Differences in Certain Non-Intellectual Traits," *Psychological Bulletin* 24 (1927): 201; and Lewis M. Terman and C. C. Miles, *Sex and Personality: Studies in Masculinity and Femininity* (New York: McGraw-Hill, 1936).

46. C. Landis and M. M. Boles, *Personality and Sexuality of the Physically Handicapped Woman* (New York: Hoeber, 1942). This study consisted of interviews of one hundred handicapped women, the majority of whom were between the ages of eighteen and twenty-five, were Catholic, and were in the state psychiatric hospital. Subjects answered 116 questions in an interview conducted by Boles. Though there is a statistical breakdown, many of the subgroups included too few subjects from which to draw conclusions. The study was supported by the CRPS.

47. F. M. Strakosch, *Factors in the Sex Life of Seven Hundred Psychopathic Women* (Utica, N.Y.: State Hospital Press, 1934). His study was done as a part of his doctoral dissertation in psychology.

48. Alfred Kinsey, Wardell Pomeroy, and Clyde Martin, *Sexual Behavior in the Human Male* (Philadelphia: Saunders, 1958), 31–34.

49. Chair of the committee was the psychiatrist Eugen Kahn of Yale University. Included as members were a number of others who had done sex research, including Dickinson, Landis, Lashley, Meyer, and Terman. Gershon Legman compiled a bibliography on homosexuality and lesbianism, which ended up at the New York Academy of Medicine; parts of it were included in Vern L. Bullough, W. Dorr Legg, Barret W. Elcano, and James Kepner, *An Annotated Bibliography of Homosexuality*, 2 vols. (New York: Garland, 1976). His collection was a haphazard and repetitive one, but it was the first serious attempt in the United States to examine what those outside of the field of psychiatry had written about homosexuality and lesbianism.

50. George W. Henry, *Sex Variants: A Study of Homosexual Patterns*, 2 vols. (New York: Hoeber, 1941), xii–xiii.

51. Ibid., 1049–65. The data were summarized in appendix 5 by Henry and two radiologists, Robert P. Ball and John R. Carty.

52. Ibid., 1034.

53. Robert Latou Dickinson, "The Gynecology of Homosexuality" [Appendix], in *Sex Variants: A Study of Homosexual Patterns*, 2 vols., by George W. Henry (New York: Hoeber, 1941), 1069–130.

54. Henry, *Sex Variants*, 1025.

55. George W. Henry, *Sex Variants: A Study of Homosexual Patterns* (reprint, 2 vols. in 1, New York: Paul B. Hoeber, 1948), vii.

56. George W. Henry, *All the Sexes* (New York: Rinehart, 1955), esp. xii–xiii.

57. W. S. Sheldon, *The Varieties of Human Physique* (New York: Harper & Brothers, 1940); and W. S. Sheldon, *The Varieties of Temperament* (New York: Harper & Brothers, 1944).

58. A Russian named Chlenov made the first inquiry into the sexual life of Moscow students in 1903–4. There is a German summary of these by F. Feldhusen, "Die Sexualenquete unter der Moskauer Studentschaft," *Zeitschrift für Bekämfung der Geschechtskrankheiten* 8 (1909): 211–24, 245–55. His study was followed by one conducted by psychiatrists Zbankov and Jakovenko on students; they were unable to publish their results until 1922. During the 1920s, there were some wide-ranging surveys by Jakobson, which some have said were the best overall surveys of sex behavior before Kinsey's. See Raymond R. Willoughby, *Sexuality in the Second Decade* (New York: Kraus, 1966), Monograph Society for Research in Child Development (vol. II, no. 3, serial no. 10, 1937), who summarized some of these. See also J. Raboch, "History of Sexology in Eastern Europe," *Experimental and Clinical Endocrinology* 98, no. 2 (1991): 53–56.

59. Much of this information is based on Wardell B. Pomeroy, *Dr. Kinsey and the Institute for Sex Research* (New York: Harper & Row, 1972). Judith Reisman has charged that Kinsey was not simply chosen for the university's new marriage course but that he had maneuvered for many years to gain approval for the course and to be able to direct it. See Judith A. Reisman and Edward W. Eichel, *Kinsey, Sex and Fraud: The Indoctrination of a People* (Lafayette, La.:

Lochinvar-Huntington House, 1990), 5. There is little evidence either way, but the accusation should not be dismissed out of hand. The one course on the Indiana campus that had anything about human sexuality in it was in the traditional mode of emphasizing the dangers of sex, particularly the risk of infection and the bad sequelae of masturbation.

60. George W. Corner, *The Seven Ages of a Medical Scientist* (Philadelphia: University of Pennsylvania Press, 1981), 314.

61. Ibid., 268.

62. Ibid., 269.

CHAPTER 7: FROM STATISTICS TO SEXOLOGY

1. Alfred Kinsey, Wardell Pomeroy, and Clyde Martin, *Sexual Behavior in the Human Male* (Philadelphia: Saunders, 1948); and Alfred Kinsey, Wardell Pomeroy, Clyde Martin, and Paul Gebhard, *Sexual Behavior in the Human Female* (Philadelphia: Saunders, 1953).

2. Wardell B. Pomeroy, *Dr. Kinsey and the Institute for Sex Research* (New York: Harper & Row, 1972), 4.

3. Pomeroy says about ten were changed over the course of the interviews. He indicates that one such question was about extramarital petting, which was added only in 1948, because "Kinsey was still a little naive on the subject" and because he resisted changing the interview questions. Ibid., 121.

4. Ibid., 121; and Wardell Pomeroy, personal communication. I sponsored a class in which Pomeroy taught the code to others.

5. See Judith A. Reisman and Edward W. Eichel, in *Kinsey, Sex, and Fraud*, ed. J. Gordon Muir and John H. Court (Lafayette, La.: Lochinvar-Huntington House, 1990). This is a badly written and poorly edited book, in which Kinsey is described as unscientific for relying on either the memory of older subjects or data gathered from a pedophile. It implies that Kinsey must have conducted experiments himself. When questioned about the data, Gebhard emphasized that sexual experiments on human infants and children were illegal and that the Kinsey group never attempted follow-up studies. See the reply from Gebhard, "Dr. Paul Gebhard's Letter to Dr. Judith Pressman Regarding Kinsey Research Subjects and Data" (March 11, 1981) [Appendix B], in *Kinsey, Sex, and Fraud*, 223.

6. Ibid., 122–3.

7. William G. Cochran, Frederick Mosteller, and John W. Tukey, *Statistical Problems of the Kinsey Report* (Washington, D.C.: American Statistical Association, 1954), 23.

8. There is a written report from the committee set up to examine this; see "Report" foundation 1, ser. 200, box 41. Rockefeller Foundation Archives, Pocantico Hills, North Tarrytown, New York.

9. George W. Corner, *The Seven Ages of a Medical Scientist: An Autobiography* (Philadelphia: University of Pennsylvania Press, 1981), 315–6.

10. Pomeroy, *Dr. Kinsey*, 464.

11. See Kinsey et al., *Sexual Behavior in the Human Female*, 28–31.

12. Kinsey et al., *Sexual Behavior in the Human Male*, 161, 610–50.

13. For more on this see Vern L. Bullough, "The Kinsey Scale in Historical Perspective," in *Homosexuality/Heterosexuality: Concepts of Sexual Orientation*, ed. David P. McWhirter, Stephanie A. Sanders, and June Machover Reinisch (New York: Oxford University Press, 1990), 3–14.

14. In the 1930s Kinsey had become deeply attached to a student of his named Ralph Voris, who died of pneumonia two years after the sexual project was launched. Pomeroy wrote that Voris was the closest friend Kinsey ever had, but there is no other indication of anything other than friendship. See Pomeroy, *Dr. Kinsey*, 46; and Paul Robinson, "Dr. Kinsey and the Institute for Sex Research," *Atlantic* 229 (May 1972): 99–102.

15. Kinsey et al., *Sexual Behavior in the Human Male*, 610, 633–6. This is the explanation advanced by Paul Robinson, *The Modernization of Sex* (New York: Harper & Row, 1976), 70–71.

16. For various aspects of this view see Kenneth Lewes, *The Psychoanalytic Theory of Male Homosexuality* (New York: Simon & Schuster, 1988).

17. Kinsey et al., *Sexual Behavior in the Human Female*, 375–408.

18. Ibid., 377, 383.

19. Edward M. Brecher, *The Sex Researchers* (Boston: Little, Brown, 1969), 124.

20. Ibid., 568.

21. Ibid., 547, 549, 559; and Kinsey et al., *Sexual Behavior in the Human Female*, 284.

22. Ibid., 186.

23. Ibid., 311.

24. Ibid., 328.

25. Ibid., 416.

26. Kinsey et al., *Sexual Behavior in the Human Male*, 585.

27. Ibid., 650–1.

28. Kinsey et al., *Sexual Behavior in the Human Female*, 450–1.

29. Ibid., 460.

30. Pomeroy, *Dr. Kinsey*, 101.

31. Kinsey et al., *Sexual Behavior in the Human Female*, 435–6.

32. See Reinhold Niebuhr, "Kinsey and the Moral Problems of Man's Sexual Life," in *An Analysis of the Kinsey Reports*, ed. Donald Porter Geddes (New York: New American Library, 1954), 62–70.

33. Actually, he had collected data on pregnancy, birth, and abortion, which appeared in Paul H. Gebhard, Wardell B. Pomeroy, Clyde E. Martin, and Cornelia V. Christenson, *Pregnancy, Birth, and Abortion* (New York: Harper, 1958).

34. A Gallup poll following the book's publication found that 58 percent of the men and 55 percent of the women thought Kinsey's research was a good thing; only 10 and 14 percent, respectively, thought it a bad thing. See Pomeroy, *Dr. Kinsey*, 283–4.

35. Dodds went so far as to meet with officials of the Rockefeller Foundation to express his unhappiness and that of Van Dusen. C. I. Barnard, who spoke with Dodds, said the controversy was not nearly as great as he expected but probably

would have been more serious if Fosdick had not closed down on it. See Memo of June 28, 1948, Foundation ser. 200, box 40, Rockefeller Foundation Archives.

36. Lionel Trilling, *The Liberal Imagination* (1950; reprint, New York: Viking, 1957), 218.

37. Pomeroy, *Dr. Kinsey*, 298–9.

38. Quoted in ibid., 367.

39. As early as 1944, there is a letter in the Rockefeller Archives indicating disapproval of this policy of Kinsey, but apparently he persisted. Robert A. Lambert to Dr. D. F. Milam, July 21, 1944, Foundation Records, group 1, ser. 200, box 40, 457, Rockefeller Foundation Archives.

40. Corner, *Seven Ages of a Medical Scientist*, 316–7. Kinsey had received about two hundred thousand dollars in royalties from the first volume, but the cost of running his institute was approximately one hundred thousand dollars a year.

41. The controversy was reported in the *New York Times*, "U.S. Customs Refuses to Pass Obscene European Photos," November 18, 1950, n18, 9:5, and in *Indianapolis Start-News*, December 8, 1950. See Foundation Records [National Research Council] ser. 200, box 41, 463, Rockefeller Foundation Archives.

42. *United States v. 31 Photographs*, 156 F. Supp. 350 (S. D. N. Y., 1957).

43. Morris L. Ernst and Alan U. Schwartz, *Censorship: The Search for the Obscene* (New York: Macmillan, 1964), 125.

44. Pomeroy, *Dr. Kinsey*, 317–9.

45. Ernst Gräfenberg, "An Intrauterine Contraceptive Method" (1931), reprinted in *Contraception: Benchmark Papers in Human Physiology*, ed. L. L. Langley (Stroudsburg, Pa.: Dowden, Hutchinson & Ross, 1973), 339–56.

46. T. Ota, "A Study on Birth Control with an Intrauterine Instrument," *Japanese Journal of Obstetrics and Gynecology* 17 (1934): 210–4.

47. See Christopher Tietze, "Intra-Uterine Contraceptive Rings: Historical and Statistical Appraisal," *Intra Uterine Contraceptive Devices: Proceedings of the Conference, April 30–May 1, 1962, New York City*, ed. Christopher Tietze and Sarah Lewitt (Amsterdam: Excerpta Medica International Congress, ser. 54, 1962), 11–18; Willi Oppenheimer, "Prevention of Pregnancy by the Graefenberg Ring Method," reprinted in *Contraception*, 357–65, and Atsumi Ishihama, "Clinical Studies on Intrauterine Rings, Especially the Present State of Contraception in Japan . . . ," reprinted in *Contraception*, 366–82.

48. "Overpopulation and Family Planning," *Report of the Proceedings of the Fifth International Conference on Planned Parenthood, 1955* (London: International Planned Parenthood Federation, 1955).

49. Jack Lippes, "PID and IUD" (paper presented at the World Congress of Gynecology and Obstetrics, Tokyo, October 1979); and Jack Lippes, personal communication.

50. See Taek Il Kim and Syng Wook Im, "Mass Use of Intra-Uterine Contraceptive Devices in Korea," in *Family Planning and Population Programs*, ed. Bernard Berelson (Chicago: University of Chicago Press, 1966), 425–32.

51. See J. Lippes, "A Study of Intra-uterine Contraception: Development of a Plastic Loop," in *Contraception*, 383–90; and L. C. Margulies, "Intrauterine Contraception: A New Approach," in *Contraception*, 391–7.

52. World Health Organization (WHO), *Mechanization of Action, Safety and Efficacy of Intrauterine Devices,* Technical Report Ser. 753 (Geneva: World Health Organization, 1987).

53. For a full discussion of this see Morton Mintz, *At Any Cost: Corporate Greed, Women, and the Dalkon Shield* (New York: Pantheon Books, 1985).

54. See International Planned Parenthood Federation, International Medical Advisory Panel (IPPF), "Statement on Intrauterine Devices," *IPPF Medical Bulletin* 16 (December 1981): 1–3.

55. H. Selye, J. S. Brown, and J. B. Collip, "Effects of Large Doses of Progesterone in the Female Rat," *Proceedings of the Society for Experimental Biological Medicine* 34 (1936): 472.

56. A. W. Makepeace, G. L. Weinstein, and M. H. Friedman, "The Effect of Progestin and Progesterone on Ovulation in the Rabbit," *American Journal of Physiology* 119 (1937): 512; and E. W. Dempsey, "Follicular Growth Rate and Ovulation After Various Experimental Procedures in the Guinea Pig," *American Journal of Physiology* 120 (1937): 126.

57. Raphael Kurzok, "The Prospects for Hormonal Sterilization," *Journal of Contraception* 2 (1937), 27–29.

58. Quoted in James Reed, *From Private Vice and Public Virtue* (New York: Basic Books, 1978): 316.

59. These figures are given by Maisel, *Hormone Quest* (New York: Random House, 1965), 44.

60. R. E. Marker, R. B. Wanger, P. R. Ulshafer, E. L. Wittbecker, D. P. J. Goldsmith, and C. H. J. Ruof, "New Sources for Sapogenins," *Journal of the American Chemical Society* 69 (1947): 2242. This was published several years after Marker had left Penn State; it lists earlier citations.

61. Quoted in Pedro A. Lehmann, Antonio Bolivar, and Rodolfo Quintero, "Russell E. Marker, Pioneer of the Mexican Steroid Industry," *Journal of Chemical Education* 50 (1973): 195–9. For an update see Pedro A. Lehmann, "Early History of Steroid Chemistry in Mexico: The Story of Three Remarkable Men," *Steroids* 57 (1992): 403–8. Marker is still alive as of this writing.

62. See U.S. Congress, Senate, Subcommittee on Patents, Trademarks, and Copyrights of the Committee on the Judiciary, *Wonder Drug Hearings on S. Res. 167,* 84th Cong., 2d sess., 1956.

63. This account is based on Maisel, *Hormone Quest,* 43–58.

64. Carl Djerassi, "Steroid Oral Contraceptives," *Science* 151 (March 4, 1966): 1055–61. See also Carl Djerassi, "Prognosis for the Development of New Chemical Birth Control Agents," *Science* 166 (October 24, 1969): 468–73, 167 (March 6, 1970), 1315–6; and Carl Djerassi, "Birth Control after 1944," *Science* 169 (September 4, 1970): 941–51. I am indebted to Djerassi for helping me trace down the full story of Marker.

65. This account is primarily based on Reed, *From Private Vice to Public Virtue,* 334–45.

66. Ibid., 357.

67. John Rock, *The Time Has Come: A Catholic Doctor's Proposals to End the Battle over Birth Control* (New York: Knopf, 1963).

68. Gregory Pincus, John Rock, and Celso Ramon Garcia, "Effects of Certain 19-Nor Steroid upon Reproductive Processes," *Annals of the New York Academy of Sciences* 71 (1958): 677.

69. G. Pincus, J. Rock, C. R. Garcia, E. Rice-Wray, M. Paniagua, I. Rodriguez, and P. Pedras, "Fertility Control with Oral Medication," in *Benchmark Papers in Human Physiology*, ed. L. L. Langley (Stroudsburg, Pa.: Dowden, Hutchinson & Ross, 1973), 413–26; see also Maisel, *Hormone Quest*, 132.

70. For some of these see Vern L. Bullough and Bonnie Bullough, *Contraception: A Guide to Birth Control Methods* (Buffalo, N.Y.: Prometheus, 1990); and for current updates of progress see the various editions of R. A. Hatcher et al., *Contraceptive Technologies*, 14th ed. (New York: Irvington, 1988). The book is updated and reissued periodically.

71. William H. Masters and Virginia E. Johnson, *Human Sexual Inadequacy* (Boston: Little, Brown, 1970), 1. The therapeutic intent is not emphasized in their first study, William H. Masters and Virginia E. Johnson, *Human Sexual Response* (Boston: Little, Brown, 1966).

72. Corner, *Seven Ages of a Medical Scientist*, 212.

73. Masters has often summarized the advice in his talks. A condensed version is in Corner, *Seven Ages of a Medical Scientist*, 213; see also Ruth Brecher and Edward Brecher, eds., *An Analysis of Human Sexual Response* (Boston: Little, Brown, 1966); and Edward Brecher, *The Sex Researchers* (Boston: Little, Brown, 1969), 285.

74. Among his articles are W. H. Masters, "The Rationale and Technique of Sex Hormone Replacement in the Aged Female and a Preliminary Result Report," *South Dakota Journal of Medicine* 4 (1951): 296–300; W. H. Masters, "Long Range Sex Steroid Replacement: Target Organ Regeneration, *Journal of Gerontology* 8 (1953): 33–39; W. H. Masters, "Sex Life of the Aging Female," in *Sex in Our Culture*, ed. C. A. Groves and A. Stone (New York: Emerson, 1955); W. H. Masters, "Endocrine Therapy in the Aging Individual," *Obstetrics and Gynecology* 8 (1956): 61–67; and W. H. Masters, "Sex Steroid Influence on the Aging Process," *American Journal of Obstetrics and Gynecology* 74 (1957): 733–46.

75. Félix Roubaud, *Traité de l'Impuissance et de la Sterilité chez l'Homme et chez la Femme* (1855; reprint, Paris: Baillière, 1876); see also Brecher, *Sex Researchers*, 288–9. There is a good summary chapter of the physiology of the sexual response and orgasm in Kinsey et al., *Sexual Behavior in the Human Female*, 594–641.

76. Ibid., 631, n. 46.

77. Joseph R. Beck, "How Do Spermatozoa Enter the Uterus?" *St. Louis Medical and Surgical Journal* 9 (September 1872): 449; and Joseph R. Beck, "How Spermazoa Enter the Uterus," *American Journal of Obstetrics and Diseases of Women and Children* (November 1874). I am indebted to Brecher, *Sex Researchers*, 291, for the reference.

78. B. S. Talmey, "Birth Control and the Physician," *New York Medical Journal* 105, no. 25 (June 23, 1917): 1185.

79. Robert Latou Dickinson, *Atlas of Sex Anatomy*, 2d ed. (Baltimore, Md.: Williams & Wilkins, 1949), 91–92.

80. G. Klumbies and H. Kleinsorge, "Das Herz in Orgasmus," *Medizinische Klinik* 45 (1950): 952–8; and G. Klumbies and H. Kleinsorge, "Circulatory Dangers and Prophylaxis During Orgasm," *International Journal of Sexology* 4 (1950): 61–66.

81. These data appeared in Kinsey et al., *Sexual Behavior in the Human Female*, 630, fig. 140.

82. Ibid., 631, n. 46.

83. Masters and Johnson, *Human Sexual Response*, 10.

84. Ibid., 300.

85. Ibid., 124.

86. Ibid., 57–61.

87. Ibid., 192.

88. Ibid., 194–5.

89. Ibid., 242.

90. Ibid., 270.

91. Masters and Johnson, *Human Sexual Inadequacy*, 3.

92. J. Wolpe, *Psychotherapy by Reciprocal Inhibition* (Stanford, Calif.: Stanford University Press, 1958).

93. Ibid., 214–8, 222–6.

94. ibid., 62.

95. Masters and Johnson, *Human Sexual Inadequacy*, 366, tab. 11B.

96. Helen Singer Kaplan, *The New Sex Therapy* (New York: Brunner/Mazel, 1974).

97. Joseph LoPiccolo, J. R. Heiman, D. R. Hogan, and C. W. Roberts, "Effectiveness of Single Therapists Versus Cotherapy Teams in Sex Therapy, *Journal of Counsulting and Clinical Psychology* 53 (1985): 287–94.

98. S. Schumacher and C. W. Lloyd, "Physiology and Psychological Factors in Impotence, *Journal of Sex Research* 17 (1981): 40-53; and B. Zilbergeld and M. Evans, "The Inadequacy of Masters and Johnson," *Psychology Today* 14 (1980): 29–43.

99. C. B. Broderick, "To Arrive Where We Started: The Field of Family Studies in the 1930's," *Journal of Marriage and Family* 50 (1988): 569–84. For potential early curriculum materials see F. M. Thurston, A *Bibliography on Family Relationships* (New York: National Council of Parent Education, 1932).

100. See, for example, Albert Ellis, *Folklore of Sex* (New York: Boni, 1951); and Albert Ellis, *Sex Beliefs and Customs* (London: Nevill, 1951–2).

101. Quoted in Pomeroy, *Dr. Kinsey*, 298.

102. See Carol Cassel, "A Perspective on the Great Sex Debate," in *Challenges in Sexual Science*, ed. Clive Davis (Lake Mills, Iowa: Graphic, 1983), 85–108.

103. See William Masters, "Phony Sex Clinics—Medicine's Newest Nightmare," *Today's Health* (November 1974): 22–26; and Robert J. Levin, "Most Sex Therapy Clinics Are Frauds," *Physician's World* 3, no. 1 (January 1975): 17–21.

CHAPTER 8: THE MATTER OF GENDER

1. John Money, *Hermaphroditism: An Inquiry into the Nature of a Human Paradox* (Ann Arbor, Mich.: University Microfilms, 1952).

2.   John Money, "Linguistic Resources and Psychodynamic Theory," *British Journal of Medical Psychology* 28 (1955): 264–6; and John Money, "Hermaphroditism, Gender and Precocity in Hyper-andrenocorticism: Psychologic Findings," *Bulletin of the Johns Hopkins Hospital* 96 (1955): 253–64.

3.   See John Money and Anke A. Ehrhardt, *Man & Woman Boy & Girl* (Baltimore, Md.: Johns Hopkins University Press, 1972).

4.   Donald Mosher, "Macho Men, Machismo and Sexuality," *Annual Review of Sex Research* 2 (1991): 199–247.

5.   J. H. Tijio and A. Levan, "The Chromosome Number of Man," *Hereditas* 42 (1956): 1–6. For background to the discovery see Malcolm Jay Kottler, "From 48 to 46: Cytological Technique, Preconception and the Counting of Human Chromosomes," *Bulletin of the History of Medicine* 48 (1974): 465–502.

6.   H. H. Turner, "A Syndrome of Infantilism, Congenital Webbed Neck, and Cubitus Valgus," *Endocrinology* 23 (1938): 566–74.

7.   H. F. Klinefelter, E. C. Reifenstein, and F. Albright, "Syndrome Characterized by Gynecomastia, Aspermatogenesis without A-Leydigism, and Increased Excretion of Follicle-Stimulating Hormone," *Journal of Clinical Endocrinology* 2 (1942): 615–27.

8.   For a more complete discussion see Money and Ehrhardt, *Man & Woman*, 23–33.

9.   Dean H. Hamer, Stella Hu, Victoria L. Magnuson, Nan Hu, and Angela M. L. Pattatucci, "A Linkage between DNA Markers on the X Chromosome and Male Sexual Orientation," *Science* 261 (July 16, 1993): 321–7.

10. Ibid., 6–7.

11. John Money, *Sex Errors of the Body* (Baltimore, Md.: Johns Hopkins Press, 1968), 41.

12. I have observed this myself. See also John Money, "Psychosexual Differentiation," in *Sex Research: New Developments*, ed. John Money (New York:, Holt, Rinehart, Winston, 1965), 3–12; and Money, *Sex Errors of the Body*.

13. A. A. Ehrhardt, R. Epstein, and J. Money, "Fetal Androgens and Female Gender Identity in the Early-Treated Adrenogenital Syndrome," *Johns Hopkins Medical Journal* 122 (1968): 160–7; and A. A. Ehrhardt, K. Evers, and J. Money, "Influence of Androgen and Some Aspects of Sexuality Dimorphic Behavior in Women with the Late-treated Adrenogenital Syndrome," *Johns Hopkins Medical Journal* 123 (1968): 115–122.

14. John Money, A. Ehrhardt, and D. Masica, "Fetal Feminization Induced by Androgen Insensitivity in the Testicular Feminizing Syndrome: Effect on Marriage and Maternalism," *Johns Hopkins Medical Journal*, 123 (1968): 105–14; D. N. Masica, J. Money, A. A. Ehrhardt, and V. G. Lewis, "IQ, Fetal Sex Hormones and Cognitive Patterns: Studies in the Testicular Feminizing Syndrome of Androgen Insensitivity," *Johns Hopkins Medical Journal* 124 (1969): 34–43; and D. N. Masica, A. A. Ehrhardt, and J. Money, "Fetal Feminization and Female Gender Identity in the Testicular Feminizing Syndrome of Androgen Insensitivity," *Archives of Sexual Behavior* 1 (1971): 131–42.

15. For further discussion see John Bancroft, "John Money: Some Comments on his Early Work," *Journal of Psychology & Human Sexuality* 4, no. 2 (1991): 1–8. See also Eli Coleman, ed. *John Money: A Tribute* (New York: Haworth, 1992).

16. John Money, J. G. Hampson, and J. L. Hampson, "An Examination of Some Basic Sexual Concepts: The Evidence of Human Hermaphroditism," *Bulletin of the Johns Hopkins Hospital* 97 (1955): 301–19.

17. Christian Hamburger, George K. Sturup, and E. Dahl-Iversen, "Transvestism: Hormonal, Psychiatric, and Surgical Treatment," *Journal of the American Medical Association* 152 (May 30, 1953): 391–6.

18. Danish Sterilization and Castration Act No. 176, May 11, 1935; and Danish Sterilization Act No. 130, June 1, 1929. There were similar laws at the time in Norway, Sweden, and Holland and certain cantons existed in Switzerland. The citations are from Hamburger et al., "Transvestism," 393.

19. Christine Jorgensen, *A Personal Autobiography* (New York: Eriksson, 1967).

20. For a more complete discussion, see Vern L. Bullough and Bonnie Bullough, *Cross Dressing, Sex and Gender* (Philadelphia: University of Pennsylvania Press, 1993). The book examines many early cases.

21. Christian Hamburger, "Desire for Change of Sex as Shown by Personal Letters from 465 Men and Women," *Acta Endocrinoligica* 14 (1953): 361–75.

22. George Wiedeman, "Transvestism" [Letters], *Journal of the American Medical Association* 152, no. 16 (1953): 1553.

23. Mortimer Ostow, "Transvestism" [Letters], *Journal of the American Medical Association* 152, no. 16 (1953): 1167.

24. David O. Cauldwell, "Psychopathia Transsexualism," *Sexology* 16 (December 1949): 274–80. Cauldwell developed the term further in two booklets: David O. Cauldwell, *Questions and Answers on the Sex Life and Sexual Problems of Trans-Sexual* (Girard, Kans.: Haldeman-Julius, 1950); and David O. Cauldwell, *Sex Transmutation—Can One's Sex Be Changed?* (Girard, Kans.: Haldeman-Julius, 1961).

25. Daniel G. Brown. "Inversion and Homosexuality," *American Journal of Orthopsychiatry* 28 (April 1948): 424–9, for example, used the term.

26. Harry Benjamin, "Clinical Aspects of Transsexual in the Male and Female," *American Journal of Psychotherapy* 18 (July 1964): 458–69.

27. Harry Benjamin, *The Transsexual Phenomenon* (New York: Julian, 1966). Richard Doctor, *Transvestites and Transsexuals* (New York: Plenum, 1988) developed a staging process by which some individuals crossed over from transvestites to transsexuals.

28. Harry Benjamin, "Nature and Management of Transsexualism," *Western Journal of Obstetrics and Gynecology* 72 (March–April 1964): 105–11.

29. Ira B. Pauley, "Gender Identity Disorders," in *Human Sexuality: Psychosexual Effects of Disease*, ed. Martin Farber (New York: Macmillan, 1985), 295–316; Ira B. Pauley and Milton T. Edgerton, "The Gender Identity Movement: A Growing Surgical-Psychiatric Liaison," *Archives of Sexual Behavior* 15 (1986): 315–26.

30. Betty Steiner, "The Management of Patients with Gender Disorders," in *Gender Dysphoria: Development, Research, and Management*, ed. Betty Steiner (New York: Plenum, 1985), 336.

31. Wendi Pierce, "Interview with Dr. Biber," *Rites of Passage: A Magazine for Female-to-Male Transsexuals and Cross Dressers* 1 (December 1989): 7.

32. "Diagnostic Criteria for 302.50: Transsexualism," in *Diagnostic and Statistical Manual of Mental Disorders*, 3d ed., rev. ed. (Washington, D.C.: American Psychiatric Association, 1980), 76.

33. Bullough and Bullough, *Cross Dressing*.

34. George W. Henry, *Sex Variants: A Study of Homosexual Patterns* (New York: Hoeber, 1941). Henry included one transvestite among his male homosexuals.

35. There have been a number of studies since Benjamin's pioneering ones, and the subject has been approached from a number of different disciplines. Included in any list would be the one by Richard Docter, *Transvestites and Transexuals*, already cited. For books, see John T. Talamani, *Boys Will Be Girls: The Hidden World of the Heterosexual Male Transvestite* (Washington, D.C.: University Press of America, 1982); Harry Brierly, *Transvestism: A Handbook with Case Studies* (Oxford, UK: Pergamon, 1979); Holly Devor, *Gender Blending: Confronting the Limits of Duality* (Bloomington: Indiana University Press, 1989); and Annie Woodhouse, *Fantastic Women* (New Brunswick, N.J.: Rutgers University Press, 1989). Among the most active researchers have been Neil Buhrich and Neil McConaghy, whose articles have appeared in a number of publications. A good bibliography of articles is in Dallas Denny, *A Bibliography of the Transvestism and Transsexualism* (New York: Garland, in press).

36. A. K. Cohen, *Delinquent Boys: The Culture of the Gang* (Glencoe, Ill.: Free Press, 1974).

37. Kenneth J. Gergen, "The Social Constructionist Movement in Modern Psychology," *American Psychologist* 40 (March 1985): 166.

38. John H. Gagnon and William Simon, *Sexual Conduct: The Social Sources of Human Sexuality* (Chicago: Aldine, 1973).

39. Anselm Strauss, "Interactionism," in *History of Sociological Analysis*, ed. Tom Bottomore and Robert Nisbet (New York: Basic, 1978), 456–98; and John P. Hewitt, *Self and Society: A Symbolic Interactionist Social Psychology* (Boston: Allyn & Bacon, 1979).

40. Kenneth Plummer, "Symbolic Interaction and Sexual Conduct: An Emergent Perspective," in *Human Sexual Relations: Towards a Redefinition of Sexual Politics*, ed. Miles Brake (New York: Pantheon, 1982), 230.

41. Jeffrey Weeks, "Discourse, Desire and Sexual Deviance: Some Problems in a History of Homosexuality," in *The Making of the Modern Homosexual*, ed. Kenneth Plummer (Totowa, N.J.: Barnes & Noble, 1981), 76–101; and Jeffrey Weeks, *Sex Politics, and Society* (London: Longman, 1981), 96–121.

42. R. C. Anderson, *Cognitive Psychology* (New York: Academic, 1975); and V. F. Guidano and G. Liotti, *Cognitive Process and Emotional Disorders: A Structural Approach to Psychotherapy* (New York: Guilford, 1983).

43. John Turner, "Toward a Cognitive Redefinition of the Social Group," in *Social Identity and Intergroup Relations*, ed. Henri Taefel (Cambridge, UK: Cambridge University Press, 1982), 18–19, 21.

44. Money and Ehrhardt, *Man & Woman*, 185.

45. Ibid., 122–6.

46. Milton Diamond "A Critical Evaluation of the Ontogeny of Human Sexual Behavior," *Quarterly Review of Biology* 40 (1965): 147–75.

47. See Vern L. Bullough, *Sexual Variance in Society and History* (Chicago: University of Chicago Press, 1976).

48. Frederick L. Whitam, "A Cross Cultural Perspective on Homosexuality, Transvestism and Trans-sexualism," in *Variant Sexuality: Research and Theory*, ed. Glen D. Wilson (London: Croom Helm, 1978).

49. Richard C. Pillard, Jeannette Pomadere, and Ruth A. Carretta, "Is Homosexuality Familial? A Review, Some Data and a Suggestion," *Archives of Sexual Behavior* 19 (1981): 465–75.

50. Henry, *Sex Variants*.

51. M. S. Margolese and O. Janiger, "Androsterone/Etiocholanolone Ratios in Male Homosexuals," *British Medical Journal* 3 (1973): 207–10.

52. Pillard et al., "Is Homosexuality Familial?"

53. Franz J. Kallman, "Comparative Twin Study on the Genetic Aspects of Male Homosexuality," *Journal of Nervous and Mental Disease* 115 (1952): 283–98. For a collection of other studies on twins, see Geoff Puterbaugh, ed., *Twins and Homosexuality: A Casebook* (New York: Garland, 1990).

54. Milton Diamond, "Bisexualities: A Biological Perspective" (paper presented at the third international Berlin Conference of Sexology, Berlin, 1990). Diamond noted that the recent findings will be published as a book by Diamond, Whitam, and Dannemiller.

55. John Bancroft, "The Relationship Between Hormones and Sexual Behavior in Humans," in *Biological Determinants of Sexual behavior*, ed. J. B. Hutchison (New York: Wiley , 1978), 494–519.

56. See Jean D. Wilson, Frederick W. George, James E. Griffin, "The Hormonal Control of Sexual Development," *Science* 211 (March 1981): 1278–84.

57. Anke A. Ehrhardt and Heino F. L. Meyer-Bahlburg, "Effects of Prenatal Sex Hormones on Gender-Related Behavior," *Science* 211 (March 1981): 1312–7.

58. Anke A. Ehrhardt, Heino F. L. Meyer-Bahlburg, Laura R. Rosen, Judith F. Feldman, Norman Veridiano, L. Zimmerman, and Bruce S. McEwen, "Sexual Orientation After Prenatal Exposure to Exogenous Estrogen," *Archives of Sexual Behavior* 14 (1985): 57–75. See also F. M. E. Slijper, H. J. van der Kamp, H. Brandenburg, S. M. P. F. de Muinck Keizer-Schrama, S. L. S. Drop, and J. C. Molenaar, "Evaluation of Psychosexual Development of Young Women with Congenital Adrenal Hyperplasia: A Pilot Study," *Journal of Sex Education and Therapy* 18 (Fall 1991): 200–7.

59. Gunter Dorner, Wolfgang Rohde, Fritz Stahl, Lothar Krell, and Wolf-Gunther Masius, "A Neuroendocrine Predisposition for Homosexuality in Men," *Archives of Sexual Behavior* 4 (1975): 1–8; Garfield Tourney, "Hormones and Homosexuality," in *Homosexual Behavior: A Modern Reappraisal*, ed. Judd Marmor (New York: Basic, 1980), 41–58.

60. Simon LeVay, "News and Comment: Is Homosexuality Biological?" *Science* 253 (August 30, 1991): 253, 257–9.

61. Milton Diamond, "Human Sexual Development: Biological Foundations for Social Development," in *Human Sexuality in Four Perspectives*, ed. Frank Beach (Baltimore, Md.: Johns Hopkins University Press, 1977), 22–61.

62. Janet Shibley Hyde, "Gender Differences in Aggression," in *The Psychology of*

Gender: Advances through Meta-Analysis, ed. Janet Shibley Hyde and Marcia C. Linn (Baltimore, Md.: Johns Hopkins University Press, 1986), 51–66.

63. For this see Vern L. Bullough, Brenda Shelton, and Sarah Slavin, *The Subordinated Sex: A History of Attitudes Toward Women*, rev. ed. (Athens: University of Georgia Press, 1988).

64. Donald L. Mosher and S. S. Tompkins, "Scripting the Macho Man: Hypermasculine Socialization and Enculturation," *Journal of Sex Research* 25, no. 1 (1988): 60–84.

65. G. Dorner, *Hormones and Brain Differentiation* (New York: Elsevier, 1976).

66. See Richard C. Pillard and James D. Weinrich, "The Periodic Table of the Gender Transposition: Part I. A Theory Based on Masculinization and Defeminization of the Brain," *Journal of Sex Research* 23 (1987): 425–54.

67. Muriel Wilson Perkins, "Female Homosexuality and Body Build," *Archives of Sexual Behavior* 10 (1981): 337–45.

68. Margolese and Janiger, "Androsterone/Etiocholanolone Ratios."

69. Brian A. Gladue, Richard Green, and Ronald E. Hellman, "Neuroendocrine Response to Estrogen and Sexual Orientation," *Science* 225 (September 28, 1984): 1496–9.

70. Garfield Tourney, "Hormones and Homosexuality," 41–58.

71. Eli Coleman, Louis Gooren, and Michael Ross, "Adversaria: Theories of Gender Transpositions: A Critique and Suggestions for Further Research," *Journal of Sex Research* 26 (November 1989): 525–38; Lynda I. A. Birke, "Is Homosexuality Hormonally Determined?" *Journal of Homosexuality* 6 (Summer 1981), 35–49; and Wendell Ricketts, "Biological Research on Homosexuality: Ansell's Cow or Occam's Razor," *Journal of Homosexuality* 9 (Summer 1984): 65–93.

72. Louis Gooren, Eric Fliers, and Keith Courtney, "Biological Determinants of Sexual Orientation," *Annual Review of Sex Research* 1 (1990): 175–96.

73. Edward O. Wilson, *On Human Nature* (Cambridge, Mass.: Harvard University Press, 1978), 142–8.

74. T. F. Hoult, "Human Sexuality in Biological Perspective," *Journal of Homosexuality* 9 (Spring 1984): 137–55; and D. J. Futuyma and S. J. Risch, "Sexual Orientation, Sociobiology, and Evolution," *Journal of Homosexuality* 9 (Spring 1984), 1576–86.

75. Frans de Waal, *Chimpanzee Politics: Power and Sex among Apes* (New York: Harper & Row, 1982), 64.

76. G. V. Hamilton, "A Study of Sexual Tendencies in Monkeys and Baboons," *Journal of Animal Behavior* 4 (1914): 295–318. Much of this was reprinted in *Mammalian Sexual Behavior*, ed. Donald A. Dewsbury, Benchmark Papers in Behavior 15 (Stroudsburg, Pa.: Dowden, Hutchinson & Ross, 1981), 79–91.

77. D. A. Goldfoot, K. Wallen, D. A. Neff, M. C. McBrair, and R. W. Goy, "Social Influences upon the Display of Sexually Dimorphic Behavior in Rhesus Monkeys: Isosexual Rearing," *Archives of Sexual Behavior* 13 (1984): 395–406.

78. F. A. Beach, "Animal Models for Human Sexuality," *Sex, Hormones and Behaviour*, Ciba Symposium No. 61, new ser. (Amsterdam: Excerpta Medica, 1979), 113–46.

79. Gooren et al., "Biological Determinants of Sexual Orientation," 178.

80. Eleanor E. Maccoby, ed., *The Development of Sex Differences* (Stanford, Calif.: Stanford University Press, 1966); Charles J. Lumsden and Edward O. Wilson, *Genes, Mind and Culture: The Evolutionary Process* (Cambridge, Mass.: Harvard University Press, 1981); and Alan P. Bell, Martin S. Weinberg, and Sue Kiefer Hammersmith, *Sexual Preference: Its Development in Men and Women* (Bloomington: Indiana University Press, 1981).

81. J. Imperato-McGinley, L. Guerrero, T. Gautier, and R. E. Peterson, "Androgens and the Evolution of Male Gender Identity among Male Pseudo-Hermaphrodites with 5-alpha Reductase Deficiency," *New England Journal of Medicine* 300 (1979): 1233–7.

82. H. F. L. Meyer Bahlburg, "Hormones and Psychosexual Differentiation: Implications for the Management of Inter-Sexuality, Homosexuality and Transsexuality," *Clinics in Endocrinology and Metabolism* 2 (1982): 681–701; and John Money, *Gay, Straight, and In-Between* (New York: Oxford University Press, 1988).

83. G. H. Herdt and J. Davidson, "The Sambia 'Turnim-Man'; Sociocultural and Clinical Aspects of Gender Formation in Male Pseudohermaphrodites with 5 alpha-reductase deficiency in Papua, New Guinea," *Archives of Sexual Behavior* 17 (1988): 1–31.

84. Barrie Thorne, *Girls and Boys in School* (New Brunswick, N.J.: Rutgers University Press, 1993).

85. For example, see Frederick Jameson, "Imaginary and Symbolic in Lacan: Marxism, Psychoanalytic Criticism and the Problem of the Subject," *The Question of Reading Otherwise*, Yale French Studies Nos. 55–56, ed. Shoshana Felman (New Haven, Conn.: Yale French Studies, 1977); see also Jacques Lacan, *The Four Fundamental Concepts of Psychoanalysis*, trans. Alan Sheridan (New York: Norton, 1982).

86. Marjorie Garber, *Vested Interests: Cross-Dressing and Cultural Anxiety* (New York: Routledge, 1992).

87. Bullough and Bullough, *Cross Dressing*.

88. Ibid.

89. Ann Constantinople, "Masculinity-Femininity: An Exception to a Famous Dictum?" *Psychological Bulletin* 80 (1973): 389–407.

90. S. L. Bem, "The Measurement of Psychological Androgyny," *Journal of Consulting and Clinical Psychology* 42 (1974): 155–62.

91. J. T. Spence and R. L. Helmreich, *Masculinity and Femininity* (Austin: University of Texas Press, 1978).

92. See Donald Mosher, "Gender: Psychological Measurements," in *Human Sexuality: An Encyclopedia*, ed. Vern L. Bullough and Bonnie Bullough (New York: Garland, 1994), pp. 237–42.

93. This theory is developed in Bullough and Bullough, *Cross Dressing*, chap. 13.

94. This was best demonstrated by a large-scale quantitative study by a number of psychoanalysts, led by Irving Bieber. The results were published in Irving Bieber, Harvey J. Dain, Paul R. Dince, Marvin G. Drellich, Henry G. Grand, Ralph H. Gundlach, Malvina W. Kremer, Alfred H. Rifkin, Cornelia B. Wilbur, and Toby B. Bieber, *Homosexuality: A Psychoanalytic Study* (New York: Basic,

1962). The study emphasizes what the social constructionists claim, namely that we see data through the blinders of our assumptions.

95. A. P. Bell, M. S. Weinberg, and S. K. Hammersmith, *Sexual Preference, Its Development in Men and Women* (Bloomington: University of Indiana Press, 1981).

96. Frederick L. Whitam and Michael Zent, "Cross-Cultural Assessment of Early Cross Gender Behavior and Familial Factors in Male Homosexuality," *Archives of Sexual Behavior* 13 (1984): 427–39.

97. Michael Newcomb, "The Role of Perceived Relative Parent Personality in the Development of Heterosexuals, Homosexuals, and Transvestites," *Archives of Sexual Behavior* 14 (1985): 147–64.

98. Richard Green, "One-Hundred Ten Feminine and Masculine Boys: Behavioral Contrasts and Demographic Similarities," *Archives of Sexual Behavior* 5 (1976): 425–46.

99. Richard Green, *The "Sissy Boy Syndrome" and the Development of Homosexuality* (New Haven, Conn.: Yale University Press, 1987).

100. Bernard Zuger, "Effeminate Behavior in Boys from Early Childhood," *The Journal of Pediatrics* 69 (1966), 1098–107.

101. Bernard Zuger, "Effeminate Behavior Present in Boys from Childhood: Ten Additional Years of Follow-up," *Comparative Psychiatry* 19 (1978): 363–9.

102. Phil S. Lebovitz, "Feminine Behavior in Boys: Aspects of Its Outcome," *American Journal of Psychiatry* 128 (April 1962), 1283–9.

103. John Money and A. J. Russo, "Homosexual Outcome of Discordant Gender Identity/Role in Childhood: Longitudinal Follow-up," *Journal of Pediatric Psychology* 4 (1979): 29–41.

104. Charles W. Davenport, "A Follow-Up Study of 10 Feminine Boys," *Archives of Sexual Behavior* 15 (1986): 511–7.

105. Neil Buhrich and Neil McConaghy, "Parental Relationships during Childhood in Homosexuality, Transvestism, and Transsexualism," *Australian and New Zealand Journal of Psychiatry* 12 (1978): 103–8; and Michael D. Newcomb, "The Role of Perceived Relative Parent Personality in the Development of Heterosexuals, Homosexuals, and Transvestites," *Archives of Sexual Behavior* 14 (1985) 147–64.

106. Vern L. Bullough, Bonnie Bullough, and Richard Smith, "A Comparative Study of Male Transvestites, Male to Female Transsexuals and Male Homosexuals," *Journal of Sex Research* 19 (August 1983): 238–57; and Vern Bullough, Bonnie Bullough, and Richard Smith, "Childhood and Family of Male Sexual Minority Groups," *Health Values* 7 (July–August, 1983): 19–26.

107. Deborah Heller Feinbloom, *Transvestites, Transsexuals: Mixed Views* (New York: Delacourte, 1967).

108. M. Saghir and W. Robins, *Male and Female Homosexuality* (Baltimore, Md.: Williams & Wilkins, 1973), 25; and Joseph Harry, "Defeminization and Adult Psychological Well-Being among Male Homosexuals," *Archives of Sexual Behavior* 12 (1983): 1–19.

109. Joseph Harry, "Defeminization and Social Class," *Archives of Sexual Behavior* 14 (1985): 1–12.

110. Frederick L. Whitam and Robin M. Mathy, "Childhood Cross-Gender Behavior of Homosexual Females in Brazil, Peru, the Philippines, and the United States," *Archives of Sexual Behavior* 20 (1991): 151–70.

111. Bullough et al., "Transvestites, Transsexual, and Homosexual Men."

112. Davenport, "Follow-Up Study."

113. Such a change is apparent in the third edition. See American Psychiatric Association, *Diagnostic and Statistical Manual*, 3d ed. (Washington, D.C.: American Psychiatric Association, 1980). See also Frederick Suppe, "Classifying Sexual Disorders: *The Diagnostic and Statistical Manual* of the American Psychiatric Association," *Journal of Homosexuality* 9 (Summer 1984): 9–28.

Chapter 9: Other Voices, Other Views

1. Betty Friedan, *The Feminine Mystique* (New York: Norton, 1963).

2. Boston Women's Health Book Collective, *Our Bodies, Ourselves* (New York: Simon & Schuster, 1971), 24.

3. Michel Foucault, *History of Sexuality: An Introduction* (New York: Pantheon, 1978). The English translation left out the French subtitle, best translated as "the will to power," which is a better key to the theme of the book.

4. See James Miller, *The Passion of Michel Foucault* (New York: Simon & Schuster, 1993). See also the interesting review of the book by Kenneth L. Woodward in *Newsweek*, February 1, 1993, p. 63.

5. D. Megill, "Foucault, Structuralism and the Ends of History," *Journal of Modern History* 51 (1979): 451–503.

6. One example of this rapidly expanding interest in the history of sex was the formation of the *Journal of the History of Sexuality*, published by the University of Chicago Press and edited by John C. Fout. It began publication in 1990.

7. See, for example, the brief discussion of rape by Robert Veit Sherwin, "Lawson Sex Crimes," in *The Encyclopedia of Sexual Behavior*, ed. Albert Ellis and Albert Abarbanel (New York: Hawthorn, 1961), 2:621–3. The subject was not even discussed in Victor Robinson, ed., *Encyclopedia Sexualis* (New York: Dingwall-Rock, 1936).

8. See Susan Brownmiller, *Against Our Will: Men, Women and Rape* (New York: Simon & Schuster, 1975), 389.

9. J. D'Emilio and E. B. Freedman, *Intimate Matters: A History of Sexuality in America* (New York: Harper & Row, 1988), 91–92.

10. H. S. Beild and L. B. Bienen, *Jurors and Rape* (Lexington, Mass.: Heath, 1980).

11. A. W. Burgess and L. L. Holmstrom, "Sexual Trauma of Children and Adolescents," *Nursing Clinics of North America* 10 (1975): 551–63. See also A. W. Burgess, A. N. Groth, L. L. Holmstrom, and S. M. Sgroi, eds., *Sexual Assault of Children and Adolescents* (Lexington, Mass.: Heath, 1978); and A. W. Burgess and L. L. Holmstrom, eds., *Rape and Sexual Assault: A Research Handbook* (New York: Garland, 1985).

12. E. J. Kanin, "Selected Dyadic Aspects of Male Sex Aggression," *Journal of Sex Research* 5 (1969): 12–28.

13. E. J. Kanin, "Date Rapists: Differential Sexual Socialization and Relative Deprivation," *Archives of Sexual Behavior* 14 (1985): 219–31.

14. K. Rappaport and B. R. Burkhart, "Personality and Attitudinal Correlates of Sexually Coercive College Males," *Journal of Abnormal Personality* 93 (1984): 216–21.

15. C. L. Muehlenhard and S. W. Cook, "Men's Self Reports of Unwanted Sexual Activity," *Journal of Sex Research* 24 (1988): 58–72; and C. L. Muehlenhard and C. L. Hollabaugh, "Do Women Sometimes Say No When They Mean Yes?" *Journal of Personality and Social Psychology* 54 (1988): 58–72.

16. Andrea Parrot, *Sexual Assertiveness Dramatization* (Ithaca, N.Y.: Cornell University Press, 1985).

17. The Women's History Research Center of Berkeley, California, attributes this quote to State Senator Bob Wilson, spring 1979. Michael D. A. Freeman, "The Marital Rape Exemption Re-examined," *Family Law Quarterly* 15 (1981), 1.

18. See Diana E. H. Russell, *Rape in Marriage* (New York: Macmillan, 1982); and David Margolick, "Rape in a Marriage Is No Longer Within Law," *New York Times*, December 23, 1984, sec. E, p. 6. For an overview see Susan Estrich, *Real Rape* (Cambridge, Mass.: Harvard University Press, 1987).

19. M. Dawn McCaghy, *Sexual Harassment: A Guide to Resources* (Boston: Hall, 1985), 1.

20. D. E. Ledgerwood and S. Johnson-Dietz, "The EECO's Foray into Sexual Harassment: Interpreting the New Guidelines for Employer Liability," *Labor Law Journal* 31 (1980): 741–4.

21. See Vern L. Bullough, "Nursing, Sexual Harassment, and Florence Nightingale: Implications for Today," in *Florence Nightingale and Her Era: A Collection of New Scholarship*, ed. Vern Bullough, Bonnie Bullough, and Marietta Stanton (New York: Garland, 1990), 168–87.

22. D. L. Mosher, *Sex Callousness Toward Women*, Technical Reports of the Commission on Obscenity and Pornography No. 8 (Washington, D.C.: U.S. Government Printing Office, 1970), 313–25.

23. See D. L. Mosher, *Scripting the Macho Man: Theory, Research, and Measurement of Hypermasculinity* (New York: Guilford, 1992); D. L. Mosher and R. A. Anderson, "Macho Personality, Sexual Aggression and Reactions to Realistic Guided Imagery of Rape," *Journal of Research in Personality* 20 (1986): 77–94; D. L. Mosher and M. Sirkin, "Measuring a Macho Personality Constellation," *Journal of Research in Personality* 18 (1984): 150–63; and D. L. Mosher and S. S. Tomkin, "Scripting the Macho Man: Hypermasculine Socialization and Enculturation" *Journal of Sex Research* 25 (1988): 60–84. For an overall summary see D. L. Mosher, "Macho Men, Machismo and Sexuality," *Annual Review of Sex Research* 2 (1991): 199–247.

24. Andrea Dworkin, *Pornography: Men Possessing Women* (New York: Putnam's, 1981).

25. M. K. Blakely, "Is One Woman's Sexuality Another Woman's Pornography?" *MS Magazine* (April 1985): 37–47.

26. Andrea Dworkin, *Intercourse* (New York: Free Press, 1987).

27. See Ira Reiss, *Journey into Sexuality: An Exploratory Voyage* (Englewood Cliffs, N.J.: Prentice-Hall, 1986).

28. V. R. Padgett, J. A. Brislin-Slutz, and J. A. Neale, "Pornography, Erotica, and Attitudes Toward Women: The Effects of Repeated Exposure," *Journal of Sex Research* 26 (1989): 479–91.

29. N. M. Malamuth and J. V. P. Check, "Debriefing Effectiveness Following Exposure to Pornographic Rape Depictions," *Journal of Sex Research* 20 (1983): 1–13. There is a vast and somewhat contradictory literature on this to which Malamuth has contributed considerably. See, for example, N. M. Malamuth and E. Donnerstine, eds., *Pornography and Sexual Aggression* (Orlando, Fla.: Academic, 1984).

30. D. Zillman, and J. Bryant, "Pornography, Sexual Callousness, and the Trivialization of Rape," *Journal of Communication* 32, no. 4 (1982): 10–21; D. Zillman and J. Bryant, "Effects of Massive Exposure to Pornography," in *Pornography and Sexual Aggression*, ed. N. M. Malamuth and E. Donnerstine (Orlando, Fla.: Academic, 1984), 115–38; D. Zillman and J. Bryant, "Pornography's Impact on Sexual Satisfaction," *Journal of Applied Social Psychology* 18 (1988): 438–53.

31. Feminist Anti-Censorship Taskforce (FACT), *Caught Looking: Feminism, Pornography and Censorship* (New York: Caught Looking, 1986); see also Varda Burstyn, ed., *Women Against Censorship* (Vancouver, B.C.: Douglas & McIntyre, 1984).

32. See William B. Lockhart, chair, *The Report of the Commission on Obscenity and Pornography* (Washington, D.C.: U.S. Government Printing Office, 1970). In fitting with its research mission, there were eight volumes of research findings published as *Technical Report of the Commission on Obscenity and Pornography* (Washington, D.C.: U.S. Government Printing Office, 1970). These reports were not reviewed or approved by the full commission, but it was on the basis of them that the decision was arrived at. Interestingly, one of the dissenting statements to the report was by Charles H. Keating, then chair of Citizens for Decent Literature; he was later convicted of defrauding the government of billions of dollars in the savings and loan scandals of the Reagan administration. His opposition was not based on any studies into the effects of pornography.

33. The report was issued as U.S. Department of Justice, *Attorney General's Commission on Pornography Final Report* (Washington, D.C.: U.S. Government Printing Office, 1986). It was not officially printed but was issued as a typescript, further emphasizing the political nature of the report. See also Martha Cornoy and Timothy Perper, "Censorship," in *Human Sexuality: An Encyclopedia*, ed. Vern L. Bullough and Bonnie Bullough (New York: Garland, 1994), pp. 91–105. See also B. Lynn, *Polluting the Censorship Debate: A Summary and Critique of the Final Report of the Attorney General;'s Commission on Pornography* (Washington, D.C.: American Civil Liberties Union, 1986).

34. Among such studies are P. Bart and M. Jozsa, "Dirty Books, Dirty Films, and Dirty Data," in *Take Back the Night: Women on Pornography* (New York: Mor-

row, 1980). The volume includes chapters by D. Russell and S. Griffin, which are pertinent. See also T. McCormack, "Machismo in Media Research: A Critical Review of Research on Violence and Pornography," *Social Problems* 25 (1978): 552–4; and D. Russell, "Research on How Women Experience the Impact of Pornography," in *Pornography and Censorship*, ed. D. Copp and S. Wendell (Buffalo, N.Y.: Prometheus, 1983).

35. Ira Reiss, *Journey into Sexuality: An Exploratory Voyage* (Englewood Cliffs, N.J.: Prentice-Hall, 1986).

36. S. Kappeler, *The Pornography of Representation* (Minneapolis: University of Minnesota Press, 1986); N. M. Malamuth and V. Billings, "Why Pornography? Models of Functions and Effects," *Journal of Communication* 34, no. 3 (1984): 117–29; N. M. Malamuth and V. Billings, "The Functions and Effects of Pornography: Sexual Communication Versus the Feminist Models in Light of Research Findings," in *Perspectives on Media Effects*, ed. J. Bryant and D. Zillman (Hillsdale, N.J.: Erlbaum, 1986); A. Soble, *Pornography: Marxism, Feminism, and the Future of Sexuality* (New Haven, Conn.: Yale University Press, 1986); and M. Valverde, *Sex, Power and Pleasure* (New York: Women's, 1986). I am indebted to Robert T. Francoeur, *Becoming a Sexual Person*, 2d ed. (New York: Macmillan, 1991), 636, for this concept of models.

37. Albert Richard Allgeier and Elizabeth Rice Allgeier, *Sexual Interactions*, 2d ed. (Lexington, Mass.: Heath, 1988), 521.

38. C. D.Coles and M. J. Shamp, "Some Sexual, Personality, and Demographic Characteristics of Women Readers of Erotic Romances," *Archives of Sexual Behavior* 13 (1984): 187–209.

39. See the various catalogs from the Sexuality Library, 1210 Valencia Street, San Francisco, CA 94110.

40. Alice Ladas, Beverly Whipple, and John Perry, *The G-Spot and Other Recent Discoveries about Human Sexuality* (New York: Holt, Rinehart & Winston, 1982). There was some criticism that the findings were rushed into print, but these complaints were made by those who do not know or understand the pressure that Whipple and Perry were under. They either had to publish their findings with the cautions they expressed or have others publish it without such cautions. They chose to publish.

41. Bonnie Bullough, Madeline Davis, Beverly Whipple, Joan Dixon, Elizabeth Rice Allgeier, and Kate Cosgrove Drury, "Subjective Reports of Female Orgasmic Expulsion of Fluid," *Nurse Practitioner* 9 (March 1984): 55–59. Whipple has other articles in press that describe the further analysis of the fluid and argue that there is analagous tissue to the prostate.

42. This brief summary only touches the surface of the massive number of studies carried out in the past thirty years. For a guide to these see Vern L. Bullough and Lilli Sentz, *Prostitution: A Guide to Sources, 1960–1990* (New York: Garland, 1992). This included 1,965 items and was a supplement to Vern L. Bullough, B. Elcano, M. Deacon, and B. Bullough, A *Bibliography of Prostitution* (New York: Garland, 1977), which had 6,494 references. The second volume is annotated.

43. The initial study was Vern Bullough and Bonnie Bullough, A *History of Prosti-*

tution (New Hyde Park, N.Y.: University, 1964); although it had a feminist perspective in terms of the questions asked, it had to be updated and amplified: Vern Bullough and Bonnie Bullough, *Prostitution: An Illustrated Social History* (New York: Crown, 1978), which in turn was slightly updated as Vern Bullough and Bonnie Bullough, *Women and Prostitution: A Social History* (Buffalo, N.Y.: Prometheus, 1987).

44. J. R. Walkowitz, "The Politics of Prostitution," *Signs* 6, no. 1 (1980): 123–35.

45. Martha L. Stein, *Lovers, Friends, Slaves: Nine Male Sexual Types: Their Psycho-Sexual Transactions with Call Girls* (New York: Putnam's, 1974).

46. Harold Greenwald, *The Call Girl: A Social and Psychoanalytic Study* (New York: Ballantine, 1958). This was revised: Harold Greenwald, *The Elegant Prostitute* (New York: Walker, 1970).

47. Edwin M. Schur, "Reactions to Deviance: A Critical Assessment," *American Journal of Sociology* 75 (November 1969): 309–22.

48. See, for example, S. Smart, "Research on Prostitution: Some Problems for Feminist Research," *Humanity and Society* 8, no. 4 (November 1984): 407–13.

49. See Eli Coleman, "The Development of Male Prostitution Activity among Gay and Bisexual Adolescents," *Journal of Homosexuality* 17 (1989): 151–84; Neil R. Coombs, "Male Prostitution: A Psychosocial View of Behavior," *American Journal of Orthopsychiatry* 44 (1974): 782–9; David F. Luckenbill, "Entering Male Prostitution," *Urban Life* 14 (1985): 131–53; Paul W. Mathews, "On Being a Prostitute," *Journal of Homosexuality* 15, nos. 3–4 (1988): 119–35; and A. J. Reiss Jr., "The Social Integration of Queers and Peers, " *Social Problems* 9 (1961): 102–20.

50. S. E. Caulkins, and N. R. Coombs, "The Psychodynamics of Male Prostitution," *American Journal of Psychotherapy* 30, no. 3 (July 1976): 441–51.

51. Paul H. Gebhard, Wardell B. Pomeroy, Clyde E. Martin, and Cornelia V. Christenson, *Pregnancy, Birth, and Abortion* (New York: Harper, 1958).

52. Ibid., 65.

53. Ibid., 213.

54. American Law Institute, *Model Penal Code*, proposed Official Draft, § 230.3.2 (Philadelphia: American Law Institute, 1962).

55. Larence Lader, *Abortion* (Indianapolis, Ind.: Bobbs-Merrill, 1966).

56. *Roe v. Wade*, 314 F. Supp. 1217 (N.D. Tex., June 17, 1970), jurisdiction postponed, 402 U.S. 941 (1971); *Doe v. Bolton*, 319 F. Supp. 1048 (N.D. Ga., 1970), jurisdiction postponed, 402 U.S. 941 (1971).

57. See Carl Djerassi, *The Politics of Contraception* (San Francisco: Freeman, 1981).

58. Dorothy Tennov, *Love and Limerence; The Experience of Being in Love* (Briarcliff Manor, N.Y.: Stein & Day, 1979).

59. E. Walster and G. Walster, *A New Look at Love* (Reading, Mass.: Addison-Wesley, 1978).

60. J. D. Weinrich, "The Periodic Table Model of Gender Transpositions: Part II. Limerent and Lusty Sexual Attraction and the Nature of Bisexuality," *Journal of Sex Research* 24 (1988): 113–29.

61. Timothy Perper, *Sex Signals: The Biology of Love* (Philadelphia: ISI, 1985); T.

Perper and D. L. Weis, "Proceptive and Rejective Strategies of U.S. and Canadian College Women," *Journal of Sex Research* 24, no. 4 (1987): 455–80.

62. H. T. Remoff, *Sexual Choice: A Woman's Decision* (New York: Dutton/Lewis, 1984).

63. M. M. Moore, "Nonverbal Courtship Patterns in Women: Context and Consequences," *Ethology and Sociobiology* 6 (1985): 237–47.

64. Perper, *Sex Signals*.

65. Erik Erikson, *Childhood and Society*, rev. ed. (New York: Norton, 1968); and E. H. Erikson, *Identity, Youth, and Crisis* (New York: Norton, 1986).

66. B. Zahacu-Chrstiansen and E. M. Ross, *Babies: Human Development During the First Year* (London: Wiley, 1975).

67. As of this writing, only partial data have been available on her long-term studies of infants. See June M. Reinisch, Leonard A. Rosenblum, Donald B. Rubin, and M. Fini Schulsinger, "Sex Differences in Developmental Milestones during the First Year of Life," *Journal of Psychology and Human Sexuality* 4, no. 2 (1991): 19–36. See also Eli Coleman, ed., *John Money: A Tribute* (New York: Haworth Press, 1991). For other work by Reinisch, see J. M. Reinisch and S. A. Sanders, "Behavioral Influences of Prenatal Hormones," in *Handbook of Clinical Psychoneuroendocrinology*, ed. C. B. Nemeroff and P. T. Loosen (New York: Guilford, 1987), 431–48; and J. M. Reinisch, "Prenatal Exposure to Synthetic Progestins Increases Potential for Aggression in Humans," *Science* 211 (1981): 1171–3.

68. J. Leo, "Cradle to Grave Intimacy," *Time*, September 7, 1981, p. 69.

69. Mary S. Calderone, [Letter], *Sexuality Today*, November 2, 1981, p. 3.

70. See Havelock Ellis, *Sexual Inversion*, vol. 2 of *Studies in the Psychology of Sex* (New York: Davis, 1901), cases xvii, xxv, xlii, app. B, pt. 1.

71. Floyd M. Martinson, *Infant and Child Sexuality: A Sociological Perspective* (St. Peter, Minn.: Martinson, 1973).

72. Ernest Borneman, *Reifungsphasen der Kindheit*. The book is presently being translated by Michael Lombardi-Nash and is scheduled for publication in 1994 by Prometheus Books.

73. D. Finkelhor, "Sex among Siblings: A Survey on Prevalence, Variety, and Effects," *Archives of Sexual Behavior* 9 (1980): 171–97.

74. J. Elias and P. Gebhard, "Sexuality and Sexual Learning in Childhood," *Phi Delta Kappan* 50 (1969): 401–5.

75. D. Finkelhor and D. E. H. Russell, "The Gender Gap among Perpetrators of Child Sexual Abuse," in *Sexual Exploitation: Rape, Child Sexual Abuse, and Workplace Harassment*, ed. D. E. H. Russell (Beverly Hills, Calif.: Sage, 1984).

76. Paul Gebhard, Wardell Pomeroy, Clyde Martin, and Cornelia V. Christenson, *Sex Offenders* (New York: Harper & Row, 1965). This is the fourth publication of the Kinsey team. The Kinsey study broke new ground in this area, as it had in others. Perhaps the most influential study before this one was Benjamin Karpman, *The Sexual Offender and His Offenses* (New York: Julian, 1954), but it was treatment oriented.

77. Gebhard et al., *Sex Offenders*.

78. Ibid.

79. D. Finkelhor, *Child Sex Abuse: New Theory and Research* (New York: Free

Press, 1984). I am indebted to Allgeier and Allgeier, *Sexual Interactions*, for guidance in this section.

80. Finkelhor, *Child Sexual Abuse*; and D. E. H. Russell, ed., *Sexual Exploitation: Rape, Child Sexual Abuse, and Workplace Harassment* (Beverly Hills, Calif.: Sage, 1984).

81. For a discussion of some of these issues see Gail Elizabeth Wyatt, "Child Sexual Abuse and Its Effects on Sexual Functioning," *Annual Review of Sex Research* 2 (1991): 249–66.

82. T. G. M. Sandfort, "Sex in Pedophiliac Relationships: An Empirical Investigation Among a Non Representative Group of Boys," *Journal of Sex Research* 20 (1984): 123–42.

83. A. C. Kilpatrick, "Childhood Sexual Experiences: Problems and Issues in Studying Long-Range Effects," *Journal of Sex Research* 23 (1987): 173–96.

84. A. C. Kilpatrick, "Some Correlates of Women's Childhood Sexual Experience: A Retrospective Study," *Journal of Sex Research* 22 (1986): 221–42.

85. P. R. Abramson, *Sarah: A Sexual Biography* (Albany, N.Y.: SUNY Press, 1984).

86. D. Finkelhor, *Sexually Victimized Children*.

87. C. Henry Kempe et al., "The Battered Child Syndrome," *Journal of the American Medical Association* 181 (July 7, 1962): 17–24.

88. A good account of the issues behind this is by Barbara J. Nelson, *Making an Issue of Child Abuse* (Chicago: University of Chicago Press, 1984).

89. See also Paul Eberle and Shirley Eberle, *The Politics of Child Abuse* (Seacaucus, N.J.: Stuart, 1986).

90. Cynthia Kisser, "Satanism as a Social Movement," *Free Inquiry* 13, no. 1 (1992–3): 54–56. See also Paul Eberle and Shirley Eberle, *The Abuse of Innocence: The McMartin Preschool Trial* (Buffalo, N.Y.: Prometheus, 1993); and Eberle and Eberle, *Politics of Child Abuse*.

91. See, for example, Vern L. Bullough and Bonnie Bullough, "Lesbianism in the 1920s and 30s," *Signs* 2 (1977): 895–904.

92. Probably the best of these remains the A. R. Bell, M. S. Weinberg, and S. K. Hammersmith, *Sexual Preference: Its Development in Men and Women* (Bloomington: Indiana University Press, 1981).

93. An early example of this was Carroll Smith-Rosenberg, "The Female World of Love and Ritual: Relations Between Women in Nineteenth-Century America," *Signs* 1, no. 1 (Autumn 1975): 19–27. Another early study was Janet Todd, *Women's Friendship in Literature* (New York: Columbia University Press, 1980).

94. Sometimes friendship patterns and identity are closely related as in Lillian Faderman, *Surpassing the Love of Men: Romantic Friendship and Love Between Women from the Renaissance to the Present* (New York: Morrow, 1981). For a discussion of lesbian historiography in the context of the history of sexuality see Martha Vicinus, "Sexuality and Power: A Review of Current Work in the History of Sexuality," *Feminist Studies* 8 (Spring 1982): 147–51; Estelle Freedman, "Sexuality in Nineteenth-Century America: Behavior, Ideology and Politics," *Reviews in American History* 10, no. 4 (1982): 196–215; and George Chauncey Jr., "From Sexual Inversion to Homosexuality: Medicine and the

Changing Conceptualization of Female Deviance," *Salmagundi* 58–59 (Fall 1982–Winter 1983): 114–46. One of the pioneering attempts at a historical study was Jeannette H. Foster, *Sex Variant Women in Literature* (London: Muller, 1958).

95. For a summary of many of the early studies see Susan Krieger, "Lesbian Identity and Community: Recent Social Science Literature," *Signs* 8, no. 1 (Autumn 1982): 91–108; and Chela Sandoval, Ann R. Bristow, and Pam Langford Pearn, "Comment on Krieger's 'Lesbian Identity and Community: Recent Social Science Literature,'" *Signs* 9, no. 4 (Summer 1984): 725–9. For a bibliography see Dolores Maggiore, *Lesbianism: An Annotated Bibliography and Guide to the Literature* (Metuchen, N.J.: Scarecrow, 1988).

96. See Gillian E. Hanscombe and Jackie Forster, *Rocking the Cradle: Lesbian Mothers* (London: Owen, 1981). This study concentrates on the UK, and the American legal system has proved somewhat more tolerant than the British.

97. E. Hooker, "The Adjustment of the Male Overt Homosexual," *Journal of Projective Techniques* 21 (1957): 18–31; E. Hooker, "Male Homosexuality in the Rorschach," *Journal of Projective Techniques* 21 (1958): 33–54; and E. Hooker, "An Empirical Study of Some Relations Between Sexual Patterns and Gender Identity in Male Homosexuals," in *Sex Research: New Developments*, ed. John Money (New York: Holt, Rinehart & Winston, 1965), 24–52.

98. Michael Schofield, *Sociological Aspects of Homosexuality* (London: Longman, 1965).

99. Bell et al., *Sexual Preference*.

100. R. E. L. Masters, *The Homosexual Revolution* (New York: Julian, 1962), 39.

101. *One, Inc. v. Olesen*, 241 F. 2nd 772. (9th Cr, 1957) and 355 U.S. 271 (1958). See also *One Institute Quarterly* 2 (1958).

102. Donald Webster Cory, *The Homosexual in America: A Subjective Approach* (New York: Greenberg, 1951).

103. See Marvin Cutler, ed., *Homosexuals Today: A Handbook of Organizations and Publications* (Los Angeles: One, Inc., 1956); Edward Sagarin [Donald Webster Cory], "Structure and Ideology in an Association of Deviants" (Ph.D. diss., New York University, 1966); and Don Teal, *The Gay Militants* (New York: Stein & Day, 1971).

104. Sir John Wolfenden, chair, *Report of the Committee on Homosexual Offenses and Prostitution* (London: Her Majesty's Stationery Office, 1957).

105. The Southern California policy was written by a committee that I headed.

106. American Friends Service Committee, *Toward a Quaker View of Sex* (London: Friends Home Service Committee, 1963).

107. National Institute of Mental Health Task Force on Homosexuality, John M. Livingood, ed., *Final Report and Background Papers* (Rockville, Md.: National Institute of Mental Health, 1972). The report was long delayed and much of it was written in 1967–8. Various versions of it, however, circulated in the gay community.

108. Thomas Szasz, *The Myth of Mental Illness* (New York: Hoeber-Harper, 1961); and Thomas Szasz, *Law, Liberty and Psychiatry* (New York: Macmillan, 1963).

109. Actually, the committee on nomenclature had reported such a change in

December 1973, and this normally would have been enough, but a referendum was demanded. This supported the nomenclature committee, but made the act official in 1974 instead of 1973. For a discussion of the issues see Ronald Bayer, *Homosexuality and American Psychiatry: The Politics of Diagnosis* (New York: Basic, 1981).

110. Martin S. Weinberg and Colin J. Williams, *Male Homosexuals: Their Problems and Adaptations* (New York: Oxford University Press, 1974), 6.

111. David P. McWhirter and Andrew M. Mattison, *The Male Couple* (Englewood Cliffs, N.J.: Prentice-Hall, 1984).

112. Denis Altman, *The Homosexualization of America; The Americanization of the Homosexual* (New York: St. Martin's, 1982).

113. David F. Greenberg, *The Construction of Homosexuality* (Chicago: University of Chicago Press, 1988).

Chapter 10: Problems of an Emerging Science

1. Erwin J. Haberle and Rolf Gindorf, *Sexology Today* (Düsseldorf, Germany, 1993) list programs from all over the world, but most are paper programs, that is individual programs that can be arranged. Many are undergraduate minors, several are one-shot courses in medical schools, some are programs of special emphasis. A total of twenty-two U.S. programs are listed, ten German ones, seven French ones, six Belgian ones, two Canadian ones, and a variety of others. Many of those listed are nonexistent or are no longer functioning.

2. Among those most up-to-date on research results, as of this writing, is Elizabeth Rice Allgeier and Albert Richard Allgeier, *Sexual Interactions*, 3d ed. (Lexington, Mass.: Heath, 1991). Written with a more conservative approach but also accurate is Janet Shibley Hyde, *Understanding Human Sexuality*, 4th ed. (New York: McGraw-Hill, 1990). Also based on current research is Robert Francoeur, *Becoming a Sexual Person*, 2d ed. (New York: Macmillan, 1991); Susan L. McCamman, David Knox, and Caroline Schacht, *Choices in Sexualities* (St. Paul: West Publishing, 1993); Robert Crooks and Karla Baur, *Our Sexuality*, 4th ed. (Redwoodham, Calif.: Benjamin Cummings, 1990). There are many others but these are those with most current publications.

3. Janice M. Irwin, *Disorders of Desire* (Philadelphia: Temple University Press, 1990), 105–34.

4. William Hartman and Marilyn Fithian, *Treatment of Sexual Dysfunction: A Bio-Psycho-Social Approach* (Long Beach, Calif.: Center for Marital & Sexual Studies, 1972), 105.

5. *SAR Guide for a Better Sex Life* (San Francisco: National Sex Forum, 1975).

6. For a skeptical view of many of those involved, including the newspaper advice columnists, see Patrick McGrady, *The Love Doctors* (New York: Macmillan, 1972).

7. Vern L. Bullough, "Problems of Research on a Delicate Topic: A Personal View," *Journal of Sex Research* 21 (November 1985): 375–86.

8. For discussion of many of these see John Money, *Love Maps* (Buffalo, N.Y.: Prometheus, 1988).

9. For a description of such activities see Thomas Weinberg and G. W. Levi Kamel, *S and M: Studies in Sadomasochism* (Buffalo, N.Y.: Prometheus, 1983).

10. Martin S. Weinberg, Colin J. Williams, and Charles Moser, "The Social Constituents of Sadomasochism," *Social Problems* 31 (1985): 379–89.

11. Paul Robinson, *The Modernization of Sex* (New York: Harper & Row, 1976), 2–3, 194–5.

12. Alex Comfort, *Joy of Sex* (New York: Simon & Schuster, 1972); Alex Comfort, *More Joy* (New York: Crown, 1974).

13. There is a growing literature on this. See, for example, F. J. Bardach and J. Goodgold, *Sexuality and Neuromuscular Disease* (New York: Institute of Rehabilitation Medicine and the Muscular Dystrophy Association, 1979); P. A. Csesko, "Sexuality and Multiple Sclerosis," *Journal of Neuroscience Nursing* 20, no. 6 (1988): 353–5; T. O. Money, T. M. Cole, and R. A. Chilgren, *Sexual Options for Paraplegics and Quadraplegics* (Boston: Little, Brown, 1975); C. S. Schuster, "Sex Education of the Visually Impaired Child: The Role of Parents," *Journal of Visual Impairment and Blindness* 80, no. 4 (1986): 675–80.; and Task Force on Concerns of Physically Disabled Women, *Toward Intimacy: Family Planning and Sexuality Concerns of Physically Disabled Women* (New York: Human Sciences, 1978). For a good overview see Dwight Dixon and Joan Dixon, "Physical Disabilities and Sex," in *Human Sexuality: An Encyclopedia*, ed. Vern L. Bullough and Bonnie Bullough (New York: Garland, 1994), pp. 450–7.

14. See, for example, Randy Shilts, *And the Band Played On: Politics, People, and the AIDS Epidemic* (New York: St. Martin's, 1987).

15. It is possible that an AIDS-related death may have occurred as early as 1969 in St. Louis.

16. See Eve K. Nichols, *Mobilizing Against AIDS* (Cambridge, Mass.: Harvard University Press, 1989), 12–13.

17. *Science* (1986): 697.

18. Morton Hunt, *Sexual Behavior in the 1970s* (Chicago: Playboy, 1974).

19. G. H. Gallup, *The Gallup Opinion Index*, Report No. 153 (Princeton, N.J.: The American Institute of Public Opinion, 1978); and G. H. Gallup, "More Today Than in 1985 Thought That Premarital Sex was Wrong," *The Gallup Report* 263 (1987): 20.

20. J. D. LeMater and P. MacCoquodale, *Pre-marital Sexuality: Attitudes, Relationships, Behavior* (Madison: University of Wisconsin Press, 1979).

21. P. W. Blumenstein and P. Schwartz, *American Couples* (New York: Morrow, 1983). There are many other studies on more limited topics that are invaluable.

22. Carol Tavris and Susan Sadd, *The Redbook Report on Female Sexuality* (New York: Delacorte, 1975).

23. L. Wolfe, *The Cosmo Report: Women and Sex in the 80s* (New York: Bantam, 1982).

24. Robert Athanasiou, Phillip Shaver, and Carol Tavris, "Sex," *Pyschology Today* 4, no. 2 (1970): 39–52. See also Robert Athanasiou, "A Review of Public Attitudes on Sexual Issues," in *Contemporary Sexual Behavior: Critical Issues in the 1970's*, ed. Joseph Lubin and John Money (Baltimore, Md.: Johns Hopkins University Press, 1973), 361–90.

25. "Playboy Readers' Sex Survey, 1983: Part 1," *Playboy*, January 1983, p. 108; "Playboy Readers' Sex Survey, 1983: Part 2," *Playboy*, March 1983, p. 90; "Playboy Readers' Sex Survey, 1983: Part 3," *Playboy*, May 1983, p. 126; "Playboy Readers' Sex Survey, 1983: Part 4," *Playboy*, July 1983, p. 130; "Playboy Readers' Sex Survey, 1983: Part 5," *Playboy*, September 1983, p. 92.

26. Shere Hite, *The Hite Report: A Nationwide Study on Female Sexuality* (New York: Macmillan, 1976); Shere Hite, *The Hite Report on Male Sexuality* (New York: Knopf, 1981); and Shere Hite, *Woman and Love: A Cultural Revolution in Progress* (New York: Knopf, 1987).

27. See, for example, Sandra Kahn, *The Kahn Report on Sexual Preferences*, with Jean Davis (New York: St. Martin's, 1981).

28. Milton Diamond, *Sex Watching: Looking into the World of Sexual Behaviour* (London: Prion, 1992), 19–20.

29. Albert D. Klassen, Colin J. Williams, and Eugene E. Levitt, *Sex and Morality in the U.S.*, ed. Hubert J. O'Gorman (Middletown, Conn.: Wesleyan University Press, 1989).

30. Samuel S. Janus and Cynthia L. Janus, *The Janus Report on Sexual Behavior* (New York: Wiley, 1993).

31. Unlike most sexual surveys, this was published in the refereed *Family Planning Perspectives*, the journal of the Alan Guttmacher Institute. See the general article by John O. G. Billy, Koray Tanfer, William R. Grady, and Daniel H. Lepinger, "The Sexual Behavior of Men in the United States," *Family Planning Perspectives* 25 (March–April 1993): 52–60. There are a number of other articles in the same issue based on survey data such as Tanfer, Grady, Kelpinger, and Billy, "Condom Use Among U. S. Men, 1991," 61–66; Grady, Kelpinger, Billy, and Tanfer, "Condom Characteristics: The Perceptions and Preferences of Men in the United States," 67–73; Klepinger, Billy, Tanfer, and Grady, "Perceptions of AIDS Risk and Severity and Their Association with Risk Related Behavior Among U.S. Men," 74–82.

32. See Traci Watson, "Sex Surveys Come Out of the Closet," *Science* 260 (April 30, 1993): 615–6.

33. F. N. Judson, "Fear of Aids and Gonorrhea Rates in Homosexual Men," *Lancet* 2 (1983): 159–60.

34. Bonnie Bullough and George Rosen, *Preventive Medicine in the United States 1900–1990* (Canton, Mass: Science History, 1992).

35. Michel Foucault, *The History of Sexuality*, trans. R. Hurley (London: Allen Lane, 1979), 1: 146. See also Michel Foucault, *Discipline and Punish: The Birth of the Prison*, trans. Alan Sheridan (London: Allen Lane, 1977); and Michel Foucault, *Power/Knowledge: Selected Interviews and Others Writings, 1972–1977* (New York: Pantheon, 1981). Foucault is a postmodernist and casts doubt on the world as objective reality and consequently on objective truth, thus discourse becomes an artifact of communal knowledge. Still, if his ideas are not pushed too far, he does offer basic insights into the changes taking place. I do not accept his explanations of the changes. For discussion of his ideas, including those about sexuality, see Mark Cousins and Athar Hussain, *Michel Foucault*, ed. Anthony Gliddens (New York: St. Martin's, 1983); Hubert

L. Dreyfus and Paul Rabinoe, *Michel Foucault: Beyond Structuralism and Hermeneutics*, 2d ed. (Chicago: University of Chicago Press, 1983); Pamela Major-Poetzl, *Michel Foucault's Archeology of Western Culture: Toward a New Science of History* (Chapel Hill: University of North Carolina Press, 1983); Jeffrey B. Minson, *Genealogies of Morals: Nietzsche, Foucault, Donzelot, and the Eccentricity of Ethics* (New York: St. Martin's, 1985); and Mark Poster, *Foucault, Marxism and History: Mode of Production Versus Mode of Information* (New York: Blackwell, 1984). Particularly important in discussing Foucault and sexuality is Jeffrey Weeks, *Sex, Politics, and Society: The Regulation of Sexuality Since 1800* (London: Longman, 1981); and Jeffery Weeks, *Sexuality and Its Discontents: Meaning, Myths, and Modern Sexualities* (New York: Routledge & Kegan Paul, 1985). Feminist scholars in particular have seized on Foucault's ideas of sex. See G. Rubin, "Thinking Sex: Notes for a Radical Theory of the Politics of Sexuality," in *Pleasure and Danger: Exploring Female Sexuality*, ed. Carole Vane (New York: Routledge & Kegan Paul, 1984), 167–319. Feminist theorists find he gives a basis for explaining women's oppression and how women become gendered and sexualized objects. Carol A. Polis, "The Apparatus of Sexuality: Reflections on Foucault's Contributions to the Study of Sex in History," *Journal of Sex Research* 23 (1987): 401–7; Ellen Ross and Rayna Rapp, "Sex and Society: A Research Note from Social History and Anthropology," *Comparative Studies in Social History* 23 (1981): 51–72; Ann Snitow, C. Stansell, and S. Thompson, eds., *Power of Desire: The Politics of Sexuality* (New York: Monthly Review, 1983); and Bryan S. Turner, *The Body and Society: Explorations in Social Theory* (New York: Blackwell, 1983). For examples of church control of sexuality, see James A. Brundage, *Law, Sex, and Christian Society in Medieval Europe* (Chicago: University of Chicago Press, 1987); and Vern L. Bullough and James Brundage, eds., *Sexual Practices and the Medieval Church* (New York: Prometheus, 1982).

36. Vern L. Bullough, *Sexual Variance in Society and History* (Chicago: University of Chicago Press, 1976), 461–503.

37. Donna Haraway, "Situated Knowledges: The Science Question in Feminism and the Privilege of Partial Perspective," *Feminist Studies* 14 (1988): 575–99.

38. For an overview of these developments, see Vern L. Bullough, *The Scientific Revolution* (New York: Holt, Rinehart & Winston, 1970).

39. Thomas S. Kuhn, *The Structure of Scientific Revolutions* (Chicago: University of Chicago Press, 1962).

40. Karl Popper, *The Logic of Scientific Discovery* (London: Hutchinson, 1959); and Karl Popper, *Conjectures and Refutations* (London: Routledge & Kegan Paul, 1963).

Index

Aberle, Sophie, 123
Abortion, 73, 116, 256–57
Abraham, Karl, 64, 68
Abramson, Paul, 263
Accumulation, 83–84
Acton, William, 26
Adam (biblical figure), 3, 11
Addison, Thomas, 124–25
Adler, Alfred, 68, 150–51
Adolescence, 76, 152, 160–61. *See also*
    Childhood sexuality; Puberty
Adrenal hyperplasia, 227–28
Adrenogenital syndrome, 216
Aggression, 229–31, 249, 252
A. H. Robins Co., 188–89
AIDS, 241, 244, 284–91. *See also* STDs
    (sexually-transmitted diseases)
Albigensians, 9
Albright, Fuller, 190
Alcoholism, 101–2, 166, 290
Algolagnia (sadomasochism), 48, 84
Alienism, 40
Alien Property Act, 193
Allen, Edgar, 127, 128
Allen, Willard M., 129, 197
Allgeier, Elizabeth, 252–53
Allgeier, Richard, 252–53
All the Sexes (Henry), 167
Altman, Denis, 271

Altruism, 41
American Academy of Sexology, 208
American Association for the Advance-
    ment of Science, 149
American Association of Sex Educators,
    Counselors, and Therapists
    (AASECT), 207–8, 275–76, 289
American Birth Control League, 146
American Board of Sexology, 208
American Chemical Society, 193
American Civil Liberties Union (ACLU),
    184, 251, 270
American Federation for Sex Hygiene,
    104
American Friends Service Committee,
    270
American Gynecological Society, 29, 109
American Law Institute, 257
American Medical Association (AMA),
    93, 105, 153, 275
American Museum of National History,
    121
American Psychiatric Association, 221,
    240, 270, 271, 280
American Psychological Association, 270
American Purity Alliance, 96–97, 104
American Social Hygiene Association,
    104, 115, 116
American Vigilance Association, 104

Ananga Ranga, 55
Anatomical Exercitations Concerning the Generation of Living Creatures (Harvey), 14
Anatomy of Plants (Grew), 11
Anima, 19, 152
Animalcula, 15
Animal studies, 9–12, 14–17, 46, 154–56, 231–32; and the association of love with pain, 84; and contraceptives, 187–88; and hormone studies, 122–26, 129–31, 133–35, 156, 187–88; and Masters and Johnson, 197
Animus, 152
Anorgasmia, 178–79
Anthropological Society of London, 55
Anthropology, 55, 96, 147, 154, 156–60, 170, 218, 271
Anthropophyteia, 58
Antibiotics, 171
Appeal to Reason, 143
Aquinas, Thomas, 13, 293, 294
Arbuthnot, Foster F., 55
ARC (AIDS-related complex), 284
Archives of Sexual Behavior, 208
Argentina, 73
Aristotle, 9–13, 124, 293
Aristotle's Masterpiece, 9–10
Armed Forces (United States), 296
Armour & Co., 126
Arrowsmith (Lewis), 133
Ascheim, Selmar, 128
Ashbee, H. S., 55, 58
Associationism, 43–44
Augustine, Saint, 2–3, 7, 18, 30
Austria, 58
Autoeroticism, 83–85, 87–88, 89, 115
Avicenna, 13

Bacon, Francis, 81
Baker, Alden H., 184
Bali, 159
Banting, Frederick G., 127
Baptism, 3
Bartelernez, G. W., 130
Bartels, Max, 56
Bartholin's glands, 200
Batelle Memorial, 287
Battered-child syndrome, 253

Battle Creek Sanitarium, 22
Baudelaire, Charles, 58
Bayliss, William M., 125
Beach, Frank A., 158–59, 232
Beam, Lura, 110, 164
Beard, George M., 23, 93
Bebel, August, 67
Beck, Joseph R., 197–98
Behaviorism, 88
Beigel, Hugo, 206
Beiträge zur Aetologie der Psychopathia Sexualis (Bloch), 56
Bell, A. P., 236, 267–68
Beneden, Eduard van, 17
Benjamin, Harry, 73, 219–20
Benkert, Karl Maria (Kertbeny), 39–40, 41, 67
Berdache, custom of, 96, 159
Berlin Psychoanalytical Society, 64
Bem, Sandra, 234
Bem Sex Role Inventory, 234–35
Bernard, Claude, 125
Bernays, Martha, 86
Berrian, William, 98
Berthold, A. A., 124
Best, Charles H., 127
Biber, Stanley, 221
Bieber, Irving, 235–36
Biedl, Arthur, 124
Binberg Bows, 187
Binet, Alfred, 42–44, 50–51, 89
Birken, Lawrence, 70
Birth control, 5, 109, 145–46; and international congresses, 73; and marriage manuals, 138. See also Contraceptives
Birth rates, 5, 50, 134, 147
Bisexuality, 65–66, 68–70, 177–78; and AIDS, 287; and biogenetic views of homosexuality, 44–45; and gender issues, 226, 234; and Henry, 165, 166; and Jung, 151–52
Bissonette, T. H., 155
Blackwell, Elizabeth, 28
Blank, Joani, 253
Bleuler, Eugen, 120
Bloch, Iwan, 56, 68, 71, 88, 273
Blüher, Hans, 66, 79
Blumer, G. Alder, 95
Blumstein, P. W., 285

Boas, Ernst P., 111
Boas, Franz, 157
Boerhaave, Hermann, 19
Bok, Edward, 104–5
Bordeu, Théophile de, 19, 20
Bornerman, Ernest, 260
Boston Women's Health Collective, 243
Bouret, Edouard, 143
Brand, Adolf, 66, 67, 79
Brave New World (Huxley), 133
Brazil, 239
Brecher, Edward M., 4, 40
Breitenstein-Hooglandt, Martha, 141
Breuer, Josef, 87
Brinkley, John H., 126–27
Britain, 15, 42–43, 49–53, 58, 60, 100
British Medical Journal, 42–43
British Museum, 58
Britten, F. H., 162–63
Bromley, Dorothy D., 162–63
Brown, John, 19–21
Brownmiller, Susan, 246
Brown-Séquard, Charles, 125–27
Bruck, Carl, 99
Brunonianism, 19
Bryant, J., 250
Bullough, Bonnie, 221, 235, 238, 255
Bureau of Social Hygiene, 113, 114–15, 118, 147
Burnham, John, 95, 148–49
Burou, George, 221
Burton, Richard, 54–58, 157
Bush, George, 290
Butenandt, Adolf, 128

Cabez de negro, 192
Calderone, Mary Steichen, 197, 207, 259, 279
Camerarius, Rudolph Jacob, 11
Captive, The (Bouret), 143
Carpenter, Edward, 78, 80, 137
Carretta, Ruth A., 226
Casonova, Giacomo, 78
Casper, Johann L., 32
Castration, 20, 26–27, 89, 124–25, 218
Category crises, 234
Catholicism, 2–3, 52. *See also* Christianity
Cattell, J. McKeen, 149

Cauldwell, David O., 219
CBS (Columbia Broadcasting System), 136
Celibacy, 2–3. *See also* Chastity
Censorship, 102, 143–45, 183–85; feminists against, 253; and pornography, 143, 249, 251
Centers for Disease Control (CDC), 171, 284
Cerf, Bennett, 145
Cervical caps, 108, 146. *See also* Contraceptives
Charcot, Jean-Martin, 43, 86, 87
Chastity, 2–3, 9, 97, 101, 119, 255, 261
Cheyenne Indians, 246
Chicago County Asylum for the Insane, 45
Child abuse, 24, 263–64, 266. *See also* Sexual abuse
Child care centers, 264–65
Child sexuality, 46–49, 121, 279–80; and associationism, 44; and Bloch, 57; and Ellis, 76, 260; and Foucault, 244; and Freud, 88–89, 259; and gender identity formation, 234–40; and Kinsey, 174–75, 184, 259; and masturbation, 87; and Reich, 152; and sex assignment, 215, 217. *See also* Adolescence; Puberty; Sexual abuse
China, 2, 73
Christianity, 76, 207, 292–94; and celibacy, 2–3; and the flesh/spirit dualism, 48–49; and marital sex, 24–25; and premarital sex, 176. *See also* Catholicism; God; Protestantism; Sin
Chromosomes, 17, 122–23, 213–14. *See also* Heredity
Chuang, Min Chueh, 194
Circumcision, 224–25
Civilization and Its Discontent (Freud), 90
Civil rights movement, 242–43, 248, 250
Civil War, 101, 108
Clark, LeMon, 111
Clarke, Edward H., 28–29
Clitorectomies, 27, 72
Cockburn, Alexander, 52

Cocoa butter, 106, 108. *See also* Contraceptives

Collier's, 134

Comfort, Alex, 283

Coming of Age in Samoa (Mead), 157

Committee for Research in the Problems of Sex (CRPS), 117, 118, 120–23, 133–35, 147, 163, 175, 181; and Albright, 190; and Bissonette, 155–56; and Ford and Beach, 158; and Kinsey, 169; and Marcuse, 154–55

Committee for the Study of Sex Variants, 165

Committee on Maternal Health, 109, 110

Committee on Obscenity and Film Censorship, 251

Communication: models of pornography based on, 252; and redefining rape, 247, 248

Communist Party, 75, 207, 268, 279

Comstock, Anthony, 52–53, 96, 105

Comstock Law, 102–3, 106

Comte, August, 292, 295

Condoms, 105, 108, 189. *See also* Contraceptives

Congress (United States), 52, 73, 102, 181-83

Congresses, sexology, 72–75, 242, 257

Connelly, Mark, 96

Consciousness raising, 243

Consent, use of the word, 297

Constantinople, Ann, 234

Constitution (United States), 184, 251

Constitutional psychology, 167

Constructionism, social, 222, 223, 271, 295

Consumers Union, 146

Contraceptives, 134–35, 185–96, 207, 218, 257; and the birth control movement, 145-47, 194; cervical caps, 108, 146; and the Comstock Law, 102; condoms, 105, 108, 189; and Davis, 116; diaphragms, 105–6, 146; and Ellis, 76, 77; and international congresses, 73; intrauterine devices (IUDs), 185–89, 195; oral, 172, 189–95, 242; statistics on the use of, 108; and Tissot, 20–21; as "unnatural," 23

Copenhagen Consecutive Prenatal Cohort Study, 259

Copernicus, Nicolaus, 293

Cornell University, 248

Corner, George W., 3, 123, 129, 131–32, 170–71, 175, 182–83, 197

Cory, Donald W., 268

Cosmopolitan, 286

Courtney, Keith, 231

Criminality, 6, 45, 51, 113–15, 295, 297

Cross-dressing, 159, 269, 282, 297; and Ellis, 85; and gender issues, 219–24, 234, 235, 237–38; and transvestism, 69, 85, 219–24

Crow Indians, 96

Custer, George Armstrong, 246

Cytology, 16

Czechoslovakia, 72, 73–74

Dalkon Shield, 188–89

Dannemiller, J. E., 227

Darwin, Charles, 5, 16, 41, 44

Das Konträre Geschlechtsgefühl (Ellis), 79

Das Weib (Ploss), 56

Daughters of Bilitis, 266

Davenport, Charles, 237

Davidson, J., 233

Da Vinci, Leonardo, 81

Davis, Hugh, 188–89

Davis, Katharine Bement, 112–21, 135, 154, 164, 254, 266

Death instinct, 88

Declaration of Independence, 21

Delisle, François, 78

Dell, Floyd, 149

Denmark, 72

Dennett, Mary W., 144

Dessoir, Max, 47, 58

Determinism, 89

Deviation, use of the term, 4, 6, 81, 92

Diabetes, 127

Diagnostic and Statistical Manual, 221, 240, 280

Diamond, Milton, 225–26, 228, 286

Diaphragms, 105–6, 146. *See also* Contraceptives

Dickinson, Robert Latou, 109–11, 119, 121, 134–35, 147, 154, 164, 175, 198, 273

Die Homosexualität des Mannes und des Weibes (Hirschfeld), 69
Die Konträre Sexualempfindung (Moll), 46
Die Transvestiten (Hirschfeld), 69
Die Zeit, 90
Die Zukunft, 63
Dioscorea (barbasco), 193
Diseases. *See* STDs (sexually-transmitted diseases)
Divorce, 73, 137, 138
Djerassi, Carl, 193
Dodds, Harold W., 180, 182
Doisy, Edward A., 127, 128, 190
Doll's House, A (Ibsen), 137
Domagk, Gerhard, 100
Dominican Republic, 233
Dorner, Gunter, 228, 229
Double standard, 96–97, 100, 203–4, 256
Douglas, Lord Alfred, 53
Druten, John van, 73
Drysdale, George, 77
Duffy, E. B., 26
Dworkin, Andrea, 249–51

Education, sexological, 276–77. *See also* Sex education
Ego, 7, 88
Egypt, 2
Ehrhardt, Anke A., 228
Ehrlich, Paul, 99–100
Elias, James, 261
Eliot, George, 58
Ellis, Edward P., 76
Ellis, Havelock, 52, 58, 60, 61, 62, 73, 75–86, 91, 206, 209, 217, 273, 282–83; and American sex research, 92, 94, 97, 105, 119, 142, 148; and endocrinology research, 120, 124, 137; and Freud, 87, 90, 91; and Hirschfeld, 73; and homosexuality, 39–40, 92, 212; and international congresses, 73; and the Jewish influence, 58, 60; and Kinsey, 170; and Malinowski, 157; and marriage manuals, 136; and modesty, 82–85; and Moll, 47; and Money, 212; and retrospective research, 260; and sexual inversion, 39-40, 94; strengths and weaknesses of, 85-86; summary of

sexology by, up to Krafft-Ebing's time, 44; and transsexualism, 217
Embryology, 65, 129, 213–15, 224, 227–28
Endocrinology, 4, 72, 85, 124–36; and animal studies, 124–26, 129–31, 133–35, 156, 187–88; and gender issues, 227–31; and quackery, 126–27; and transsexualism, 219. *See also* Hormones
Enduring Passion (Stopes), 139
Engels, Friedrich, 91
Englemann, George J., 29
English Eugenics Society, 50
Enovid, 195
Eonism, 84, 85
Epilepsy, 40, 95
Equal Opportunity Commission, 248
Erasmus, Desiderius, 81
Erickson, Reed, 291
Erickson Educational Foundation, 291
Erikson, Erik, 259
Ernst, Morris, 145
Erotic Symbolism (Ellis), 84
Essentialism, sexual, 212
Eugenics, 5, 49–52, 169
Eulenburg, Albert, 46–48, 58–59
Eulenberg, Philip Fürst zu, 63–64, 71
Evans, Herbert M., 130
Evans, Wainwright, 138
Eve (biblical figure), 3, 11
Evil, 53, 76
Evolution, 5, 6, 44, 46, 69–70
Excitability, 19–20
Exhibitionism, 84, 180, 261, 263, 297
Exner, Max J., 111–12, 115, 116
Extramarital sex, 4, 180

Falloppio, Gabriele, 13
Fantasies, 198–99
Fay, Erica, 139
Federal Bureau of Investigation (FBI), 279
Federal Child Abuse Act, 264
Feminine Mystique, The (Friedan), 242
Femininity, 163–64, 167, 210–40. *See also* Gender
Feminism, 4, 7, 213, 241–44, 280, 290; and abortion, 256–57; and eugenics,

Feminism (continued)
    51; and Freud, 90; and homosexual-
    ity, 265; and Mosher, 108; and
    pornography, 249–54; and prostitu-
    tion, 255, 256; and rape, 245–46, 247,
    248; and renewed interest in children,
    259; and social constructionism,
    271–72. See also Women, subordi-
    nation of
Fermentation, discovery of, 11
Fetishism, 41–43, 56, 69, 218; and Ellis,
    84; and Freud, 89, 151–52
Fielding, William J., 142
Finkelhor, David, 260–61, 263
Finland, 72
First International Congress on Hypno-
    tism, 44
Fithian, Marilyn, 203, 254, 278
Flaubert, Gustave, 58
Fleming, Alexander, 100
Fleming, Donald, 133
Flexner, Abraham, 114
Flexner, Simon, 117
Fliers, Eric, 231
Fliess, Wilhelm, 149
Food and Drug Administration (FDA),
    153, 189, 195
Foot, Edward B., 106
Ford, Clellen S., 158–59
Forel, August, 73
Forensic medicine, 32
Fornication, 23
Fosdick, Harry E., 182
Fosdick, Raymond B., 114
Foucault, Michel, 244–45, 250–51, 292
Fournier, Jean-Alfred, 103
Fraenkel, Ludwig, 129
France, 31–32, 72, 74, 103, 128, 197
Free-association method, 165
Free-love movement, 101
Freemartins, 122–23
Free speech. See Censorship
Freud, Sigmund, 47, 57, 82, 83, 84,
    86–91, 105, 254, 271; and American
    sex research, 97, 148–55; and associa-
    tionism, 43–44; concept of libido, 84,
    153; ego and superego in, 7, 88; and
    Ellis, 82, 83, 84, 86, 120; and Erikson,
    259; and Hirschfeld, 64, 70–71, 72,

86; on hysteria, 49, 68, 72, 83, 87; and
    the Jewish influence, 59–60; and
    Kinsey, 169, 170; and marriage manu-
    als, 136; and the mystique of mother-
    hood, 97; Oedipus complex in, 70–71,
    82, 88, 89; and science, 149–55; social
    and sexual desire in, 70–71
Friedan, Betty, 242
Friede, Donald, 143
Friedländer, Benedict, 66, 79
Frigidity, 178–79

Gagnon, John H., 222
Galen, 12–14
Galileo, 294
Gallagher, T. F., 129
Gallup Poll, 285
Galton, Francis, 5, 50, 51
Gamble, Clarence J., 134–35, 147
Garber, Marjorie, 234
Garden of Eden, 3, 11
Gates, Reginald R., 138
Gay, Jan, 165
G. D. Searle (company), 134
Gebhard, Paul, 173, 257, 261–62
Gender, 210–40, 274; and the concept of
    critical periods, 224; identity, 211–12,
    223–25; and the nature/nurture con-
    troversy, 225–27; "scripts," 222–23,
    239; use of the term, 210, 211. See
    also Femininity; Masculinity
Genetics. See Chromosomes; Heredity
Geniuses, 50
Geocentric theory, 293–94
George W. Henry Foundation, 167
Germany, 6, 32–49, 58–59, 72–73, 128.
    See also Nazism
Getting Married (Shaw), 137
Ghosts (Ibsen), 137
Gladue, Brian, 230
Glide Foundation, 276
God, 3, 12, 35, 36, 212, 293
Goddesses, 36
Gonadotropin, 130
Gonorrhea, 94–95, 99–100, 110, 289,
    290. See also STDs (sexually-
    transmitted diseases)
Good Housekeeping, 164
Goodman, John, 29

Good Vibrations, 253
Gooren, Louis, 231
Graaf, Regnier de, 15
Gräfenberg, Ernst, 186, 187, 254
Graham, Sylvester, 22, 101
Grant, Ulysses S., 102
Great Depression, 134
Green, Richard, 208, 230, 236, 237
Greenberg, David, 271–72
Greenwald, Harold, 255
Gregg, Alan, 181
Grew, Nehemiah, 11
G spot, 254
Guttmacher, Alan F., 186, 194, 197, 207
Guze, Henry, 206
Gynecology, 4, 13–14, 25, 62, 93; and
    Dickinson, 109, 111; and Kinsey, 184;
    and Masters and Johnson, 196, 197,
    205; and the specialization of science,
    5; and transsexualism, 218

Haeckel, Ernst, 69–70
Haire, Norman, 74
Haldeman-Julius, Emanuel, 142–43
Hall, G. Stanley, 24
Hall, Marguerite R., 143
Ham, Johan, 15
Hamburger, Christian, 217–18
Hamilton, Allan M'Lane, 95
Hamilton, Gilbert V., 117–21, 154, 164
Hammersmith, S. K., 236, 267–68
Hampson, J. G., 217
Hampson, J. L., 217
Hand, Augustus N., 144, 145
Hand, Learned, 145
Handicapped individuals, 164
Haraway, Donna, 293
Harden, Maximilian, 63–64, 67
Harden–Eulenburg–von Moltke affair,
    63–64, 67
Harman, William, 203
Harris, Frank, 80
Harry, Joseph, 239
Hartman, William, 254, 278
Harvard Medical College, 28, 29
Harvey, William, 14–15, 18
Hata, Sahachiro, 99
Hay, Harry, 268
Hayes, Wayne L., 183

Health insurance, 281
Heliocentric theory, 293–94
Hellman, Ronald, 230
Helmreich, R. L., 235
Helms, Jesse, 288
Henry, George, 165–67, 226
Herdt, Gilbert, 233
Heredity, 5, 40, 122–23, 166; and gender
    issues, 216–17, 231–32, 240; and herm
    aphrodites, 216–17. See also Chromo-
    somes; Eugenics
Hermaphrodites, 2, 6, 36, 81, 93, 159;
    and gender issues, 211–14, 216–17,
    224; and homosexuality, 214, 231–32;
    plants as, 11; pseudo-, 233
Herrick, Robert, 137–38
Hertwig, Oscar, 17
Hicklin decision, 52
Hieronymus Fabricus of Aquapendente,
    14
Hill, Anita, 248
Hinduism, 55, 142
Hinton, James, 77
Hirschfeld, Hermann, 64
Hirschfeld, Magnus, 57, 58, 60, 61,
    62–75, 80, 81, 82, 83, 85–86, 91, 92,
    179, 209, 273, 280; early years of,
    64–65; and Ellis, 76, 80–83, 85–86;
    and endocrinology research, 124, 126,
    127, 135; and Freud, 91; and homo-
    sexuality, 92, 176, 212, 268; and inter-
    national sexual congresses, 72–74;
    and the Jewish influence, 58, 60; and
    Kinsey, 170; and Moll, 47, 74–75; and
    Money, 212; and the Nazis, 74–75;
    and politics, 66–67; as a researcher,
    67–72; and Stekel, 151; strengths and
    weaknesses of, 85–86; theory of, sum-
    mary of, 65–66; and transsexualism,
    217, 218, 219
History of European Morals (Lecky), 97
HIV (human immunodeficiency virus),
    284–89. See also AIDS; STDs
    (sexually-transmitted diseases)
Hoagland, Hudson, 134
Hoffman, Friedrich, 19
Hoffmann, Erich, 99
Holder, A. B., 96
Holland, 72

Homosexuality, 4, 162, 165–67, 266–71, 280, 298; and aggression, 229–30; and AIDS, 284–85, 287–89; in anthropological studies, 158, 159, 160; biogenetic views of, 44–45; and cross-dressing, 222, 237-38; and Davis, 266; and Ellis, 78–82, 83-85; and eugenics, 49–52; and Foucault, 245; and Freud, 87, 89–90; and gender issues, 212, 213, 218, 226–27, 235, 240; the German sexological movement, 6–7, 32-49, 63–64, 65, 66–70, 75; among the Greeks, 160; and Haldeman-Julius, 142; and heredity, 214, 231–32; and hormones, 228–31; and industrialization, 30–31; and Kinsey, 171, 174–78, 184, 266; and Krauss, 58; and Lydston, 93; medicalization of, 38; and mental illness, 95; and military service, 296; and prostitution, 256; and Reich, 152; and sexological education, 276–77; and sexual inversion, 79, 80–81, 94, 95, 164; and sexually-transmitted diseases (STDs), 284–91; and social constructionism, 271–72; and twin studies, 227; and the two-stage theory of sexual development, 46–47; and Ulrichs, 274; as "unnatural," 23; use of the term, 39, 69, 79, 160, 269; and Wilde, 53–54

Hooker, Evelyn, 267, 270, 286

Hormones, 70, 72, 122–23; and contraceptives, 188, 189–95; and gender issues, 218, 220, 230–31, 233, 240; and homosexuality, 228–31; isolation of, 128-29; and theories of menstruation, 127–28; and transsexualism, 218. See also Endocrinology

Hughes, C. H., 93–94

Humanitarian Committee, 71

Human Relations Area Files, Inc., 158

Humboldt, Baron von, 81

Hungary, 72

Hunt, Morton, 285

Huxley, Aldous, 133

Hyde, Janet, 229

Hypnosis, 44, 48

Hypogonadism, 220

Hysteria, 14, 49, 68, 72, 83, 87, 244

Ibsen, Henrick, 137

Ideal Marriage (Van de Velde), 140–41

Illness. See STDs (sexually-transmitted diseases)

Incest, 42, 257. See also Sexual abuse

India, 54–55

Industrialization, 30–31

Indus Valley, 2

Inquisition, 294

Institute for Advanced Study of Human Sexuality, 276

Institute for Sex Research, 257

Insurance, health, 281

Intelligence tests, 50–51

Interactionism, 223, 226

International Committee for Sexual Equality, 268

International Committee for the Taxonomy of Viruses, 285

International Conference of Sexual Reform Based on Sexual Science, 72–73

International Congress of Sex Research, 74

International Professional Surrogates Association (IPSA), 278

International Society for Sexual Research, 74

Inversion, sexual, 79, 80–81, 94, 95

IQ tests, 51

Ishihama, Atsumi, 186

Italy, 68, 72, 78

IUDs (intrauterine devices), 186–89, 195

Jacobi, Abraham, 23–24

Jacobs, Aletta, 106

Jahrbuch für sexuelle Zwischenstufen, 67

Janiger, O., 226, 230

Janus Report, 287

Japan, 58, 73, 186

John Birch Society, 207

Johnson, Lyndon B., 251

Johnson, Virginia, 111, 172, 185, 195–210, 273–74, 278; criticism of SAR sessions, 279; and the Freudian myth of vaginal orgasm, 254; and Kinsey, 196; and sexology fads, 281; and the sexual response cycle, 200–202

Jorgensen, Christine, 213, 217, 218, 219, 221
Journal of Homosexuality, 208, 271
Journal of Human Fertility, 193
Journal of Marriage and the Family, 136
Journal of Sex Education Therapy, 208
Journal of Sex Research, 206
Journal of the American Medical Association, 93
Journal of the History of Human Sexuality, 208
Joyce, James, 145
Joy of Sex, The (Comfort), 283
Judaism, 3, 58–60, 134, 179
Judson, F. N., 289
Jung, C. G., 151–52

Kallman, Franz J., 227
Kama Shastra Society, 55
Kama Sutra, 55
Kanin, E. J., 247
Kaplan, Helen Singer, 203–4, 278
Kaposi's sarcoma, 285
Karsch-Haack, F., 58
Kellogg, John Harvey, 22
Kelly, Howard, 93
Kempe, C. Henry, 263
Kennedy, John F., 181
Kertbeny (Karl Maria Benkert), 39–40, 41, 67
Key, Ellen, 137
Kiernan, James G., 45
Kinsey, Alfred, 116, 119, 136, 164, 168–71, 172–85, 206–10, 219, 274; and abortion, 257; and AIDS, 285, 286; and bisexuality, 234; and Brecher, 4; and censorship, 183–85; data on sexual abuse, 261–62; and homosexuality, 171, 174–78, 184, 267; interview technique of, 173–75; and Masters and Johnson, 196, 199, 200, 203–4; and retrospective research, 260; and scientific detachment, 280; seven-point scale of, 176–77, 227; and sex therapy, 279. *See also* Kinsey Institute
Kinsey Institute, 283, 286
Kirkendall, Lester, 207
Kisch, Enoch Heinrich, 58

Kleinsorge, H., 198–99
Klinefelter's syndrome, 214
Klumbies, G., 198–99
Knauer, Emil, 124
Kneeland, George J., 114
Koch, Fred C., 129
Kölreuter, Josef Gottlieb, 11–12
Krafft-Ebing, Richard von, 6–7, 37–39, 40–45, 46, 48, 49, 92; Bloch's criticism of, 56–57; and Freud, 90; and Hirschfeld, 62, 75; and homosexuality, 40–42, 65, 92; influence of, summary of, 40–43; and the Jewish influence, 58; and Kinsey, 169–70, 280; and Ulrichs, 37–38
Krauss, Friedrich S., 57–58, 68
Krupp, Friedrich, 63
Kuhn, Thomas, 294
Kuzrok, Raphael, 189–90

Laboratorios Hermona, 192
Ladder, 266, 269
Ladder, Lawrence, 257
Ladies' Home Journal, 104–5
Landis, Carney, 162, 164
Landon, Alfred M., 126–27
Lamater, J. D., 285
Laquer, Ernst, 190
Lashley, K. S., 122
Laumann, Edward, 288
Leach, Fred A., 126
Lebovitz, Phil S., 237
Lecky, William E. H., 97
Lees, Edith, 78
Leeuwenhoek, Anton van, 15, 18
Lehfeldt, Hans, 206
Lehman, Federico, 192
Lesbianism, 21, 66, 78, 110–11, 165–67, 254; and AIDS, 241, 289–90; and censorship, 143; and Davis, 115–16; and Ellis, 82; and gender issues, 213, 229–30, 236–40, 268; and Kinsey, 116, 176–77; "political," 266; and pornography, 253. *See also* Homosexuality
Letter on the Sex Life of Plants (Camerarius), 11
Leunbach, J., 73, 74
LeVay, Simon, 228
Levene, P. A., 191

Lewis, Denslow, 93
Lewis, Sinclair, 133
Lillie, Frank R., 122–23, 130
Limerance, 257
Lindsey, Benjamin B., 138
Linnaean Society, 16
Linnaeus, Carolus, 16
Lippes, Jack, 187
Lippes Loop, 187–88
Little Blue Books, 142, 143
Living, 136
Locke, John, 91
Loeb, Jacques, 133
Lombroso, Cesare, 5–6, 45, 68
Lord Campbell's Act, 52
Love, 77, 84, 255; and erotic chemotro-
   pism, 70; and limerance, 258; sexual-
   ization of, 136–43
Lydston, G. Frank, 45, 92–93, 95

MacCarthy, Desmond, 73
McClung, Clarence E., 122
McCormick, Katharine D., 194
MacCorquodale, P., 285
McIlvenna, Ted, 276
MacKinnon, Catharine, 249–50
Macleod, J., 127
McMartin case, 264
McWhirter, David, 271
Magnus, Albertus, 13
Mail fraud, 126
Malinowski, Bronislaw, 156–57, 161, 170
Malpighi, Marcello, 15
Man and Woman (Ellis), 78
Mann, Frederika, 64
Mantegazza, Paolo, 68
Manuals: Diagnostic and Statistical
   Manual, 221, 240, 280; marriage,
   136–43; sex, 23
MAPS (Make-A-Picture-Story test), 267
Marcuse, Herbert, 153–55
Marcuse, Max, 59
Margolese, M. S., 226, 230
Margulies, Lazar C., 186, 187
Margulies Spiral, 187
Markee, J. E., 130
Marlowe, Christopher, 81
Marriage: companionate, 138, 242;
   "complex," 101; counseling, 72,

203–5; and the free-love movement,
   101; manuals, 136–43; marital rape,
   247; objective of, and prostitution,
   255; and sex therapy, 203–5
Marrian, Guy F., 129
Married Love (Stopes), 139, 145
Martin, Clyde, 173
Martinson, Floyd M., 260
Marx, Karl, 91
Marxism, 91, 152, 252. See also Commu-
   nist Party
Masaryk, Jan, 73–74
Masculinity, 163–64, 167, 210–40; and
   aggression, 229–31; and sexual harassment, 249. See also Gender
Masochism, 69, 84, 88, 282; and the
   German sexological movement, 41,
   42, 44, 48; and Kinsey, 180
Masters, R. E. L., 268
Masters, William, 111, 172, 185,
   195–210, 273–74, 278, 283; and
   Corner, 3, 196–97; criticism of SAR
   sessions, 279; and the Freudian myth
   of vaginal orgasm, 254; and Kinsey,
   196; and sexology fads, 281; and the
   sexual response cycle, 200–202; three
   principles of, 197
Masturbation, 1, 87, 162, 202, 265,
   283–84, 297; in anthropological studies, 158; condemnation of, 3, 19,
   22–23; and Davis, 116; and Ellis, 87;
   and the German sexological move-
   ment, 40, 42, 46, 71–72; and homo-
   sexuality, 21, 24, 79–80, 82, 83; and
   Kinsey, 169, 174, 176; and Lydston,
   93; and pornography, 52; survey sta-
   tistics on, 110–11; and Tisso, 20–21
Mathy, Robert, 239
Mattachine Review, 268–69
Mattachine Society, 268
Mattison, Andrew M., 271
Meade, Margaret, 157–58
Medical Review of Reviews, 144
Medical systems, origins of, 18–19
Meese, Edwin, 251, 252
Meese Commission, 251
Mencken, H. L., 133, 157
Menninger, Karl, 181
Menopause, 26, 108, 202

Mensinga, W. P. J., 105, 106, 146
Menstruation, 27–29, 123, 141; cycle, study of, 127–28, 130–32; and Mosher, 108; and sexual identification, 215
Mental illness, 6, 20, 164; and the "dangers of sex," 22–23; and eugenics, 51; and homosexuality, 166
Mesmer, Friedrich A., 34
Mexico, 127, 191–92, 193
Meyer, Adolf, 169
Meyer-Bahlburg, Heino, 228
Michelangelo, 81
Miles, C. C., 163–64
Milky Way Galaxy, 294
Millet, Kate, 90
Milnes, Colin, 16
Mitchell, John P., 114
Modesty, 82–85
Moench, Mary, 165
Moll, Albert, 39, 46–48, 89; and Ellis, 83; and Freud, 87, 91; and Hirschfeld, 64, 67, 70, 74, 75; and the Jewish influence, 58, 60; and Kinsey, 170
Moltke, Kuno von, 63–64, 67
Mondale, Walter, 263 Money, John, 210–17, 224–27, 237, 274, 281-82, 290
Monism, 70
Moore, C. R., 155
Moore, Monica, 258
Morality, 18, 58, 112, 274, 287; and censorship, 143–44; and child abuse, 265; "civilized," 96–97, 98; and the Comstock Law, 102; and contraception, 105; and eugenics, 51, 52; and the free-love movement, 101; and Kinsey, 170, 181; and Krafft-Ebing, 41; and pornography, 53, 249–50; and premarital sex, 163; and prostitution, 97; and sex as a symbol of evil, 30
More Joy (Comfort), 283
Morrow, Prince A., 103–5
Mosher, Clelia, 107–9, 111, 119
Mosher, Donald, 212, 229, 235, 248–52, 282
Mosovich, Abraham, 199
Muehlenhard, C. L., 247
Muret, Marc-Antony, 81
Murphy, Starr J., 113

Murray, George R., 125
My Secret Life (Fraxi), 58, 161

Napier, Charles, 54
Napoleonic Code, 33, 37
National Association for the Repeal of Abortion Laws, 257
National Center for Child Abuse and Neglect, 263
National Committee Maternal Health, 147
National Council on Family Relations, 136, 205–6
National Endowment for the Humanities (NEH), 291
National Institute of Child Health and Human Development, 287
National Institutes of Health (NIH), 123, 216, 270
National Institutes of Mental Health (NIMH), 286, 287, 291
National Opinion Research Center, 286–87, 288
National Organization for Women (NOW), 242
National Research Council (NRC), 116–18, 120–22
National Sex Forum, 276, 279
Nature/nurture controversy, 44, 225–27
Naturgesetze der Liebe (Hirschfeld), 69
Nazism, 58–60, 74–75, 192, 268, 282; and endocrinology research, 132, 135; and Freud, 91
Neisser, Albert, 99
Neosalvarsan, 100
Netherlands, 128, 232
Neurasthenia, 23, 93
Newcomb, Michael, 236
New Guinea, 157–58
New Left movement, 243
Newly Revealed Mystery of Nature in the Structure and Fertilization of Flowers (Sprengel), 12
Newsweek, 243
New York Medical Society, 103
New York Times, 133
New York University, 276
Niemoller, A., 142
Nietzsche, Friedrich, 64, 78

Nobel Prize, 127, 128
Nonconformist, use of the label, 212, 238
Norplant inserts, 195
North German Confederation, 39
Norway, 72
Noyes, John H., 101
Noyes, William, 94
Nutrition studies, 155

Objectivity, 295
Oedipus complex, 71, 82, 88, 89
Onanism, 20–21, 23
One (magazine), 268, 269
On Generation, 12
Oppenheimer, Willi, 186
Orchis Extract, 126
Organo Product Co., 126
Orgone box, 153
Ostow, Mortimer, 219
O'Sullivan, Thomas C., 112–13
Ota, Tenrei, 186
Our Bodies, Ourselves, 243

Packers Product Co., 126
Paidicka, 208
Palmieri, Edmund L., 184
Papanicolaou, George N., 127
Parent-Duchâtelet, Jean Baptiste, 31
Parke, J. Richardson, 95
Parke-Davis, 128, 191, 192
Parran, Thomas, Jr., 135
Pasteur, Louis, 11
Patent Office (United States), 27, 105
Pathological, use of the label, 4, 38
Patriarchy, 266
Pearson, Karl, 49–50, 78
Pedophilia, 42, 175, 260–63. See also
    Sexual abuse
Penicillin, 100, 171
Periodicity, sexual, 83–85
Perkins, Muriel W., 230
Perper, Timothy, 258
Perry, John, 254
Pershing, John J., 107
Peru, 239
Perverse, use of the label, 4, 30–31, 51,
    93, 95
Pflüger, E. F. W., 27–28
Pfost, Gracie, 183

Philippines, 236, 239
Pillard, Richard C., 226, 227, 229
Pilpel, Harriet, 184
Pinchot, Gertrude M., 147
Pincus, Gregory, 132–36, 193, 194
Planned Parenthood, 51, 106, 146, 193,
    194, 207, 257, 289
Plant studies, 9, 11–12, 16, 191–92
Plato, 13, 36
Playboy, 243, 249, 285, 286
Ploss, Hermann, 56
Polygamy, 101
Pomadere, Jeannette, 226
Pomeroy, Wardell, 173, 174–75, 205–6,
    276, 278
Popper, Karl, 294–95
Popular Mechanics, 164
Population Council, 147
Pornography, 52–53, 94, 101–2, 265, 285,
    298; and censorship, 143, 249, 251;
    and feminism, 249–54; and homosex-
    uality, 160; and sexual histories, 161
Postmodernism, 292–93, 295
Post Office Department (United States),
    146
Pregnancy, 1, 26, 131, 180; and anatomy,
    14, 109; and hormones, 215; tests,
    128. See also Abortion
Premarital sex, 4, 162–63, 176, 180, 285
President's Commission on Obscenity
    and Pornography, 251–52
Prince, Virginia, 222, 223, 238
Productos Esteroides, 194
Progress in Hormonal Research, 193
Progressive era, 102, 107
Prostitution, 4, 6–7, 30, 97–98, 254–56,
    297–98; and Bloch, 57; and Davis,
    114-15; in France, 31–32; and Kinsey,
    168, 174, 184; and Masters and John-
    son, 199; and the military, 112; and
    the public health movement, 103;
    and sexually-transmitted diseases
    (STDs), 98, 100
Protestantism, 18, 58. See also Chris-
    tianity
Psychoendocrinism, 72
Psychology and Human Sexuality, 208
Psychology Today, 286
Puberty, 76, 21, 123; and gender issues,

215–16, 225, 233, 235–36; and the two-stage theory of sexual development, 46-47. See also Adolescence
Public health movement, 103–5
Public Health Service (United States), 135
Puerto Rico, 195
Puritans, 18
Purity crusade, 96–97, 104, 119

Quackery, 126–27
Quakers, 167

Racism, 50–52, 93–94, 242. See also Civil rights movement
Radiant Motherhood (Stopes), 139
Random House, 141, 145
Rape, 245–48, 257
Reagan, Ronald, 290
Real Encyclopädie für Medizin (Eulenburg), 47–48
Reconstruction, 101
Redbook, 286
Redi, Francesco, 10
Reece, B. Carroll, 182–83
Reich, Wilhelm, 152–53
Reinish, June, 259, 283
Reiss, Ira, 250
Religion, 101, 296. See also Christianity; Hinduism; Judaism; Protestantism
Remof, H. T., 258
Renaissance, 78, 81
Rendell, W. J., 106
Republican Party, 182, 183
Ribaud, Félix, 197
Rice, Thurman, 169
Rice-Wray, Edris, 195
Ricord, Philip, 21, 99
Robie, W. F., 111
Robinson, Paul, 4, 83, 282–83
Robinson, William, 73, 138
Rock, John, 194, 195
Rockefeller, John D., Jr., 112–21, 133, 135
Rockefeller, John D., III, 147
Rockefeller Foundation, 121, 135, 180–83, 290
Rockefeller Institute, 191
Roe, Humphrey V., 138

Rohleder, Hermann, 48, 57, 68, 71
Roosevelt, Franklin D., 126
Roper, Joseph C., 165
Rorschach test, 267
Roubaud, Felix, 111
Rush, Benjamin, 21–22
Rusk, Dean, 181–82
Russell, Bertrand, 73, 157
Russo, A. J., 237
Ruzicka, Leopold, 130

Sacher-Masoch, Leopold von, 42
Sade, Marquis de, 42
Sadism, 41, 42, 44, 48, 84, 88, 180
Sadomasochism, 48, 84, 241, 253, 282
Salvarsan, 99–100
Sandfort, Theodor, 262–63
Sanger, Margaret, 51, 73, 109, 134, 138, 140, 142, 146, 193–94
Sanger, William, 32
Sapogenin, 191–92
Sappho and Socrates (Hirschfeld), 65, 66
SAR (Sexual Attitude-Reassessment), 279
Satterthwaite, Adeline P., 195
Schaudinn, Fritz, 99
Schering Corp., 193
Schering-Kahlbaum, 128
Schilder, Paul, 120
Schleiden, Matthias J., 16
Schofield, Michael, 267
Schopenhauer, Artur, 33
Schrenck-Notzing, Albert von, 42–44, 46–48
Schwann, Theodor, 16
Schwartz, P., 285
Schweitzer, Johann Baptist von, 35
Scientific Society for Sexology and Forensic Sexological Expertise, 168
Searles (company), 193, 194
Secretin, 125
Sex and Character (Weininger), 137
Sex change surgery, 213, 218–22
Sex education, 28–29, 172, 203–5, 207, 275-76; and contraception, 146–47; and Davis, 116; and Dickinson, 109; and eugenics, 51; and Exner, 111–12; and international congresses, 73, 74; and the public health movement,

Sex education (continued)
    103–5; and Rice, 169; and sexually-
    transmitted diseases (STDs), 135-36,
    289
Sexology, coining of the term, 26
Sex Side of Life (Dennett), 144
Sex therapy, 142, 172, 203–8, 275,
    278–84
Sexual abuse, 26, 160–61, 184, 257, 259-
    65; statistics on, 112, 175
Sexual harassment, 248–49
Sexual inversion, 79–81, 94, 95
Sexual Inversion (Ellis), 79–80
Sexual Life of Our Time and Its Relations
    to Modern Culture (Bloch), 57
Sexual Life of Savages, The (Malinowski),
    157
Sexually-transmitted diseases. See STDs
    (sexually-transmitted diseases)
Sexual Pathology, 71–72
Sexual Probleme, 69
Sexual response cycle, 200–202
Shakespeare, William, 143
Shaw, George Bernard, 80, 137
Sheldon, W. S., 167
Sherwin, Robert, 206
Sibling rivalry, 150
SIECUS, 207, 259, 289
Simon, William, 222
Sin, 35, 292; original, 3, 11; and prostitu-
    tion, 97
Single Woman, The (Dickinson), 110
Slavery, 101, 242
Social Democratic Party (Germany),
    35n, 67
Social Disease and Marriage (Morrow),
    103–4
Social hygiene societies, 104, 116
Society for Sex Therapy and Research,
    208
Society for the Scientific Study of Sex
    (SSSS), 206–7, 251, 257, 277, 280,
    289, 299
Society of Sanitary and Moral Prophy-
    laxis, 104
Sociobiology, 231–32
Sollins, Irvin V., 194
Somlo, Emeric, 192
Soranus of Ephesus, 13

Soul, 49, 59
Soviet Union, 72
Speke, John Hanning, 55
Spence, J. T., 235
Spermatozoa, 15, 16, 17
Spermicides, 106, 135. See also Contra-
    ceptives
Sprengel, Christian K., 12
SRS (sex reassignment surgery), 219, 221
Stahl, Georg Ernst, 19
Stalin, Josef, 168
Stanley, Lawrence, 280
Starling, Ernest H., 125
Star Wars defense plan, 298
STDs (sexually-transmitted diseases), 4,
    31, 32, 98–100, 289–91; association
    of, with sex itself, 21; and Bloch, 57;
    in corrective institutions, 94–95; and
    Dennet, 144; and international con-
    gresses, 73; and the military, 106–7,
    112; prevention of, 100, 103–5,
    135–36; and prostitution, 98, 100;
    statistics on, 110; treatment of,
    98–100, 171, 289–90. See also AIDS
Stein, Martha L., 255
Steinach, Eugen, 125–27
Stekel, Wilhelm, 68, 151, 169, 280
Sterilization, 72, 218
Sternberg, Wilhelm, 68
Steroids, 128–29, 134, 189, 191, 194, 195
Stockard, Charles R., 127
Stöcker, Helene, 68
Stone, Abraham, 140, 193, 194
Stone, Hannah, 140
Stonewall Rebellion, 269
Stopes, Marie, 138–40, 142, 144–145,
    242
Strakosch, F. M., 164
Strindberg, August, 137
Studies in the Psychology of Sex (Ellis),
    75–76
Studies on Hysteria (Breuer and Freud),
    87
Suggestion therapy, 44
Sulfanilamides, 100
Sulloway, Frank, 44, 87, 150
Sunday Express, 143
Superego, 7, 88
Supreme Court, 145, 248, 257

Surgery, sex change, 213, 218–22
Sweden, 72
Symbolic interactionism, 223
Symonds, John Addington, 78–80, 81
Symposium (Plato), 36
Syntex Sociedad Anónima, 192–93
Syphilis, 21, 57, 98–100, 103, 289–90. See also STDs (sexually-transmitted diseases)
Szasz, Thomas, 270

Taboos, 157, 259–60
Talmey, B. S., 198
TAOS (The Association of Sexologists), 208
TAT (Thematic Apperception Test), 267
Tennov, Dorothy, 257
Tennyson, Alfred Lord, 58
Terman, Lewis M., 162, 163–64, 234
Thatcher, Margaret, 288
Therapy, sex, 142, 172, 203–8, 275, 278–84
Thomas, Clarence, 248
Thorne, Barrie, 233
Thousand Marriages, A (Dickinson), 110
Thousand Nights and a Night, The, 54
Tigris-Euphrates Valley, 2
Tissot, S. A. D., 20–21, 32, 49, 292
Toenggal, Syng Hyan, 159
Tourney, Garfield, 230
Transsexualism: in anthropological studies, 159; and gender issues, 217–21, 224, 229–30, 234–35, 237–38, 240
Transvestism, 69, 142, 180, 241; and aggression, 229–30; in anthropological studies, 159; and cross-dressing, 219, 220, 221–24; and Ellis, 84, 85; and gender issues, 217–24, 229–30, 236, 238, 240
Trilling, Lionel, 180–81
Trobriand Islanders, 157
Tumescence, 83–84
Turner's syndrome, 213–14
Twin studies, 227
Two-seed doctrine, 12–13
Two-stage theory of sexual development, 46-47

Ulrichs, Karl H., 33–43, 45, 51, 64, 234, 274

Ulysses (Joyce), 145
Université de Québec, 277
University of California at Los Angeles, 237
University of Chicago, 288, 291
University of Hawaii, 286
University of Minnesota Hospital, 237
University of Pennsylvania, 276
University of Wisconsin, 29
Urbanization, 30
Urning, 36, 37, 39
U.S. v. 31 Photographs, 184

Van de Velde, Theodore H., 138, 140–42
Van Dusen, Henry P., 180
Vatican, 294
Veblen, Thorstein, 133
Venereal disease. See STDs (sexually-transmitted diseases)
Vesalius, Andreas, 13
Veterans' Hospitals, 107
Vice Versa, 269
Victorianism, 53, 77, 84, 97, 161, 265
Vietnam War, 296
Villa, Pancho, 112
Virginity, 157, 242, 246
Vitalism, 19
Vitamins, 155
Voting rights, 137
Voyeurism, 169, 180, 297

Waal, Franz de, 232
Walkowitz, J. R., 255
Warburg, Paul N., 113
War Department (United States), 107
Warhol, Andy, 53
Warren, A. J., 181
Wasserman, August, 99
Wasserman testing, 136
Weimar Congress of the Psychoanalytical Association, 64
Weinberg, M. S., 236, 267–68, 270, 282
Weininger, Otto, 137
Weinrich, James, 229, 258
Wellcome Foundation, 288
Well of Loneliness, The (Hall), 143
Wells, Herman, 169
Wertham, Frederic, 167
Wesley, John, 76

Westheimer, Ruth, 283
Westphal, Carl, 38, 39, 43
What Every Girl Should Know (Sanger), 142
Wheatley, Susannah, 76
Wheeler, R. V., 138
Whipple, Beverly, 254
Whitam, Frederick L., 226, 227, 236, 239
Wiederman, G. H., 219
Wilde, Oscar, 53–54, 64, 81
Wilhelm II, 63
Wilkins, Lawson, 216
Willard, Elizabeth, 26
Williams, Colin, 270, 282
Williams Committee, 251
Wilson, Edward O., 231
Wilson, Woodrow, 106
Winslow, Randolph, 94–95
Wise Parenthood (Stopes), 139
Wissler, Clark, 121
Wittle, Fritz, 59
Wolfenden Commission, 270
Wolffe, Charlotte, 70
Women, subordination of, 101, 223; biological justifications for, 17–18, 27-29, 210; and eugenics, 49; and prostitution, 97–98. See also Feminism
Women Against Pornography, 249, 251, 252–53
Woodhull, Victoria, 101
Wooley, Helen B., 121
Woolsey, John M., 145

Woolston, Howard, 114
Worcester Foundation, 193
Working Women's Institute, 248
World War I, 71, 105, 116, 138, 160–61; American army during, 106, 111–12; attitudes toward sex education after, 104; birth control after, 109; civilized morality before, 96, 97; publications on sex during, 67; Reich's work after, 152; voting rights after, 137
World War II, 100, 149, 160, 161, 193, 242, 268; attitudes toward sex education after, 104; birth control before, 105–6; homosexuality and the military during, 164-66; making of penicillin during, 171. See also Nazism

Yale Cross-Cultural Survey, 158
Yerkes, Ada W., 155–56, 170–71
Yerkes, Robert M., 117, 118, 121, 122
Yolles, Stanley, 287, 290
Young Men's Christian Association (YMCA), 111, 115

Zeitschrift für Sexualwissenschaft (Hirschfeld), 57, 68, 69, 71
Zent, Michael, 236
Zillman, D., 250
Zinn, Earl F., 116, 117, 120
Zondek, Bernhard, 128
Zuger, Bernhard, 237
Zuni Indians, 96